# Mathematical Models in Agriculture

# Mathematical Models in Agriculture

A Quantitative Approach to Problems in Agriculture and Related Sciences

**J. France,** PhD
Senior Scientific Officer, Biomathematics Division,
Grassland Research Institute, Hurley, Maidenhead

**J.H.M. Thornley,** MA, DPhil
Head, Biomathematics Division,
Grassland Research Institute, Hurley, Maidenhead

**Butterworths**
London Boston Durban Singapore Sydney Toronto Wellington

First published 1984



**British Library Cataloguing in Publication Data**

France, J.
Mathematical models in agriculture.
1. Agriculture—Economic aspects—Mathematical models
I. Title II. Thornley, J.H.M.
338.1′0724 HD1411

ISBN 0–408–10868–1

**Library of Congress Cataloging in Publication Data**

France, J.
Mathematical models in agriculture.

Bibliography: p.
Includes index.
1. Agriculture—Mathematical models. I. Thornley, J.H.M. II. Title.
S494.5.M3F72 1984 630′.724 83-19010
ISBN 0–408–10868–1

Typeset by Scribe Design Ltd, Gillingham, Kent
Printed and bound in England by Redwood Burn Ltd.

# Preface

Agriculture and the related scientific disciplines are in a state of rapid evolution. Quantitative methods of experimentation and of thought are increasingly taking over. For describing results and matching these to our ideas and theories, it is now widely accepted that mathematics is not only the most appropriate tool but provides the only way we know of doing this. In the language of those who study the philosophy and methodology of science, it seems that the agricultural sciences are becoming 'mature', and are entering a phase where the 'normal' mode of scientific research will dominate. Normal science is characterized by a reasonable consensus about the body of theory (working hypotheses) which may be used to explain and predict the observed phenomena. The engine of progress is a continual and highly productive interaction between experiment and hypothesis, observation and theory, always accompanied by increasing precision, generality and explanatory power. Taking our cue from events in physics, chemistry, molecular biology and biochemistry, we are entering an exciting era, which promises steady and significant progress for many years, and which will change the face of agriculture.

This book sets out to be a textbook on this topic, and thereby to facilitate the processes just described. It is not a textbook of agriculture or biology, and few new agricultural or biological concepts are presented. Our primary aim is to teach agricultural scientists when and how to attempt to express their ideas mathematically, how to solve the resulting mathematical model and compare its predictions with experimental data. This is an area where the universities have not so far been notably successful.

All concerned with agricultural research and teaching should find this book of value, including scientists involved with agricultural extension work—as the computing revolution continues, suitably programmed mathematical models provide an efficient vehicle for making research results available to the adviser and farmer. The book should be essential reading for graduate students in agriculture and related sciences, and the better undergraduates could also derive much benefit from its study.

Because this is a textbook in a subject where little formal tuition is available, we have attempted to make it suitable for self-study. Where possible, problems are provided, together with outline solutions. The level of mathematics is uneven because the topics covered differ widely in their stage of development and the type of mathematics required. The advent of computers has meant that the long and complicated mathematical analyses that were common last century and in the early part of this century are no longer needed. Problem formulation is the difficult, essential and most creative step in the process; problem solution may be tedious but is relatively straightforward. A quite modest knowledge of algebra, calculus and ordinary differential equations should suffice for most of the book. Where more advanced techniques are employed, an introductory account is given, either in the text or in the Glossary.

The scope of the book is best seen in the Contents. While we have tried to cover the areas that we judged most relevant to the agricultural scientist, inevitably our own experience and competence have greatly influenced the choice of topics and their depth of treatment. Statistical methods, despite their importance, are not given proper coverage as several excellent texts on the subject are already available to the agricultural scientist. There are five general chapters which should be read before any of the subsequent material is tackled. Of the remaining eight chapters, some stand fairly well alone and can be read out of sequence, whereas others are more integrated in small groups; where this occurs should be clear from the chapter titles.

We are indebted to Alec Lazenby, former Director of the Grassland Research Institute, for allowing and encouraging us to undertake this project. We are also grateful to many colleagues, especially in the Biomathematics Division, for permitting us to use ideas and material that have resulted from joint work. Our thanks are also due to Pam Berridge for patiently typing and editing a long and difficult manuscript.

Lastly, since the writing of a textbook in a rapidly changing area is not without its difficulties, we would appreciate it if any suggestions for improvement, of appropriate examples and problems, notable omissions, errors or inconsistencies could be brought to our notice.

J. France
J.H.M. Thornley

# Contents

Chapter 1

# Role of mathematical models in agriculture and agricultural research

## Introduction

The aim in this introductory chapter is to present one or two simple mathematical models with the basic language and ideas of modelling. It is important to understand the relationship of mathematics both to the practice of agriculture and to agricultural research, which is concerned with improving agricultural practice. Models can contribute in different ways both to research and to production, and the approach adopted will largely depend on one's objectives.

## What is a mathematical model?

A mathematical model is an equation or set of equations which represents the behaviour of a system. For example, *Figure 1.1* shows the sort of growth curve that might result from an experiment in which an animal is supplied with food at

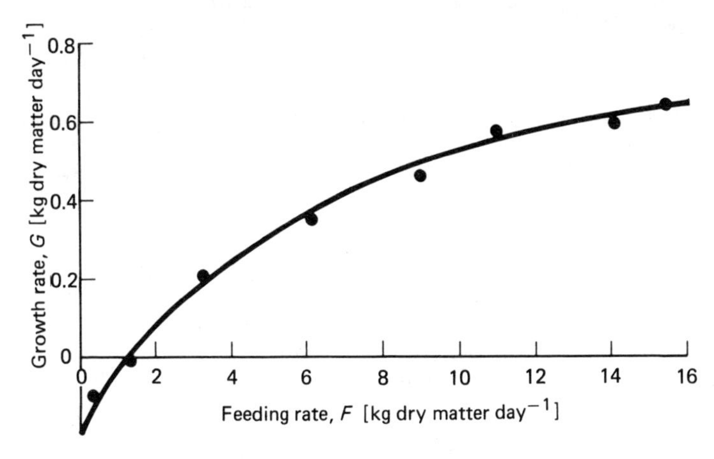


*Figure 1.1* A hypothetical experiment to show the effects of varying feeding rate, $F$, on the growth rate of an animal, $G$. Experimental data: •; the solid line represents a fitted equation (equation (1.1))

different rates. For various reasons, one may wish to represent the data points of *Figure 1.1* by a mathematical equation, such as

$$G = G_1 \frac{F}{K + F} - G_2 \tag{1.1}$$

where $F$ is the rate at which food is supplied to the animal, and $G$ is the growth rate of the animal. Both $F$ and $G$ are *variables*, and they take different numerical values. $F$ is known as an *independent variable*, since the experimenter fixes $F$ at certain values—the range of levels of feeding rate he wishes to cover; $G$ is known as a *dependent variable*, since it is not under the direct control of the experimenter, and its value is a consequence of the level of $F$ that has been chosen. The quantities $G_1$, $G_2$ and $K$ are called *parameters*, and they take a fixed value for a curve such as that drawn in *Figure 1.1*. All the three parameters, $G_1$, $G_2$ and $K$ are easily recognizable attributes of the curve in *Figure 1.1*: if the feeding rate $F$ is zero, then the growth rate $G = -G_2$, and the animal is losing weight; if the feeding rate $F$ is very large, then $F/(K + F)$ approaches unity, and the growth rate $G$ approaches $G_1 - G_2$; the parameter $K$ reflects the steepness of the curve in *Figure 1.1*, and $K$ is equal to the value of $F$ when $G$ lies half-way between its minimum value (at $F = 0$) and its maximum value ($F$ is very large). More briefly, parameters $G_1$, $G_2$ and $K$ describe the asymptote, intercept, and half-maximal response of equation (1.1).

Equation (1.1) is often referred to as a rectangular hyperbola. This is because it has two asymptotes which are at right-angles to each other. These are

$$F \to \infty \qquad G \to G_1 - G_2 \tag{1.2a}$$

and

$$F \to -K \qquad G \to -\infty \tag{1.2b}$$

The two straight lines $G = G_1 - G_2$ and $F = -K$ are at right-angles.

Responses of the diminishing-returns type as in *Figure 1.1*, are quite common in biology and elsewhere. Some examples are the nutritional response of crops to fertilizers; the photosynthetic response of crops, plants and leaves to light; and the rate of an enzyme reaction in response to substrate concentration.

It may be noted that the line in *Figure 1.1* does not go exactly through the experimental data points, so that the mathematical model of equation (1.1) only gives an approximate representation of the data, with a certain degree of error. Considered on its own, equation (1.1) does no more than redescribe and summarize the experimental data, and it does not add anything new to our knowledge about the situation. Later, the important distinction between description and understanding (or empiricism and mechanism) is considered at length.

To produce the solid line in *Figure 1.1*, equation (1.1) has been 'fitted' to the experimental data—that is, the three parameters of the equation, $G_1$, $G_2$ and $K$, have been adjusted so that the line and the data points coincide as closely as possible, in some defined way. If the experiment is performed on animals of different age, or on different species of animals, one would expect that the parameters which summarize the response (assuming that equation (1.1) still describes the data in each case) to be possibly different. Such parameter differences may be of great interest to the scientist, and the use of an empirical model, as in *Figure 1.1*, may be a valuable way of uncovering these effects.

**A simple dynamic model**

Equation (1.1) is a *static* model, in that it does not contain the time variable $t$, and there are many useful models of this type. However, there is a second very important category of models, known as *dynamic* models; these contain the time variable $t$ within them, and are often used to describe a time-course of events.

An example of a simply dynamic model is

$$W = W_0 + bt \tag{1.3}$$

where $W$ is the weight of an organism (animal or plant), $t$ is time, and $W_0$ and $b$ are parameters. $W_0$ is the value of $W$ at zero time ($t = 0$), and $b$ is the slope of the growth curve. Equation (1.3) is drawn in *Figure 1.2*.

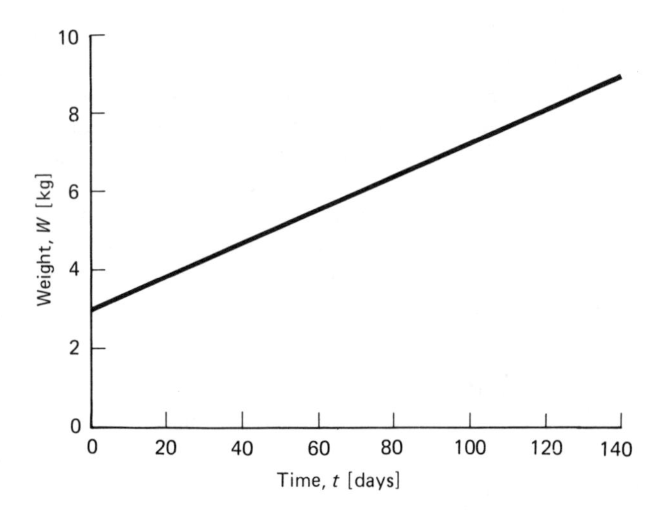


*Figure 1.2* A simple dynamic growth model. Equation (1.3) represents the weight of an organism, $W$, and its dependence on time $t$

However, it is more common to represent dynamic models by their *differential* form, that is, by means of one or more first-order differential equations. Differentiating both sides of equation (1.3) with respect to time $t$, therefore

$$\frac{dW}{dt} = b \tag{1.4}$$

In words, equation (1.4) states that the growth rate, $dW/dt$, is equal to a constant, in fact the parameter $b$.

The model of equations (1.3) and (1.4) is not very realistic, and most of the useful dynamic models result in a differential equation (or differential equations) that cannot be directly integrated to give something equivalent to equation (1.3) (an *analytic* expression), but must be integrated numerically, for which purpose a computer and purpose-built software are invariably used.

## Agriculture and agricultural research

The etymology of the word agriculture (L. *ager*, field) still conveys an accurate idea of what agricultural activity is about, although the boundary between some agricultural and non-agricultural activities is becoming increasingly difficult to define. For present purposes, agriculture is taken to mean the production activities that take place on farms, with field cultivation or use being a considerable component at the beginning of the chain or chains of activities, resulting in the production of materials which are used elsewhere (off the farm) for a variety of purposes (food, clothing, industrial processes, etc.).

The practice of agriculture is partly based on knowledge, a formal knowledge contained in books and papers, giving a rationale for taking decisions. It is partly based on an inherited folk-wisdom, where things are done because it is known that they work to a certain degree, although it is not understood precisely why, or whether better results might be obtained by doing things a little differently. Finally, it is partly based on conjecture, where the situation is novel and there is no guidance from knowledge or tradition on what to do, but one has to do something.

These three components of the practice of agriculture can be summarized as knowledge, tradition, and conjecture. The purpose of agricultural research is to increase the percentage of the first component, knowledge, at the expense of the second and third components, and thereby to increase the efficiency of agricultural production. The current efficiency of agricultural production forms a baseline from which, other things remaining equal, it is possible to move only forwards. Increased knowledge does not necessarily lead to increased efficiency, but it can uncover options which lead to higher efficiency. The farmer is always able to continue with present practices and present efficiency—again, other things remaining the same. Equally important is the fact that increased knowledge allows a more rational response when other things no longer remain equal and the environment, both natural and man-made, in which the farmer operates, changes significantly.

### Agricultural research—the nature of science

It has already been mentioned that agricultural research is concerned with adding to that part of knowledge which is connected to the principles and practice of agriculture. Scientific knowledge is not just about observational data, but it includes having a conceptual scheme or hypothesis that corresponds with those data, and it is the continual interaction between hypotheses (how we think things work) and observational data (how they actually do work) that leads to progress. This is illustrated in *Figure 1.3*. As time goes on, measurements become more extensive and more exact; similarly, ideas about what is happening become more precise and more articulated. In measuring hypotheses against observational data, one attempts to connect ideas to nature at as many points as possible and with as much precision as possible. This is relevant to mathematical modelling because, if the observations are numerical, then the hypotheses should be expressed numerically so that a proper connection can be made. This means using mathematics as a language for expressing one's ideas.

Thus, any branch of science, as it progresses from the qualitative to the more quantitative, may one day reach the point where the use of mathematics, for connecting theory to experiment, is increasingly essential and fruitful. It cannot be too strongly emphasized that the ideas themselves, the hypotheses, are not

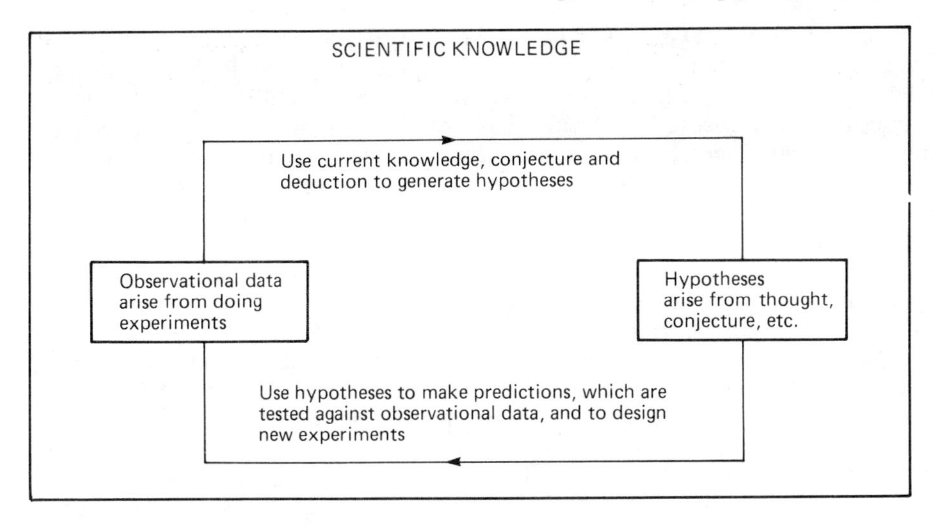


*Figure 1.3* The nature of scientific knowledge, and how it progresses

contributed by mathematics. Mathematics is merely a tool enabling the biologist to express his biological hypotheses so that quantitative prediction is possible, which can then be compared with the real world.

## Basic research, applied research and development

The way in which a modeller approaches a problem will depend upon his objectives, and these will depend on the nature of the problem he is addressing. For this reason, it is necessary to discuss the three categories of activity: basic research, applied research, and development.

*Figure 1.4* suggests how R & D (an abbreviation for basic research, applied research, and development) fits into the framework of our economy (Doyle and

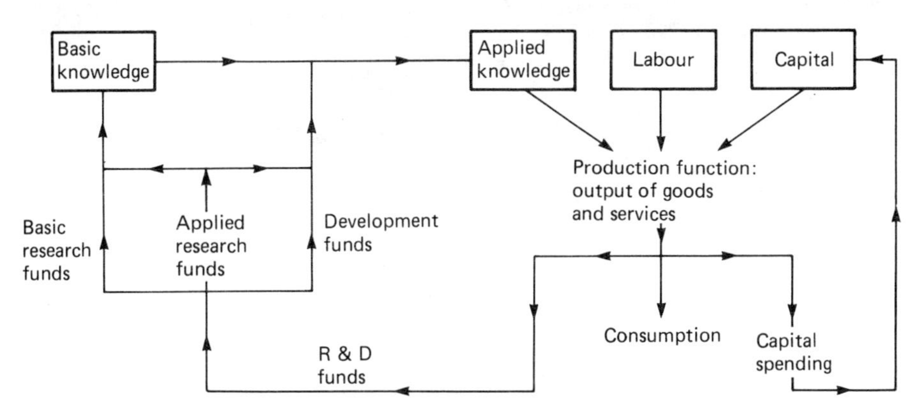


*Figure 1.4* The role of R & D expenditure in the economy. The quantities in the boxes are state variables, which define the state of the system at a particular time. The other items are flows or processes with dimensions of per unit of time. Capital spending includes expansion, updating and maintenance of the capital stock. (After Doyle and Thornley, 1982)

Thornley, 1982). R & D is an activity or process which occurs at a certain rate—£ or $ per year of expenditure on R & D. The product of the R & D process is knowledge. Knowledge can be represented as a 'state variable' whose level (it is assumed) can be measured, and which has a particular value at a particular point in time. When goods and services are produced, resources are used; at a simple level, these resources can be considered as comprising labour, capital and raw materials. However, the nature of the capital resources reflects the level of applied knowledge or technology, as does the manner in which labour makes use of the capital resources to transform the raw materials. The level of applied knowledge affects the rate at which these transformations take place.

Thus, as shown in *Figure 1.4*, scientific knowledge is divided into two categories: basic knowledge $K_b$, and applied knowledge $K_a$. Applied knowledge is that part of basic knowledge (suitably transformed) of which use is made in the production processes producing goods and services. Some of these goods and services are used to support R & D, and this in turn increases the stock of basic and applied knowledge, giving a feedback loop. The differences between basic research ($R_b$), development ($D$), and applied research ($R_a$), and the ways in which these contribute to basic knowledge ($K_b$) and to applied knowledge ($K_a$) can now be spelt out more exactly, and this is illustrated in *Figure 1.5*.

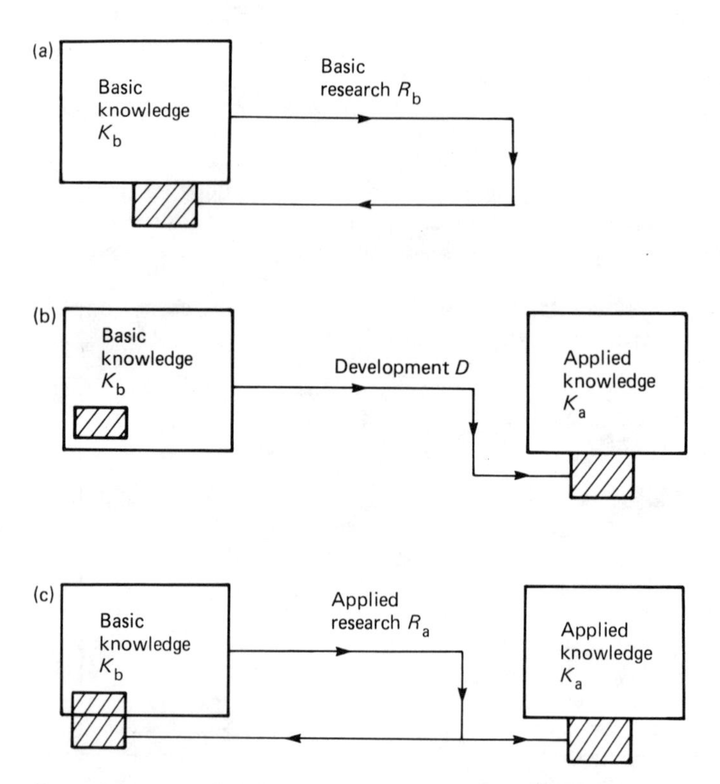


*Figure 1.5* The contributions of basic research (a), development (b), and applied research (c) to basic knowledge and applied knowledge, $K_b$ and $K_a$. In (a), the shaded area indicates an addition to $K_b$, $\Delta K_b$, resulting from $R_b$. In (b), a part or subset of $K_b$ is transformed by $D$ into an increment $\Delta K_a$ to $K_a$. In (c), elements of (a) and (b) are combined to give increments both to $K_b$ and to $K_a$

BASIC RESEARCH, $R_b$

Contrary to popular opinion, basic research is often an easier activity than applied research, and possibly this explains the widespread tendency of researchers to gravitate towards 'fundamental' problems. In basic research, the researcher chooses a problem he thinks is soluble, given the current state of knowledge ($K_b$) and the resources available to him ($R_b$). The important factor is that the scientist himself sets the boundaries of the system he is investigating. Science is 'the art of the soluble', and little credit is obtained from attempting work that is too far beyond the current state of knowledge.

A dynamic model to predict how basic knowledge changes with time might be written

$$\frac{dK_b}{dt} = f_b(R_b, K_b, K_a) \tag{1.5}$$

where $f_b$ denotes some functional dependence upon $R_b$, $K_b$ and $K_a$. Unfortunately, not enough is known to make much progress in defining equation (1.5), and there are reasons for thinking that such an equation cannot be constructed (*see* p. 8).

DEVELOPMENT, $D$

Development is concerned with attempting to make good some perceived deficiency in technology (applied knowledge), but where it is thought that the current level of basic knowledge is sufficient to solve the problem. In development work, one is transforming some part of basic knowledge into a form suitable for application. Analogous to equation (1.5), a suitable equation describing the process might be

$$\frac{dK_a}{dt} = f_d(D, K_b, K_a) \tag{1.6}$$

with $f_d$ denoting some other functional dependence upon $D$, $K_b$ and $K_a$. Note that, although the boundaries of the system being investigated are determined by the problem, and not the scientist, those evaluating the project consider that an extension of basic knowledge is not needed for solution of the problem.

APPLIED RESEARCH, $R_a$

Applied research (*Figure 1.5c*) can be viewed as combining some of the characteristics of basic research and of development (*Figure 1.5a,b*). Again (as with development), the boundaries of the system being investigated are determined by the problem and not by the scientist, although in this case, it is considered that an extension of basic knowledge will be needed before the problem can be solved. Before taking on board an applied research project, it is essential that a critical assessment be made as to whether the increment in basic knowledge required can in fact be achieved, given the current state of knowledge, duration of the project and level of support.

The modelling problem now involves the simultaneous solution of equations (1.5) and (1.6), integrating them until a final time $t = t_f$ is reached, where the level of applied knowledge $K_a(t = t_f)$ now encompasses the required technology.

## Benefits and costs of research

As shown in *Figure 1.4*, R & D activities are supported by the sections of society which are producing goods and services, and as a consequence, there is a continuing need for what is called 'accountability'. It is important that there should be an effective dialogue between those engaged in the different parts of the agricultural R & D spectrum and those concerned with agricultural production, and that this dialogue should be based on an appreciation of their symbiotic relationship. The salient features are summarized in *Table 1.1* (after Doyle, 1981).

**TABLE 1.1. Benefits and costs of research**

| *Type of research* | *Objective* | *Benefits* | *Costs* |
|---|---|---|---|
| Basic | To extend knowledge | Unquantifiable, except by historical evidence | Elastic |
| Applied | To solve a problem requiring an extension to knowledge | Can be estimated | Can be guessed, but subject to error |
| Development | To solve a problem using existing knowledge | Can be estimated | Can be estimated |

Basic research has been of great value to mankind, although it is often the least expected contributions that are of maximum benefit. That basic research will continue to serve us well, must always be an act of faith, although one for which there is abundant historical evidence. It has been shown (Popper, 1982) that the extent to which the findings of basic research can be predicted is very limited indeed; thus, a deterministic equation (1.5) does not exist. Popper also discusses the limitations which must apply to even a stochastic equivalent of equation (1.5). It is encumbent upon research administrators who pursue a 'dirigiste' approach to demonstrate that Popper's analysis is erroneous and that, historically, directed basic research has been more productive than undirected basic research; unless this is done (and some believe that it cannot be done), *dirigisme* will continue to lack credibility and should be resisted.

Both applied research and development are about problem-solving, so the benefits of solving the particular problem can usually be estimated. With development the costs can generally be assessed, but with applied research, it is usually difficult to do more than guess the costs (depending upon the size of the 'basic research' and 'development' components of the applied research), and costs can easily escalate to a point where the project has to be stopped.

## Economic evaluation and implementation of research results

Where research has produced an answer to an agricultural problem, an economic analysis can help determine whether the innovation is of value to the farmer. A new technology can only be successfully marketed to the farming community if a careful consideration of the costs and benefits to the farmer shows a clear advantage in its adoption. Such an assessment should take account of any extra investment required, or any changes in the risks to which the farmer is exposed. Since economic evaluation is a quantitative exercise, mathematical models can make a valuable contribution in this area.

Several examples of this sort of modelling appear in this book. For instance, research into nutrition, both of plants and of animals, has led to improved recipes, either of fertilizer treatments for crops, or of feeding regimes for animals. The calculations involved may be quite complex and well beyond the scope of many farmers and horticulturists. However, a mathematical model programmed for an inexpensive computer can make these research results widely accessible.

The relative economics of different dairying systems—for instance, low input–low output versus high input–high output—are a result of many interacting factors. Again, a mathematical model may be a good means of assembling the results of the relevant research programmes, and reaching a conclusion that can be put before the farming community.

## Description and understanding. Hierarchy: systems and subsystems

This section comprises several closely related components.

### Description and understanding

The scientist aims to extend current knowledge—possibly by improving the observational data, or alternatively by extending the corresponding ideas or concepts (*Figure 1.3*). The former is more concerned with description, whereas the latter is more concerned with our understanding of what is going on. Referring back to *Figure 1.1*, and to equation (1.1), which approximately describes the data shown in *Figure 1.1*, the curve gives a description of the growth of the animal that may be useful for many purposes. However, it must be emphasized that the curve simply redescribes the experimental data, and it does not contain any additional information that is not contained in the data. What the curve does not tell us, and what one may wish to know, is why the response is as it is—both in terms of its general shape and also in terms of the particular numerical values attaching to the initial slope and asymptote. For instance, one might wish to know what physiological process limits the growth rate of the animal at very high feed rates, or what processes determine how quickly the animal loses weight when it is fasting. To obtain this sort of understanding, it is necessary to relate the description of whole-animal response to the descriptions of other phenomena—phenomena which lie at a lower level in the organizational hierarchy of the animal.

### Hierarchy: systems and subsystems

The term hierarchy, or more fully organizational hierarchy, is best explained by giving a typical hierarchy for agriculture:

| *Level* | *Description of level* |
|---|---|
| ... | ... |
| $i + 1$ | Collection of organisms (herd, flock, crop) |
| $i$ | Organism (animal, plant) |
| $i - 1$ | Organs |
| ... | Tissues |
| ... | Cells |
| ... | Organelles |
| ... | Macromolecules |
| ... | ... |

The idea of hierarchical organization is, of course, not peculiar to agriculture, but is very widespread, some examples being businesses, computer programs, social groupings and electronic equipment. The growth response curve of *Figure 1.1* gives a description of behaviour at the organism level, which is labelled by the index $i$. Although this description of behaviour of the animal at level $i$ is of value in its own right, it is finding some scheme that successfully relates this description to other descriptions of behaviour at lower levels in the hierarchy, that provides an understanding.

Any level of the above diagram can be viewed as a system, composed of subsystems lying at lower levels, or as a subsystem of higher level systems.

Hierarchical systems have three important properties:

1. Each level has its own language, concepts or principles. For example, the terms animal production or crop productivity have little meaning at the cell or organelle level. In relation to a gas or a liquid, the terms pressure, volume and temperature are bulk properties that mean little at the atomic or molecular levels.
2. Each level is an integration of items from lower levels. Discoveries or descriptions made at a given level can be connected to the next higher level in an explanatory (mechanistic) scheme. Thus, a description at level $i$ can provide an understanding (mechanism, explanation) to phenomena at level $i + 1$.
3. The relationship between levels is not symmetrical. Successful operation of the higher level requires the lower level to function effectively, but not vice versa. If a table is ground up into wood chips, it will no longer work as a table, but the molecular interactions are hardly altered.

For the whole-animal response curve in *Figure 1.1*, a possible subsystem model is given in *Figure 1.6*. This particular model consists of compartments or pools that

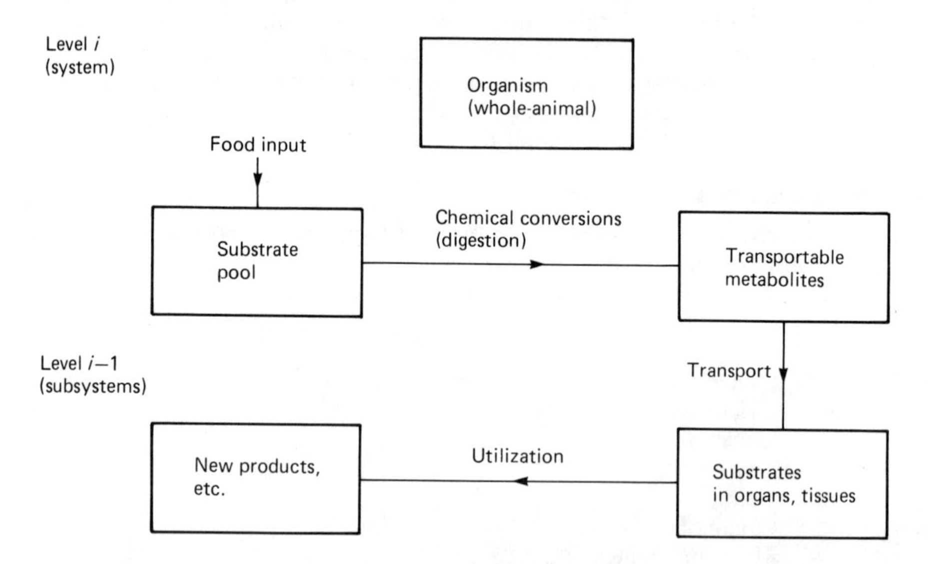


*Figure 1.6* Systems and subsystems: a possible model to 'explain' the response of *Figure 1.1*

are joined together by processes that occur with rates that can be evaluated. With a given problem, the appropriate subsystems tend to choose themselves, but a general aim is to select components of the system that are relatively self-contained, with strong interactions within each subsystem, and weak interactions between them. The relationship between the whole system and the assembly of subsystems exists, of course, in the real world, as well as in the world of the scientist's ideas, or in the mathematical model.

It is usually important that among the subsystems that may be considered, there are not too many of them that are only poorly described or poorly understood. A model is often only as good as its worst submodel. If the subsystems are too ill-defined, then in examining whole-system response and comparing this with observational data, it may be difficult to ascertain which aspects of the subsystems are giving rise to the whole-system response. The larger the model (size being measured in terms of the number of subsystems), the more critical one must be, especially in assessing current knowledge in relation to what one hopes to achieve. It is easy and can be very expensive to construct a large model which is of little value to the scientist (it does not give any new insights), or to the farm manager (it does not give sufficiently accurate predictions for decision-taking). Effective science is 'the art of the soluble', and it can be wasteful in many ways to attempt to solve problems for which current knowledge and techniques fall too far short.

## Models for research and models for management

In an earlier section and in *Figure 1.4*, a distinction was drawn between basic knowledge and applied knowledge. In either of these two categories, some of the knowledge can be represented by mathematical models. The question which is addressed in this section is this: Is there a difference between a management model (representing applied knowledge) and a research model (representing basic knowledge)?

Since applied knowledge can be viewed as a subset or derivative of basic knowledge, in terms of the finished product or finished model, the answer is *no*, there is no absolute difference between a research model and a management model. However, in terms of what modellers do, that is, the modelling process, there is a very definite difference. With a research model, especially with a model that is aiming to improve understanding, as much as if not more progress can be made with a model that is wrong—it fails to describe correctly what happens—as with a model that more correctly describes behaviour. In fact, part of the enjoyment of modelling is that one can speculate inexpensively and try something out. On the other hand, if a management model is to be successful (which means being used), it must give predictions or information that are clearly better, in some way, than existing practice.

All models, whether built for research or management purposes, are based on a mix of data, knowledge and conjecture. It is permissible, even desirable, for a research model to have a high proportion of conjecture. The management-oriented models must not only have a smaller proportion of guesswork, but should also be based on data and knowledge that are relatively secure. As with large models versus small models, unless objectives are clearly defined, there is a danger of doing work that is neither of value to the researcher nor of use as a management tool.

## Types of model

Mathematical models were briefly introduced in the second section of this chapter, and two simple models are described there. In this section, a more systematic consideration is given of the different types of model that one is likely to meet.

### Empirical and mechanistic models

An empirical model sets out principally to describe, whereas a mechanistic model attempts to give a description with understanding. Using the terminology of p. 9, the empiricist works only at a single level (say level $i$) in the organizational hierarchy, where he is constructing equations connecting level $i$ attributes alone. For example, equation (1.1) and *Figure 1.1* show an empirical model. The mechanist tries to describe the behaviour of the level $i$ attributes in terms of level $i - 1$ attributes, as in *Figure 1.6*. The two levels are connected by a process of analysis and resynthesis, accompanied by hypotheses or assumptions. The description of level $i - 1$ behaviour may be purely empirical, not containing any elements that refer to level $i - 2$, or it may be partly empirical and partly mechanistic, with some reference to $i - 2$ or lower levels. Every mechanistic model is eventually based on empiricism.

It is always possible to find an empirical model that gives a better fit to a given set of data than a mechanistic model. This arises because the empirical model has fewer constraints, whereas a mechanistic model can be very constrained by its assumptions, even if it contains more adjustable parameters.

### Static and dynamic models

A static model is a model that does not contain time as a variable, as in equation (1.1). Any specifically time-dependent components of behaviour of the system are ignored. Since all aspects of the world do change at some rate or other, a static model is always an approximation, although it may be a very good approximation, perhaps because the system is sufficiently close to equilibrium, or the time constant of the system is short compared with the time constant at which the environment is changing, so that the environment can be considered constant. It should be noted that an instantaneous picture or snapshot of a system cannot always be represented by a static model, since the system may be far from equilibrium (for example, a ball in mid-air is not easily handled by means of a static model).

An example of a static model is an analysis of the loads on the structural members of a farm building. Most dynamic models of crop growth incorporate what are basically static models of light interception by the crop canopy, although for crops whose leaves can respond to the light pattern, a dynamic model would be required.

A dynamic model, on the other hand, contains the time variable $t$ explicitly, as in equation (1.3), and also the differential equation form of equation (1.4). Most dynamic models are first expressed as differential equations, as in

$$\frac{dy}{dt} = f \tag{1.7}$$

where $y$ denotes an attribute or property of the system (possibly animal liveweight), $t$ is the time variable, and $f$ is some function, possibly of $y$, $t$ and some parameters.

Equation (1.7) must, of course, be integrated to give the actual behaviour of the system over time. This integration may be analytical, resulting in an expression as in equation (1.3), or, more frequently, only a numerical integration is possible, and this is usually performed on a computer.

It is possible that at some value of time, the rate of change of the system becomes zero, so that

$$\frac{dy}{dt} = 0 \quad \text{and} \quad f = 0$$

The equation $f = 0$ is now a static model.

### Deterministic and stochastic models

A deterministic model is one that makes definite predictions for quantities (such as animal liveweight, crop yield or rainfall), without any associated probability distribution. This might be acceptable in some instances, but for very variable quantities, such as rainfall, may be completely unsatisfactory. A stochastic model, on the other hand, contains some random elements or probability distributions within the model, so that, not only can it predict the expected value of a quantity such as animal liveweight $w$, $E(w)$, but also the variance of $w$, $V(w)$.

The greater the uncertainty in the behaviour of a system, the more important it may be to construct a stochastic treatment. Processes such as birth, migration, death, radioactive decay and chemical conversion tend to be random, and it is only where large numbers are involved that behaviour is apparently deterministic. Problems in epidemiology, population dynamics, biological control, etc. are sometimes approached with stochastic models.

Stochastic models tend to be technically difficult to handle and can quickly become very complex indeed. It is usually worthwhile attempting to formulate and solve the corresponding deterministic problem first, to see if this might yield the desired results with less effort, before attacking what may be a difficult stochastic problem.

## Contribution of models to research and to management: a summary

In the preceding sections of this chapter, the reader has been introduced to the basic ideas of modelling and to much of the terminology. In this section, a summary is given of the many different ways in which modelling can contribute to the broad spectrum of agricultural activities.

1. Hypotheses expressed in mathematics can provide a quantitative description and understanding of biological problems.
2. The requirement of a model for mathematical completeness can provide a conceptual framework which may help pinpoint areas where knowledge is lacking, and might stimulate new ideas and experimental approaches.
3. A mathematical model can be a good way of providing a recipe by which research knowledge is made available in an easy-to-use form by the farm manager.
4. The economic benefits of methods suggested by research can often be investigated and high-lighted by an agro-economic model, thus stimulating the adoption of improved methods of production.

5. Modelling may lead to less *ad hoc* experimentation, as models sometimes make it easier to design experiments to answer particular questions, or to discriminate between alternative mechanisms.
6. In a system with several components, a model provides a way of bringing together knowledge about the parts, to give a coherent view of the behaviour of the whole system.
7. Modelling can help provide strategic and tactical support to a research programme, motivating scientists and encouraging collaboration.
8. A model may provide a powerful means of summarizing data, and also a method for interpolation and cautious extrapolation.
9. Data are becoming increasingly precise, but also more expensive to obtain; a model can sometimes make more complete use of data.
10. The predictive power of a successful model may be used in many ways: priorities in research and development, management and planning. For instance, a model can be used to indicate the answers to 'what if ...?' questions: What would be the consequences of halving the maintenance requirements of ruminants on animal production? What would be the consequences for crop yields of halving the within-plant transport resistances?

## Exercises

### Exercise 1.1

The equation of the rectangular hyperbola in equation (1.1) can be written $y = ax/(b + x)$, with variables $x$ and $y$, and constants $a$ and $b$. Derive an expression for the slope, $dy/dx$. Verify that the initial slope $dy/dx(x = 0)$ and the asymptote $y(x \to \infty)$ are $a/b$ and $a$ respectively. Derive an expression for the rate of change of slope, $d^2y/dx^2$, and verify that for $x \geqslant 0$, the magnitude of this is maximum at $x = 0$.

### Exercise 1.2

The results of an experiment are well-fitted by a growth equation $W = at/(b + t)$, where $W$ denotes weight, $t$ is time, and $a$ and $b$ are constants. Differentiate to obtain a dynamic growth model of form $dW/dt = g(t)$, where $g$ denotes a function of $t$. Use the original growth equation to eliminate $t$ in favour of $W$ to obtain a dynamic growth model in the more useful form of $dW/dt = h(W)$, where $h$ denotes a function of $W$.

Chapter 2

# Techniques: dynamic deterministic modelling

## Introduction

An outline was given in the last chapter of possible contributions of a mathematical model to a research programme, and also of the different kinds of model that can be constructed. It is important to match the problem, the research objectives, and the type of modelling to be attempted. In much of agriculture, and perhaps especially in plant and animal physiology, our needs are often met by the construction of dynamic deterministic models, and therefore in this chapter, the techniques required for this type of modelling are presented. In addition, the use of a high-level computer simulation language to solve such models is illustrated with specific reference to CSMP, the Continuous System Modelling Program.

## Variables

In a dynamic model, variables are quantities that change with time, and they can be considered under four categories: state variables, rate variables, auxiliary variables and driving variables.

### State variables

A state variable is a quantity that defines, or helps to define, the state of the system at a given point in time. For example, in a plant model, leaf area $A$ and dry weight $W$ are typical state variables. As far as possible, one should choose state variables for the model that correspond to quantities or properties of the system that can be measured or are of particular interest.

Let the system be defined by $q$ state variables $X_1, X_2, X_3, \ldots, X_q$. These variables are independent, that is, it is not possible from a knowledge of the numerical values of some of them (say $X_2, X_3, \ldots, X_q$) to derive any of the others (say $X_1$). If leaf area $A$ and dry weight $W$ were related by

$$A = kW$$

where $k$ is a constant, then $A$ and $W$ would not be independent state variables, and one would use either $A$ or $W$ as the independent state variable.

The numerical values of the $q$ state variables $X_1, X_2, \ldots, X_q$ specify uniquely the state of the system at a given time $t$. The dynamic deterministic modelling problem largely consists of constructing differential equations which predict how these variables change with time; this is taken up in the next but one section (p. 18).

### Rate variables

A rate variable is a quantity that defines some process within the system at a given point in time. Rates always have dimensions of quantity per unit time; they cannot be measured instantaneously (as can a state variable), but only over an increment of time $\Delta t$. The processes in the system occur at given rates, and it is these rates which determine how the state variables $X_1, X_2, \ldots, X_q$ change with time. Examples are photosynthetic rate, respiration rate, substrate utilization rate, transport rates, protein degradation rates, and growth rates. It may be noted that all processes can be categorized as either conversion (usually chemical conversion in biology) or transport processes.

Given a knowledge of the state variables of the system, and the other parameters and constants of the model, the rates of the processes occurring can be calculated. Often the most important assumptions of the model are concerned with how the rates of processes depend on the state variables. Thus we can write

rate of process = function of state of system

and simple examples of this are in equations (1.4), (5.9) and (8.121) on p. 3, p. 78 and p. 168.

### Auxiliary variables

In most dynamic models, one wishes to obtain certain extra variables (additional to the state variables which alone define the system completely), usually to assist one in understanding the system and perhaps for comparison with measurement. For example, if leaf area $A$ and dry weight $W$ are independent state variables, then the leaf area ratio, $F_A$, can be obtained by

$$F_A = \frac{A}{W}$$

This we call an auxiliary variable, since it also varies with time, but it is only obtained for the modeller's convenience. Another example can be taken from a root–shoot partitioning model, where $W_r$ and $W_s$ are the dry weights of root and shoot, and are the two independent state variables of the model. One would almost certainly wish to know the total dry weight of the plant $W$, given by

$$W = W_r + W_s$$

$W$ is an auxiliary variable.

Sometimes, the auxiliary variables are rates. The specific or relative growth rate $R_W$ is obtained by

$$R_W = \frac{1}{W}\frac{dW}{dt}$$

where $t$ denotes time. The quantity $R_W$ is much used by biological scientists, and involves a variable (total dry weight $W$) and a rate (total growth rate $dW/dt$, a process).

### Driving variables

Driving variables are data inputs to a model which vary autonomously with time. The growth of animals, plants and many other organisms is driven by the environment. In controlled studies, this may sometimes be constant, although it is more common for the environment to vary with time. The quantities specifying the environment may have the nature of state variables (properties), for example, temperature and humidity; or alternatively they may be like rate variables (processes), such as nutritional inputs, rainfall, radiation and wind, with dimensions of quantity per unit time.

Symbolically we may write

$$E \equiv E(t)$$

to indicate that the environment, $E$, is a function of time, $t$. The environment is usually determined externally to the model, with the assumption that, although the environment affects the growth of an organism, the growth of the organism does not affect the environment appreciably.

## Parameters and constants

Parameters and constants are quantities appearing in the equations of a model that do not vary with time.

It is customary to describe as constants quantities that have reliably and accurately determined values, which remain the same when the experimental conditions are varied, or, for example, when the model is applied to different genotypes, or to different parts of an organism. Some typical constants are the relative molecular mass of glucose, the density of water, and the number of seconds in a day.

The term parameter is usually applied to quantities whose value is less certain, but are kept constant throughout a run of the model. Typical parameters might be a Michaelis–Menten constant in enzyme kinetics, photosynthetic efficiency, the maintenance respiration coefficient, and the resistance to transport of a sugar out of an organ. The values of parameters often vary with experimental conditions, the way in which the organisms are prepared (or grown), genotype, or other factors. Some or all of the parameters may be approximately known from work carried out at the next lower level in the organizational hierarchy (p. 9), which is the level at which the assumptions of the model are made. Sometimes, certain parameters are particularly poorly known, and it may be wished to fit the output of the model to experimental data by adjusting these parameters and looking at the goodness-of-fit; this process is often referred to as model calibration, and is discussed later in this chapter.

The symbol $P$ will be used to denote the set of parameters and constants:

$$P = \text{parameters} + \text{constants}$$

## Differential equations

The $q$ state variables $X_1, X_2, \ldots, X_q$ define the system at time $t$. Typically, the dynamic deterministic model consists of first-order differential equations which describe how these state variables change with time, and the same number of differential equations is required as there are state variables:

$$\frac{dX_1}{dt} = f_1(X_1, X_2, \ldots, X_q; P; E)$$

$$\frac{dX_2}{dt} = f_2(X_1, X_2, \ldots, X_q; P; E) \tag{2.1}$$

$$\cdots\cdots\cdots\cdots\cdots\cdots$$

$$\frac{dX_q}{dt} = f_q(X_1, X_2, \ldots, X_q; P; E)$$

$f_1, \ldots, f_q$ denote functions of the state variables, $X_1, \ldots, X_q$, of the parameters $P$, and of the environment $E$.

The fact that we have written

$$f_1(X_1, X_2, \ldots, X_q; P; E)$$

does not mean in practice that the function $f_1$ (and also the other functions $f_2, f_3, \ldots, f_q$) has to contain all the state variables, parameters and environmental variables. For example, consider a leaf growth model, which, in addition to variables such as area and fresh weight, has the quantities of sucrose $S_u$ and starch $S_t$ in the leaf as state variables. The differential equation for the amount of sucrose in the leaf might take the form

$$\frac{dS_u}{dt} = P_g + kS_t - (S_u - C)/r \tag{2.2}$$

where $k$, $C$ and $r$ are parameters, $P_g$ is the photosynthetic rate of the leaf (this depends on the light environment which may be constant or represented by a driving variable), the second term denotes starch breakdown to sucrose, and the last term transport of sucrose out of the leaf. It is usual for each differential equation to contain only a few of the state variables, parameters and environmental variables. If the response of leaf growth to temperature is of particular interest, then in equation (2.2), one could make $k$, the starch degradation parameter, temperature-dependent, possibly using the Arrhenius equation (p. 155).

Each function $f_i$ on the right side of equations (2.1) will generally consist of several terms, each of which is a process with a rate, as illustrated in equation (2.2), where there are the three processes of photosynthesis, degradation and transport. Again, note that the instantaneous rate of each process can always be calculated from the values of the state variables, parameters and environmental quantities.

Since the system is defined by the $q$ state variables $X_1, X_2, \ldots, X_q$ it is sometimes helpful to think of a $q$-dimensional space, called the system space, and the state of the system is defined by the position of the system point in this space. Equations (2.1) then determine the velocity of the system point as it moves through this space, and the system point will trace out a trajectory through the system space. These pictorial images are useful for developing ideas about bifurcation (alternative paths of development) and catastrophe (rapid shift from one type of trajectory to another); a very readable introduction to catastrophe theory and bifurcation is given in Wilson and Kirby (1980, Chapter 10).

**Explicit time dependence**

In equations (2.1), the time variable $t$ does not appear explicitly on the right side, apart from its possible presence in a driving function in $E$ which denotes the quantities specifying the environment. Discounting the latter, the presence of $t$ on the right side of the equations specifying the model is often an *ad hoc* device which can lead to difficulties and is best avoided if possible. Essentially, a system cannot 'know' what the time is; it is defined only by its state variables, although one or more of these may function as an internal clock. Explicit use of the time variable may put an external constraint on the dynamics of the system which is unacceptable. Sometimes, the introduction of an extra state variable (or variables) can help in this respect. For example, the Gompertz growth equation may be written (p. 82)

$$\frac{dW}{dt} = \mu_0 W e^{-Dt} \quad \text{with} \quad W = W_0 \quad \text{at } t = 0$$

where $W$ is the state variable (usually denoting weight) with initial value $W_0$, and $\mu_0$ and $D$ are parameters. This single equation can be replaced by two equations (two state variables)

$$\frac{dW}{dt} = \mu W \quad \text{and} \quad \frac{d\mu}{dt} = -D\mu \quad \text{with} \quad W = W_0 \quad \text{and} \quad \mu = \mu_0 \quad \text{at } t = 0$$

A reformulation of the problem in this manner might lead to an improved understanding of the nature of the system. Thus, equations of the general type

$$\frac{dX}{dt} = g(X; t; P; E)$$

where $g$ denotes a function containing $t$ explicitly, should if possible be replaced by equations of the type in equations (2.1) (Exercise 2.1).

**Memory functions and delays**

Equations (2.1) can be represented more simply as

$$\frac{dX}{dt} = f(X)$$

Sometimes it is also wished to make the current value of $dX/dt$ depend upon the value of $X$ time $\tau$ ago, $X(t - \tau)$, so that the above equation is replaced by

$$\frac{dX}{dt} = f[X(t), X(t - \tau)]$$

This is called a discrete lag. It may be more realistic, but is much more complicated to use a distributed lag by writing

$$\frac{dX}{dt} = f(X, Y)$$

where

$$Y = \int_0^\infty w(\tau)X(t - \tau)\,\mathrm{d}\tau$$

$w(\tau)$ is a normalized weighting function. For a discrete lag, $w(\tau)$ is simply a very sharp spike of unit area occurring at a single value of $\tau$. Such problems and their consequences are more extensively discussed on pp. 183–185, where some examples are given.

Problems involving delayed variables can always be redescribed using current variables by expanding the set of state variables, although a discrete delay requires an infinite set of ordinary differential equations. The purist can correctly maintain that the use of lagged variables (as with the explicit time dependence discussed in the last section) implies an inadequate definition of the problem, which should be reformulated. The pragmatic modeller may decide to avoid the ensuing complexities and to use lagged variables, but is well aware that alternative approaches do exist.

## Numerical integration

Given the parameter values $P$, the environment $E$ and the initial condition $X_i(t = 0)$, $i = 1, 2, ..., q$, the model can be solved by integrating equations (2.1) to give a table of predicted values of the form:

| | | *State variables* |
|---|---|---|
| | | $X_1, X_2, ......, X_q$ |
| | 0 | |
| | 1 | |
| *Time* | 2 | |
| | 3 | |
| | 4 | |
| | . | |
| | . | |
| | . | |

It is unusual that the differential equations in equations (2.1) can be solved analytically, and almost always it is necessary to use numerical techniques to obtain their solutions. It is the intention here to introduce the reader to a few of the simpler methods for numerical integration and to point out some of the possible pitfalls. A more comprehensive introduction to numerical integration can be found in any standard numerical analysis textbook (for example, Fox and Myers, 1968; Gear, 1971).

For ease of exposition, equations (2.1) are simplified to a single-state-variable problem with the state variable $x$, and the parameters $P$ and the environment $E$ are no longer explicitly mentioned, giving

$$\frac{\mathrm{d}x}{\mathrm{d}t} = f(x,t) \tag{2.3}$$

where $f$ denotes a function of $x$ and $t$. The numerical integration methods to be outlined apply in just the same way to the $q$-state-variable situation. The problem of integration may be indicated symbolically by

$$x(t) \xrightarrow[\text{equation (2.3)}]{\text{using}} x(t+\Delta t)$$

which means, given a value of $x$ at time $t$, and given the function $f$ of equation (2.3), how does one calculate the value of $x$ at time $t + \Delta t$, where $\Delta t$ denotes an incremental change in the time variable $t$? This prescription can then be applied iteratively starting from $x$ at $t = 0$ to any value of time $t$.

**Euler's method**

This is the simplest method, and can be derived by considering the definition of $\mathrm{d}x/\mathrm{d}t$. By definition,

$$\frac{\mathrm{d}x}{\mathrm{d}t} = \lim_{\Delta t \to 0} \left[ \frac{x(t+\Delta t) - x(t)}{\Delta t} \right] \tag{2.4}$$

Substituting equation (2.4) into (2.3) gives

$$\lim_{\Delta t \to 0} \left[ \frac{x(t+\Delta t) - x(t)}{\Delta t} \right] = f[x(t), t]$$

Therefore, for small values of $\Delta t$, the following approximation holds

$$\frac{x(t+\Delta t) - x(t)}{\Delta t} \approx f[x(t), t]$$

Rearranging this equation gives Euler's formula

$$x(t+\Delta t) = x(t) + \Delta t f[x(t), t] \tag{2.5}$$

Values of $x$ may be determined at time $t = 0, \Delta t, 2\Delta t, 3\Delta t$, etc. by using equation (2.5) iteratively.

Euler's method is a linear or first-order method, meaning that the error is of order $(\Delta t)^2$. This can be shown by considering a Taylor series expansion of $x(t + \Delta t)$, namely

$$x(t+\Delta t) = x(t) + \Delta t \frac{\mathrm{d}x}{\mathrm{d}t} + \frac{1}{2!}(\Delta t)^2 \frac{\mathrm{d}^2x}{\mathrm{d}t^2} + \frac{1}{3!}(\Delta t)^3 \frac{\mathrm{d}^3x}{\mathrm{d}t^3} + \text{higher-order terms} \tag{2.6}$$

By comparing equations (2.5) and (2.6), and as $\mathrm{d}x/\mathrm{d}t = f(x,t)$, it can be seen that the error $\epsilon$ in equation (2.5) is

$$\epsilon = \frac{1}{2}(\Delta t)^2 \frac{\mathrm{d}^2x}{\mathrm{d}t^2} + \text{higher-order terms}$$

This error is known as the truncation error. For a straight line which has no curvature, the second and higher derivatives are zero, and Euler's method gives an exact result. Note that halving the interval $\Delta t$ reduces the error by a factor of four.

Despite the availability of many more sophisticated methods, this method is extremely useful in practice. This is because of its inherent simplicity, as it is easily checked out in every detail and one can be sure of just what is going on in the calculation.

Before going on to second- and higher-order methods, we show how unstable oscillations can arise, which would cause the failure of a computer program. Suppose equation (2.3) takes the form

$$\frac{dx}{dt} = -kx \tag{2.7}$$

which, by analytical integration, has the solution

$$x = x(t = 0)e^{-kt} \tag{2.8}$$

In this exact solution, $x$ approaches zero asymptotically. Now generate a solution to equation (2.7) using Euler's method of equation (2.5). With $x(t = 0) = 1$, $\Delta t = 1$, and $k = ½$, we obtain

| $t$ | 0 | 1 | 2 | 3 | 4 | ... |
|---|---|---|---|---|---|---|
| $x$ | 1 | $\frac{1}{2}$ | $\frac{1}{4}$ | $\frac{1}{8}$ | $\frac{1}{16}$ | ... |

However, suppose $k = 3$. Now the application of equation (2.5) gives

| $t$ | 0 | 1 | 2 | 3 | 4 | ... |
|---|---|---|---|---|---|---|
| $x$ | 1 | –2 | 4 | –8 | 16 | ... |

For this example, it is easily shown that

$k\Delta t < 1$ gives asymptotic stability
$1 < k\Delta t < 2$ gives oscillations of decreasing amplitude

and

$2 < k\Delta t$ gives oscillations of increasing amplitude

The smaller $\Delta t$, the greater the stability of the numerical solution, and the smaller the truncation error $\epsilon$, although the computing time increases and there is an increasing possibility of rounding errors due to the fact that the computer can only store quantities to a certain number of decimal digits. The $\Delta t$ used will be a compromise between these conflicting criteria. Generally, if it is wished to reduce the truncation error, it is more efficient to use a higher-order method, than to reduce the size of $\Delta t$. However, some problems cannot be easily solved with higher-order methods as they require the calculation of the derivatives within the time interval (Exercise 2.2).

**Second-order methods: the trapezoidal method**

Euler's method (equation (2.5)) can be regarded as the first two terms of a Taylor's series (cf. equation (2.6)). In determining $x(t + \Delta t)$, it is usually not practicable to calculate the higher derivatives of the Taylor's series expansion, unless these are particularly simple, which is seldom the case. Essentially, the higher-order methods are constructed by evaluating the first derivatives at different time points within the integration interval, and calculating the higher derivatives from the way in which the first derivatives change.

Consider the second derivative in equation (2.6). By definition,

$$\frac{d^2x}{dt^2} = \lim_{\Delta t \to 0} \left[ \frac{\dot{x}(t + \Delta t) - \dot{x}(t)}{\Delta t} \right]$$

where $\dot{x} \equiv dx/dt$. Therefore, for small $\Delta t$

$$\frac{d^2x}{dt^2} \approx \frac{\dot{x}(t + \Delta t) - \dot{x}(t)}{\Delta t} \tag{2.9}$$

As $dx/dt = f(x,t)$, equation (2.9) becomes

$$\frac{d^2x}{dt^2} \approx \frac{f[x(t + \Delta t), t + \Delta t] - f(x,t)}{\Delta t} \tag{2.10}$$

Euler's formula (equation (2.5)) gives

$$x(t + \Delta t) = x(t) + \Delta t f(x,t) = x_1$$

Substituting $x_1$ for $x(t + \Delta t)$ in equation (2.10), we have

$$\frac{d^2x}{dt^2} \approx \frac{f(x_1, t + \Delta t) - f(x,t)}{\Delta t} \tag{2.11}$$

The Taylor's series expansion for $x(t + \Delta t)$ is given by equation (2.6), namely

$$x(t + \Delta t) = x(t) + \Delta t \frac{dx}{dt} + \frac{1}{2}(\Delta t)^2 \frac{d^2x}{dt^2} + \text{higher-order terms}$$

Using equations (2.3) and (2.11) to substitute for $dx/dt$ and $d^2x/dt^2$ in the above gives

$$x(t + \Delta t) \approx x(t) + \Delta t f(x,t) + \tfrac{1}{2}(\Delta t)^2 \left[ \frac{f(x_1, t + \Delta t) - f(x,t)}{\Delta t} \right]$$
$$+ \text{higher-order terms}$$

Ignoring higher-order terms and simplifying, gives the well-known *trapezoidal method*, with

$$x(t + \Delta t) = x(t) + \tfrac{1}{2}\Delta t[f(x_1, t + \Delta t) + f(x,t)] \tag{2.12}$$

The trapezoidal method is an example of a predictor–corrector method, as Euler's method is used to predict $x_1 = x(t + \Delta t)$ and then equation (2.12) is used to improve or 'correct' this estimate of $x(t + \Delta t)$. This is a second-order method, and the error is now third order and is given by

$$\epsilon = \tfrac{1}{6}(\Delta t)^3 \frac{d^3x}{dt^3} + \text{higher-order terms}$$

so that the accuracy is higher than with Euler's method (Exercise 2.3).

There are several other second-order methods, in which the error may be smaller than with the trapezoidal method, although the latter has the advantage of simplicity.

The agricultural modeller sometimes has a special problem, which can result in his choosing Euler's method in preference to the trapezoidal or higher-order methods, all of which involve evaluating the first derivatives in equations (2.1) more than once per time step. This arises because equations (2.1) contain environmental variables (denoted by $E$); these are quantities such as temperature, radiation or carbon dioxide concentration, which are not available as continuous variables, but only as averages over some time interval. For example, the mean temperature may be 18 °C on one day and 20 °C on the next. These discontinuities in the environment cause discontinuities in the first differentials of equations (2.1). It may then not be clear how one should calculate the first derivatives within or at the two ends of a time interval. The practical solution that most modellers employ is to measure all the environmental variables with respect to a particular time interval (or refer them to a common time interval), and then use this time interval for integration with Euler's method.

### Higher-order methods

Many higher-order methods exist, such as Simpson's rule, Runge-Kutta methods and Milne's method. Invariably, these involve computing the first derivatives at several points within the time interval, which can lead to the difficulties discussed earlier. However, these methods can perform very well for problems that are mathematically well-behaved and if any driving variables in the problem are given as continuous functions.

As an example of a frequently used fixed-step method, we outline the classical fourth-order Runge-Kutta method which takes the form

$$x + \Delta x = x + \tfrac{1}{6}(\Delta x_1 + 2\Delta x_2 + 2\Delta x_3 + \Delta x_4)$$

where

$$\Delta x_1 = \Delta t f(x, t)$$

$$\Delta x_2 = \Delta t f(x + \tfrac{1}{2}\Delta x_1, t + \tfrac{1}{2}\Delta t)$$

$$\Delta x_3 = \Delta t f(x + \tfrac{1}{2}\Delta x_2, t + \tfrac{1}{2}\Delta t)$$

and

$$\Delta x_4 = \Delta t f(x + \Delta x_3, t + \Delta t)$$

The error term, $\epsilon$, is of order $(\Delta t)^5$.

## Stiff equations

In equation (2.7), the constant $k$ is known as a rate constant and has dimensions of $(\text{time})^{-1}$, so that $1/k$ has the dimensions of time. $1/k$ is often referred to as the time constant or relaxation time of the equation. In equation (2.8), over a time interval of $1/k$, $x$ falls to $1/\mathrm{e}$ of its original value.

Stiff equations may be encountered in modelling if the rate constants (or time constants) in the differential equations differ greatly. These equations then have solutions which have exponents which are greatly different. In many biological models, there may be processes that occur in a few seconds, or even much less, and others that take several days. An integration interval that suits the slow processes is likely to give rise to unstable oscillations of increasing amplitude with the fast processes. A short interval that gives stable integration with the fast processes may be very expensive to run over the time period of interest. A plant or animal model may typically represent conversions between small molecules (milliseconds or less), macromolecules (minutes), the construction of new tissues (hours and days), and the production of new organs (days and weeks). It may prove difficult to restrict the range of time constants in a model without discarding physiologically significant components.

The chemical reaction

$$z \xrightarrow{k} \text{products} \tag{2.13}$$

results in equation (2.7). Suppose (2.13) is replaced by

$$y \underset{k_2}{\overset{k_1}{\rightleftharpoons}} x \xrightarrow{k_3} \text{products} \tag{2.14}$$

where $k_1$, $k_2$ and $k_3$ are rate constants. Expression (2.14) represents a two-state variable problem ($x$ and $y$) with differential equations

$$\frac{dx}{dt} = -(k_2 + k_3)x + k_1 y$$

$$\frac{dy}{dt} = -k_1 y + k_2 x$$

Consider the case where $k_3 = 1$, and $k_1 = k_2 = 1000$, so that $x$ and $y$ are in rapid equilibrium and $x$ is slowly converted to other products. The solutions, starting from several initial values, are illustrated in *Figure 2.1*. Note that the part of the solution of interest is usually that part which depends on the slower processes.

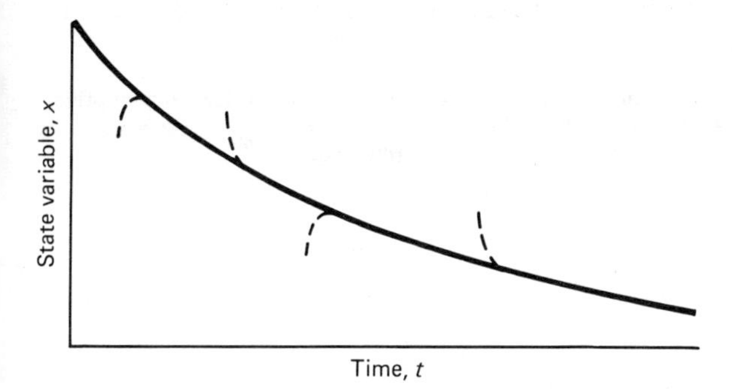


*Figure 2.1* The stiff differential equation problem. The continuous line shows the slowly decaying process, represented by $k_3$ of equation (2.14). The dashed lines show the rapid process, $k_1$ and $k_2$ of equation (2.14)

There are three ways in which one can attempt to solve the stiff differential equation problem:

1. Rapidly exchanging pools can be brought together. In the example above, (2.14) can be considered as equivalent to (2.13) with

$$z(t = 0) = x(t = 0) + y(t = 0) \quad \text{and} \quad k = \left(\frac{k_1}{k_1 + k_2}\right)k_3$$

2. Rate constants may sometimes have very little effect on the solutions. For instance, in the above problem, the slow solutions are much the same for $k_1$ and $k_2$ values of 100, 1000 or 10 000.
3. In some problems, fluxes may be of much more interest than absolute pool sizes. It may be possible to slow down the fast rates of change by 10 or 100 by increasing a pool size by 10 or 100, by increasing a volume, but maintaining concentrations and active biochemical volumes (where the reactions occur) the same.

In recent years, a number of algorithms have been developed to cope with stiff problems (Gear, 1971). These are available at many computer installations, and as a part of some simulation languages; for example, a variable-step integration method called STIFF is available in CSMP. The details of their operation are not readily accessible to the non-specialist and will not be discussed here. An understanding of the nature of stiff equations and how they arise should enable the modeller to tackle the computing problem more effectively. Stiff equation algorithms are of no benefit if there are high-frequency (fast) components in the driving functions.

## Choice of method

In choosing an integration method, the modeller should be concerned with obtaining sufficient accuracy without using excessive computer time. The various integration methods differ quite markedly in complexity and consequently require different amounts of computing time. We have compared five fixed-step methods by running a CSMP program on an ICL 4-72 computer in which integration for 100 time units was performed. The results obtained are shown in *Table 2.1*. It can be seen that, for this problem and this machine, for all but Euler's method, rounding errors dominate for $\Delta t < 0.01$. The best error/execution time ratio is obtained using the highest-order method with the longest integration interval.

**TABLE 2.1. Execution times (in time units of about 2 s) and % integration error for different integration methods and integration intervals $\Delta t$ using CSMP and a 32-bit wordlength. $dx/dt = x$, $x(t = 0) = 1$; integration until $t = 100$. % error = $100[x(t = 100, \text{numerical}) - \exp(100)]/\exp(100)$.**

| | *Integration method* | | | | | | | | | |
|---|---|---|---|---|---|---|---|---|---|---|
| *Integration interval, $\Delta t$* | *Euler* | | *Adams second-order* | | *Trapezoidal* | | *Simpson's rule* | | *Fixed-step fourth-order Runge-Kutta* | |
| | *Time* | *Error* | *Time* | *Error* | *Time* | *Error* | *Time* | *Error* | *Time* | *Error* |
| 0.1 | 1 | −99 | 1 | −32 | 1 | −14 | 2 | −11 | 2 | −0.028 |
| 0.01 | 3 | −39 | 3 | −0.6 | 4 | −0.33 | 5 | −0.29 | 7 | −0.17 |
| 0.001 | 17 | −6 | 19 | −1.6 | 29 | −1.6 | 43 | −1.6 | 58 | −1.6 |
| 0.0001 | 161 | −15 | 177 | −15 | 288 | −15 | 414 | −14.9 | 566 | −14.9 |

It is easier than is often realized to write a computer program that will compile and execute, and yet whose results are nonsense. It is essential to be a sceptic and to try and ensure, to a reasonable degree, that the results are independent of method and of integration interval. This need not be too onerous a task.

Euler's method is easily undervalued, and we advocate that all models should be run initially using this method with a range of integration intervals $\Delta t$, and that the model should be thoroughly checked out. If $\Delta t = 1$ is the interval of interest, then using $\Delta t$'s in the range 0.005 to 1 is often sufficient. The compromise between truncation and rounding errors is shown in *Table 2.1* and *Figure 2.2*, the aim being to work roughly in the minimum region.

Some of the problems of using higher-order methods have already been mentioned, and sometimes, with discontinuous environmental variables, it may only be possible to use Euler's method. However, this can be an advantage since Euler's method does not have the potential of the higher-order methods for generating spurious results.

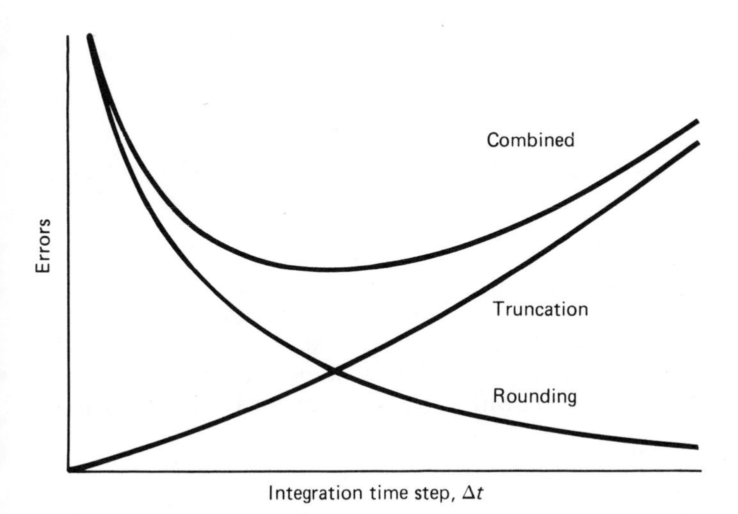


*Figure 2.2* Truncation and rounding errors in numerical integration by computer—an illustration of the compromise involved in choosing a time interval $\Delta t$ for integration

The higher-order methods can, each in its own way, exhibit unstable behaviour, although again, this is usually dependent on the interval $\Delta t$ and can be checked by varying $\Delta t$. However, if the problem and the computing facilities allow it without too much difficulty, it is desirable to try several methods and time intervals, and to check the results for consistency.

## A fitting procedure

Fitting a model usually means adjusting the parameter values ($P$ of equations (2.1)) and initial conditions ($X_i(t = 0)$, $i = 1, 2, \ldots, q$) so that the model behaves more like the real world, but retaining the same structure and basic equations (equations (2.1)). To some purists, this process, which is sometimes described as 'tuning' or 'calibrating' the model, is anathema; they feel that the parameters should be known from independent investigations, possibly investigations at the level of the assumptions. This view may neglect the empirical content and practical objectives which are associated with many models. Fitting can be an important part of model evaluation, sometimes giving useful insights. In this section, a simple fitting procedure is outlined, which, while it does have a number of shortcomings, has been found to be of practical value.

**Experimental data**

Consider the case where a single attribute $y$ of the real system is measured at $m$ time points

$$t_1, t_2, \ldots, t_m$$

to give the set of numbers

$$y_1, y_2, \ldots, y_m$$

Let us further assume that the values $y_i$, $i = 1, 2, \ldots, m$ are means over any replicates that were taken.

**Predicted data**

The state variables of the model are $X_i$, $i = 1, 2, \ldots, q$, and the model predicts at each time $t_j$

$$X_i(t_j; P; E) \quad (i = 1, 2, \ldots, q) \tag{2.15}$$

In equations (2.15), the notation of equations (2.1) is retained to indicate the dependence of the $X_i$ on time ($t$), the parameters and initial conditions ($P$) and the environment ($E$). A state variable of equations (2.15) might correspond directly with the experimentally measured attribute $y$, or it might be necessary to derive an auxiliary variable that does correspond with $y$. These predicted values are denoted by

$$Y_1, Y_2, \ldots, Y_m$$

for the $m$ time points $t_i$, $i = 1, 2, \ldots, m$. For a given model, these $Y$-values depend on the parameters and initial conditions $P$ and on the environment $E$. The problem is kept simple by assuming that only a single environment has been considered. If there were two environmental treatments $E_1$ and $E_2$, then there would be two sets of measured numbers of the type $y_i$, $i = 1, 2, \ldots, m$ and also two corresponding sets of predicted values of type $Y_i$, $i = 1, 2, \ldots, m$. Since the model is deterministic, the $Y_i$ are given without any associated probability distribution.

**Calculation of a residual**

A residual $r_i$ can be calculated according to

$$r_i = y_i - Y_i \quad \text{or} \quad r_i = \ln(y_i/Y_i) \tag{2.16}$$

or using some other appropriate measure, and these can be summed to give a residual sum of squares $R$ by

$$R = \sum_{i=1}^{m} g_i r_i^2 \tag{2.17}$$

using a weighting factor $g_i$ if required. The residual sum of squares $R$ is a measure of the lack of fit of the model, and it depends on the parameter values, so that we can write

$$R \equiv R(P)$$

Consider the simplified case where the model has just a single parameter $P$. Best fit is obtained by adjusting $P$ so that $R$ is a minimum, giving

$$\frac{dR}{dP} = 0 \quad \text{and} \quad \frac{d^2R}{dP^2} > 0$$

In *Figure 2.3*, examples of a sensitive parameter and an insensitive parameter are illustrated. In curve S, the value of $d^2R/dP^2$ is much larger, and this serves to define the value of $P$ for best fit more closely. It is desirable that $R$ should be reasonably sensitive to all the parameters of the model, and if there are some parameters to which $R$ is completely insensitive, this can indicate areas in which the model might be simplified.

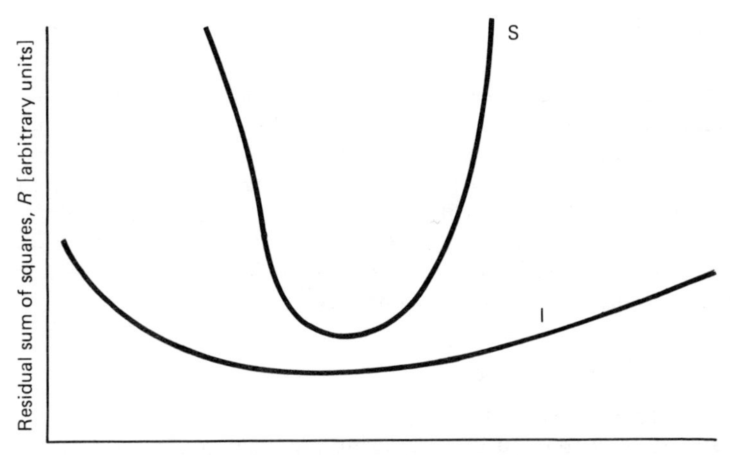


*Figure 2.3* Sensitive (S) and insensitive (I) model parameters. In curve S, $R$ varies rapidly with parameter $P$, whose value is therefore critical to the realism of the model. In curve I, the value of $P$ has little effect on $R$; if $P$ can be obained by experiment or calculation, only an approximate value is required; fitting the model is not able to define $P$ closely

Ignoring truncation and rounding errors (which are always present but are usually insignificant), the predicted values $Y_i = 1, 2, \ldots, m$ are computed without error. However, the experimental data $y_i$, $i = 1, 2, \ldots, m$ are subject to error, and this puts a lower limit to the value of $R$ that can be achieved by parameter adjustment. The residual sum of squares $R$ can be divided into two components

$$R = R_\ell + R_e \tag{2.18}$$

where $R_\ell$ is the part of the residual due to the lack of fit on the model, and $R_e$ is due to error in the experimental data. $R_e$ has an expected value of

$$R_e = (m - n)\sigma^2$$

where $m$ is the number of data points, $n$ is the number of parameters in $P$, and $\sigma^2$ is the error variance. The parameters $P$ only affect $R_\ell$, but if $R_e$ is very large (adding a large constant term to the curves in *Figure 2.3*), then $R$ may be rendered much less sensitive to the $P$ values. If the experiment resulting in the data $y_1, y_2, \ldots, y_m$ has replication, then $R_e$, the error term, may be known. Otherwise, an upper limit to

$R_e$ (and therefore $\sigma^2$) is given by the minimum value of $R$ obtained by adjusting the parameters $P$.

Suppose that instead of the single set of data $Y_i$, $i = 1, 2, \ldots, m$, we have two sets, perhaps plant dry weight and leaf area, giving

$$W_i \quad i = 1, \ldots, m_W \quad \text{and} \quad A_i \quad i = 1, \ldots, m_A \tag{2.19}$$

Following the recipe of equations (2.16) and (2.17), two residual sum of squares can be calculated, that with respect to dry weight $R_W$, and that with respect to leaf area $R_A$. It is necessary to combine these in order to carry out parameter adjustment; this can be achieved by means of

$$R = \frac{R_W}{\sigma_W^2} + \frac{R_A}{\sigma_A^2} \tag{2.20}$$

where the $\sigma_W^2$ and $\sigma_A^2$ are the respective error variances. The parameter adjustment can now be done with respect to the combined residual $R$. If the error variances are not known, then clearly equation (2.20) cannot be constructed, although for fitting, it is still necessary to combine the residuals in some way. If the second of equations (2.16) is used to calculate the residuals $r_i$, then $R_W$ and $R_A$ are independent of the dimensions of $W$ or $A$, and of any scaling factors. If it is further assumed that dry weight $W$ and area $A$ have the same coefficient of variation, often a reasonable assumption, then the combined residual can be written

$$R = R_W + R_A \tag{2.21}$$

and the parameters can be adjusted so that equation (2.21) is minimized.

The goodness-of-fit can be estimated by comparing the residual sum of squares due to lack of fit of the model ($R_\ell$ of equation (2.18)) with the error residual sum of squares ($R_e$) using the $F$-test (*see* Glossary). An estimate of the error term is only available if a replicated experiment has been performed; if this is not the case, only a qualitative and subjective assessment of the fit of the model to the data may be possible.

**Confidence intervals for fitted parameters**

The dependence of the residual sum of squares, $R$ (or some other measure of the lack of fit), on the $n$ parameters $P_i$, $i = 1, 2, \ldots, n$ can be expressed by

$$R = R(P_1, P_2, \ldots, P_n)$$

Computer methods are generally used to search for the minimum of $R$ with respect to the $P_i$. The use of these computer methods is often explained using a terminology that might be unfamiliar to the agricultural modeller. The function $R$, whose minimum one is trying to find, is known as the objective function. The gradient vector is the set of first partial derivatives (*see* Glossary) of $R$ with respect to the parameters $P_i$, $i = 1, 2, \ldots, n$, namely $\partial R/\partial P_1$, $\partial R/\partial P_2$, $\ldots$, $\partial R/\partial P_n$. At a minimum,

$$\frac{\partial R}{\partial P_1} = \frac{\partial R}{\partial P_2} = \ldots = \frac{\partial R}{\partial P_n} = 0$$

The matrix of second partial derivatives of the objective function is called the Hessian matrix, whose elements take the form

$$H_{ij} = \frac{\partial^2 R}{\partial P_i \partial P_j}$$

At the minimum, all the eigenvalues (*see* Glossary) of the Hessian matrix must be positive. Essentially, this means that the second derivatives with respect to the parameters are positive, and in whatever direction one moves from the point defined by $\partial R/\partial P_i = 0$, $i = 1, 2, \ldots, n$, $R$ increases.

Let the matrix $\boldsymbol{G}$ with elements $G_{ij}$ be the inverse of the Hessian matrix $\boldsymbol{H}$ with elements $H_{ij}$. This can be represented symbolically by

$$\boldsymbol{G} = \boldsymbol{H}^{-1}$$

although to actually obtain the inverse matrix, it is usual to use a numerical computer procedure.

Conventionally, the number of degrees of freedom is denoted by $\nu$. From $m$ data points and $n$ parameters, $\nu$ is given by

$$\nu = m - n \tag{2.22}$$

although if there are two data sets as in equations (2.19), then $\nu$ is given by

$$\nu = m_W + m_A - n$$

An unbiased estimate of the variance of parameter $P_i$ is

$$V(P_i) = \frac{R}{\nu} G_{ij} \tag{2.23}$$

and an unbiased estimate of the covariance of parameters $P_i$ and $P_j$ is

$$C(P_i, P_j) = \frac{R}{\nu} G_{ii}$$

If $P_i^*$ is the actual (true) value of the parameter $P_i$, then the $100(1 - \beta)$ per cent confidence interval for $P_i^*$ is

$$P_i \pm [V(P_i)]^{1/2} t_{\beta,\nu}$$

where $t_{\beta,\nu}$ is the $100\beta$ percentage point of the $t$-distribution with $\nu$ degrees of freedom (*see* Glossary).

### A computer-oriented summary

This process of model solution and fitting can be summarized in six procedures.

I MODEL

The structure of the model is contained in

procedure *model* $(P, E; Y)$

Specification of the parameters and initial conditions ($P$), and the environment ($E$) permits calculation of predicted values ($Y$). This module describes the equations of the model (equations (2.1)) and does the iterative integration (usually by calling standard numerical integration procedures) and produces a table of predicted values (p. 20).

II RESIDUAL

The module

procedure *residual* $(P, y, model; R)$

computes the residual or objective function, $R$, which is the criterion for fitting. One must specify to *residual* the numerical values of the parameters $P$ and the experimental data points $y$; *residual* will call *model* to evaluate the predicted values $Y$ for the given $P$, and then use $y$ and $Y$ to calculate the residual sum of squares $R$ (equation (2.17)).

III MINIMIZE

Most computer installations provide a range of methods for doing unconstrained optimization. Basically, these are of the type

procedure *minimize* $(P_{\text{initial}}, residual; P_{\text{optimum}}, R_{\text{minimum}})$

The residual $R$ depends on the parameter values $P$. One must supply *minimize* with an initial guess at the parameters, $P_{\text{initial}}$; with these starting values, *minimize* will vary the parameters, calling *residual* recurrently to see if the residual $R$ is increasing or decreasing as the parameters are changed. It will search for a minimum value of the residual, $R_{\text{minimum}}$, corresponding to an optimum (best fit) set of parameter values $P_{\text{optimum}}$.

IV DIFFERENTIATE TWICE

This procedure

procedure *differentiate twice* $(P_{\text{optimum}}, residual; \boldsymbol{H})$

performs numerical differentiation. It takes the parameters $P_{\text{optimum}}$, it varies them by small amounts, and by calling *residual* as it changes the parameters, it can calculate the rates of change of the residual $R$. It thus obtains the elements of the Hessian matrix, $\partial^2 R/\partial P_i \partial P_j$. Sometimes III and IV are conveniently carried out by a single procedure.

V INVERT

This is a straightforward matrix inversion routine, so that

procedure *invert* $(\boldsymbol{H}; \boldsymbol{G})$

takes the Hessian matrix $\boldsymbol{H}$ calculated by *differentiate twice*, and inverts it to give the matrix $\boldsymbol{G}$.

VI CONFIDENCE INTERVALS

procedure *confidence intervals* $(v, \boldsymbol{G}, R, P; P_{\pm})$

carries out the calculations described in the final paragraph of the previous section. It is necessary to provide *confidence intervals* with the degrees of freedom $v$, the matrix $\boldsymbol{G}$ (obtained by *invert*), the residual sum of squares $R$ at the point of best fit, which is assumed is a conservative estimate of the error sum of squares (some of $R$

might in fact be due to lack of fit rather than error; see equation (2.18)), and the value of the parameter $P$. The upper and lower confidence limits are returned in $P_{\pm}$.

## A programmed example

High-level computer simulation languages can be ideal for solving compartmental models of the type described in this chapter, which comprise a system of differential equations. In this section, the power and usefulness of such a language are illustrated by programming a modified logistic growth function

$$\frac{\mathrm{d}W}{\mathrm{d}t} = \mu W\left(1 - \frac{W^2}{W_f^2}\right) \tag{2.24}$$

in CSMP. A derivation of the logistic function itself is given in Chapter 5 (p. 80).

CSMP is FORTRAN-based and is application-orientated, which allows the modeller to concentrate on the problem in hand rather than on the mechanics of its solution. CSMP is a 'non-procedural' language, which means that the program statements can be written in any order (with some restrictions). The CSMP translator contains a sorting algorithm, which puts statements in the correct order for execution. This non-procedural facility allows the modeller to order his statements to give a meaningful biological definition of the problem, and is of great value. Basically, a CSMP program comprises three segments: INITIAL, DYNAMIC and TERMINAL. The INITIAL segment appears first, is optional and is used for calculations that only have to be performed once at time zero during the

```
TITLE   AN ANIMAL GROWTH MODEL
*  CSMP PROGRAM OF THE MODIFIED LOGISTIC GROWTH FUNCTION,
*  DESCRIBING THE GROWTH OF A CALF FROM BIRTH TO
*  MATURITY.
*
*
*  THE CALF WEIGHS 40 KG AT BIRTH.
INCON  W0 = 40
*
*  PARAMETER 'MU' IS SET AT 0.01 (PER DAY), AND THE
*  MATURE WEIGHT 'WF' AT 600 (KG).
PARAMETER  MU = 0.01,  WF = 600
*
DYNAMIC
*  THE DIFFERENTIAL EQUATION OF THE MODEL (EQN (2.24)).
      DWDT  =  MU * W * (1 - W * W / (WF *  WF))
*  INTEGRATION OF DWDT.
      W  =  INTGRL (W0,  DWDT)
*
*  EULER'S METHOD OF INTEGRATION (RECT).
METHOD RECT
*
TIMER  DELT = 0.1,  PRDEL = 25.0,  FINTIM = 1000.0
*  INTEGRATION INTERVAL IS DELT;  RESULTS ARE PRINTED EACH
*  INTERVAL OF PRDEL;  THE SIMULATION RUNS FOR FINTIM TIME UNITS.
PRINT  W
PRTPLT  W
END
STOP
ENDJOB
```

*Figure 2.4* A CSMP program of a simple animal growth model

```
 TIME              W
0.0000E  00      4.0000E  01    +
2.5000E  01      5.1279E  01    -+
5.0000E  01      6.5678E  01    --+
7.5000E  01      8.3994E  01    ---+
1.0000E  02      1.0716E  02    -----+
1.2500E  02      1.3617E  02    --------+
1.5000E  02      1.7198E  02    -----------+
1.7500E  02      2.1515E  02    ---------------+
2.0000E  02      2.6539E  02    --------------------+
2.2500E  02      3.2097E  02    -------------------------+
2.5000E  02      3.7846E  02    ------------------------------+
2.7500E  02      4.3323E  02    -----------------------------------+
3.0000E  02      4.8086E  02    ---------------------------------------+
3.2500E  02      5.1874E  02    ------------------------------------------+
3.5000E  02      5.4659E  02    ---------------------------------------------+
3.7500E  02      5.6585E  02    ----------------------------------------------+
4.0000E  02      5.7856E  02    ------------------------------------------------+
4.2500E  02      5.8670E  02    ------------------------------------------------+
4.5000E  02      5.9181E  02    -------------------------------------------------+
4.7500E  02      5.9497E  02    -------------------------------------------------+
5.0000E  02      5.9691E  02    -------------------------------------------------+
5.2500E  02      5.9810E  02    -------------------------------------------------+
5.5000E  02      5.9882E  02    -------------------------------------------------+
5.7500E  02      5.9926E  02    -------------------------------------------------+
6.0000E  02      5.9953E  02    -------------------------------------------------+
```

*Figure 2.5* Extract from the results from the program in *Figure 2.4*

```
TITLE    AN ANIMAL GROWTH MODEL
*  CSMP PROGRAM OF THE LOGISTIC GROWTH FUNCTION,
*  DESCRIBING THE GROWTH OF A CALF FROM BIRTH TO
*  MATURITY.
*
*  THE CALF WEIGHS 40 KG AT BIRTH.
INCON   W0 = 40
*
*  PARAMETER 'MU' IS SET AT 0.01 (PER DAY), AND THE
*  MATURE WEIGHT 'WF' AT 600 (KG).
*  IN THIS DEMONSTRATION OF THE FITTING PROGRAM, THE PARAMETERS
*  ARE SET TO HALF THE ABOVE VALUES  (CF. FIGURE 2.4).
PARAMETER    MU = 0.005,   WF = 300
*
DYNAMIC
*  THE DIFFERENTIAL EQUATION OF THE MODEL (EQN(2.24)).
      DWDT   =  MU * W * (1 - W / WF)
*  INTEGRATION OF DWDT.
      W   =   INTGRL(W0, DWDT)
*
*  EULER'S METHOD OF INTEGRATION (RECT).
METHOD RECT
*
TIMER    DELT = 0.1,   PRDEL = 25.0,   FINTIM = 1000.0
*  INTEGRATION INTERVAL IS DELT;   RESULTS ARE PRINTED EACH
*  INTERVAL OF PRDEL;   THE SIMULATION RUNS FOR FINTIM TIME UNITS.
PRINT  W
*
*  FITTING MODULE:
    STORAGE  WEX(10),  TIMEX(10)
    TABLE  TIMEX(1-10) = 25, 50, 100, 150, 200, 400, 600,   ...
           800, 900, 1000,   ...
       WEX(1-10) = 50.4, 63.2, 97.5, 145, 207, 477, 580, 597, 599,  600
    OPTIMISE   MU,  WF
    FIT W(10, WEX, TIMEX)
*  END OF FITTING MODULE.
*
END
STOP
ENDJOB
```

*Figure 2.6* Animal growth model program of *Figure 2.4*, modified to use parameter-adjustment data-fitting procedures

F I T T I N G   C O M M E N C E S
= = = = = = = = = = = = = = = = = = =

| ITN. | SIMS. | RSS | MU | WF |
|---|---|---|---|---|
| 0 | 3 | 5.3290D 00 | 5.0000E-03 | 3.0000E 02 |
| 1 | 6 | 4.4581D-01 | 9.8907E-03 | 4.4315E 02 |
| 2 | 9 | 7.9833D-03 | 9.9223E-03 | 5.7752E 02 |
| 3 | 12 | 7.0784D-06 | 9.9881E-03 | 5.9982E 02 |
| 4 | 16 | 4.5575D-06 | 9.9882E-03 | 6.0026E 02 |
| 4 | 16 | 4.5575D-06 | 9.9882E-03 | 6.0026E 02 |

****** FITTING COMPLETED

TABLE OF RESIDUALS

MU = 9.9882E-03 WF = 6.0026E 02

| VARIABLE | TIME | OBSERVED | CALCULATED | RESIDUAL (LOG) |
|---|---|---|---|---|
| W | 2.5000E 01 | 5.0400E 01 | 5.0386E 01 | -2.7180E-04 |
| | 5.0000E 01 | 6.3200E 01 | 6.3166E 01 | -5.3501E-04 |
| | 1.0000E 02 | 9.7500E 01 | 9.7421E 01 | -8.1539E-04 |
| | 1.5000E 02 | 1.4500E 02 | 1.4523E 02 | 1.6079E-03 |
| | 2.0000E 02 | 2.0700E 02 | 2.0687E 02 | -6.4754E-04 |
| | 4.0000E 02 | 4.7700E 02 | 4.7702E 02 | 4.6730E-05 |
| | 6.0000E 02 | 5.8000E 02 | 5.7983E 02 | -2.8706E-04 |
| | 8.0000E 02 | 5.9700E 02 | 5.9730E 02 | 5.0259E-04 |
| | 9.0000E 02 | 5.9900E 02 | 5.9908E 02 | 1.4210E-04 |
| | 1.0000E 03 | 6.0000E 02 | 5.9975E 02 | -4.1294E-04 |

TOTAL RESIDUAL SUM OF SQUARES = 4.55750E-06

*Figure 2.7* Extract from the results from the program in *Figure 2.6*

simulation. The DYNAMIC segment forms the core of the program and contains statements describing the dynamics of the model, such as the differential equations, that need to be evaluated at every iteration. TERMINAL is the final segment, is optional and is used for calculations that are to be performed on completion of the iterated DYNAMIC section. The statements appearing in a CSMP program may be grouped into three categories: *data* statements, *structure* statements, and *control* statements. Data statements specify such things as the numerical values of parameters and constants, and the initial conditions of the problem. Structure statements describe the relationships between the variables and parameters of the problem. The control statements specify such things as the integration step and method, the time of the simulation, and the format of the output data.

A CSMP program for solving equation (2.24) is shown in *Figure 2.4*. The logistic function programmed describes the growth of a calf from birth (time $t = 0$) to maturity. The program contains sufficient comment statements (those with an * in the first column) to make it fairly self-explanatory. The time course of the simulation is Day 0 to Day 1000 and equation (2.24) is to be integrated using Euler's method with a step length of $\Delta t = 0.1$ day. The resultant values of $W$ are printed and plotted every 25 days, and are given in *Figure 2.5*.

In *Figure 2.6*, which programs the logistic function, an extra segment has essentially been added to the program of *Figure 2.4*, to show how a parameter adjustment and fitting method has been implemented in practical terms. In this case, the usual CSMP routines have been supplemented by a pre-processor which analyses the statements in the fitting module, and adds on the required FORTRAN statements and subroutines to the program (Lainson, Sweeney and Thornley, personal communication). Some of the output from this program is shown in *Figure 2.7*.

Many CSSL (Continous System Simulation Languages) packages have been written, and CSMP is an archetype of the species. They are very similar, and with the aid of a programming manual and the brief account that has been given here, the modeller should be able to tackle some of the simpler dynamic models. There is a number of texts on the topic: for example, Brockington (1979) provides a readable primer on programming dynamic models aimed at the agriculturalist, and Speckhart and Green (1976) give a comprehensive introduction to all facets of modelling with CSMP.

## Exercises

### Exercise 2.1

The exponential quadratic growth curve (p. 89) has the equation $x = \exp(a_0 + a_1 t + a_2 t^2)$, where $x$ is the dependent variable (typically the fresh or dry weight of an animal or plant), $t$ denotes time, and $a_0$, $a_1$ and $a_2$ are constants. Show that this 'empirical' equation is equivalent to a two-state-variable problem: $\mathrm{d}x/\mathrm{d}t = x(a_1 + 2y)$, $\mathrm{d}y/\mathrm{d}t = a_2$; with $\ln x = a_0$ and $y = 0$ at time $t = 0$. $y$ is the second state variable. Suggest a biological interpretation of this reformulated problem.

### Exercise 2.2

Solve the differential equation $\mathrm{d}y/\mathrm{d}t = -y$, with $y = 1$ at time $t = 0$ using Euler's formula (equation (2.5)) for a few time steps using $\Delta t = 0.1$. Integrate the differential equation, and use the analytic solution to check your results.

### Exercise 2.3

Re-do Exercise 2.2 using the second-order trapezoidal method of equation (2.12). Note the increased accuracy.

### Exercise 2.4

A model is fitted to 84 data points by adjustment of four parameters to minimize the log residual sum of squares $R$ (equations (2.16) and (2.17)), giving an $R$(minimum) of 0.8. Estimate the mean residual sum of squares, and estimate the average relative error or lack of fit of prediction and experiment.

Suppose the error residual has been found to be 0.4 with 50 degrees of freedom. Is the model an acceptable fit to the data at the 10% probability level?

### Exercise 2.5

Write a CSMP program (as in *Figure 2.4*) for simple exponential growth $\mathrm{d}x/\mathrm{d}t = kx$ with $x(t = 0) = 1$ and $k = 1$, integrating by Euler's method.

Chapter 3

# Techniques: mathematical programming

## Introduction

Mathematical programming is a general term covering several variants of what are essentially linear programming. The emphasis of this chapter will be therefore on linear programming, though appropriate non-linear and other variations will be identified and discussed. The main purpose of this chapter is to explain what is meant by linear programming and how linear programming problems are solved on a computer. It is important to understand that the word *programming* as used in this context is synonymous with *planning*, and should not be confused with *program* as used in the context of a *computer program.* Mathematical programming generally provides a deterministic, static and empirical approach to modelling.

Linear programming is a branch of mathematics which has been developed over the past thirty or so years to deal with complex planning and investment problems arising in industry and government. It is used to allocate resources, which are restricted in some way, in the best possible manner with respect to some chosen objective. Typical applications of linear programming in agriculture include: allocating land to particular crops; deciding on the amounts of different fertilizers to apply; calculating least-cost rations for farm livestock; and planning a farm's labour and machinery needs. Much of the theoretical development of the subject is due to the mathematician George Dantzig, and the reader is referred to Dantzig (1963) for a description of the origins of linear programming.

There are four basic steps in solving a problem by linear programming. These are

1. the formulation of the problem in words and the collection of the necessary information and data;
2. the translation of the problem into the mathematical conventions of linear programming;
3. the application of mathematical rules and procedures to the problem in order to obtain a solution; and
4. the interpretation of the solution and its explanation to interested parties, such as scientists or farmers.

Steps 1, 2 and 4 are implemented mainly by the modeller; step 3 is almost invariably implemented on a computer using a purpose-built computer program. The emphasis of this chapter is on step 2—how a problem is translated into the mathematical conventions of linear programming. It is assumed that a computer package is available to implement step 3. The use of a typical mathematical programming package is illustrated with specific reference to ICL LP400 and the mathematical concepts embodied in such a package are explained, though not in great detail. The manipulation of the data into the form required for a linear programme is also discussed. It is assumed throughout that the problem to be solved has already been formulated in clear and concise English and that the relevant basic data are available.

## Mathematical formulation

A linear programming problem has three quantitative aspects: an objective; alternative courses of action for achieving the objective; and resource or other restrictions (for a full description see Heady and Candler, 1958). These must be expressed in mathematical terms so that a solution may be calculated. Thus, all mathematical formulations contain the following three essential components:

1. There are *decision variables* whose values are to be chosen. A decision variable describes a particular course of action (e.g. the amount of land to allocate to each particular crop).
2. There is an *objective function* of decision variables to be optimized, which describes the objective (e.g. the farmer wishes to maximize his profits from the crops he grows).
3. There are restrictions on the decision variables called *constraints* (e.g. the sum of the amounts allocated to each crop cannot exceed the total land available, the time needed to harvest the crops must not exceed the man-hours available).

In addition, these three components must be of a special form if the problem is to be suitable for solution by linear programming. First, the objective function and constraints must lend themselves to expression in a *linear* form (hence the term *linear* programming), and must be completely *deterministic* (i.e. contain no random elements). And secondly, the decision variables must be *continuous* and *non-negative*; which means that, within the limits of the constraints imposed upon them, they can assume any integer or fractional non-negative values.

The process and conventions of mathematical formulation are best illustrated by means of an example.

### Example

A farmer wishes to determine the amounts of pig slurry and compound fertilizer to spread on 20 ha of grassland so that his total fertilizer costs are minimized. The cost and chemical composition of both the slurry and fertilizer are shown below.

| | *Cost* [£ $t^{-1}$] | *Chemical composition* [kg $t^{-1}$] | | |
|---|---|---|---|---|
| | | *Nitrogen* | *Phosphate* | *Potash* |
| Pig slurry | 2.5 | 6 | 1.5 | 4 |
| Compound fertilizer | 130 | 250 | 100 | 100 |

The farmer wishes to apply at least 75 $kg\,ha^{-1}$ of nitrogen, 25 $kg\,ha^{-1}$ of phosphate and 35 $kg\,ha^{-1}$ of potash to his grassland. He can apply pig slurry at the rate of 8 $t\,hour^{-1}$ and fertilizer at the rate of 0.4 $t\,hour^{-1}$, and has 25 hours available in which to complete the job.

To formulate the linear programme, we first have to isolate the three essential components of the problem, namely, the decision variables, the objective function and the constraints on the decision variables and then check that they are of the special form needed for solution by linear programming.

### DECISION VARIABLES

The farmer wishes to determine the amounts of pig slurry and compound fertilizer to spread on his grassland. The decision variables, therefore, are the amount of each fertilizer to use. Let

$x_1$ = the quantity [t] of pig slurry spread

and

$x_2$ = the quantity [t] of compound fertilizer used

Incidentally, linear programming problems usually involve many decision variables. Practical problems often have hundreds, sometimes thousands of variables, and so a subscripted $x$ is generally used to denote a decision variable.

### OBJECTIVE FUNCTION

The farmer's objective is to minimize the total cost of the fertilizers he uses. Pig slurry costs him 2.5 £ $t^{-1}$ and compound fertilizer 130 £ $t^{-1}$, so his fertilizer costs are given by $2.5x_1 + 130x_2$. Let $z$ [£] denote fertilizer costs. The objective function is therefore

$$\text{minimize} \quad z = 2.5x_1 + 130x_2 \tag{3.1}$$

Some linear programming packages are based on minimization of the objective function, whilst others are based on maximization. In fact, the two approaches are equivalent because minimizing a function is equivalent to maximizing the negative of that function. Thus the optimal values of the decision variables obtained by minimizing the objective function are the same as those obtained by maximizing the negative of the same objective function, and

$$\text{minimum } z = -\text{ maximum } (-z)$$

Therefore, the objective function in our example could alternatively be written

$$\text{maximize} \quad (-z) = -2.5x_1 - 130x_2$$

### CONSTRAINTS

The constraints on the decision variables are the farmer wishes to apply at least 75 $kg\,ha^{-1}$ of nitrogen, at least 25 $kg\,ha^{-1}$ of phosphate and at least 35 $kg\,ha^{-1}$ of potash to his grassland, and he has no more than 25 hours in which to complete the work. First, let us consider the constraint on nitrogen. The pig slurry contains

6 kg t$^{-1}$ of nitrogen and the compound fertilizer 250 kg t$^{-1}$, so the amount of nitrogen applied from the slurry is $6x_1$ kg and from the fertilizer $250x_2$ kg. Thus, the total amount of nitrogen applied to the grassland is $6x_1 + 250x_2$, and this total must not be less than 75 kg ha$^{-1}$, or 1500 kg as the farmer has 20 ha of grassland. The constraint on nitrogen can therefore be expressed as

$$6x_1 + 250x_2 \geqslant 1500 \qquad (3.2)$$

Similar reasoning gives the phosphate constraint as

$$1.5x_1 + 100x_2 \geqslant 500 \qquad (3.3)$$

and the potash constraint as

$$4x_1 + 100x_2 \geqslant 700 \qquad (3.4)$$

Finally, there is the time constraint. The farmer can apply the slurry at the rate of 8 t hour$^{-1}$ and the fertilizer at the rate of 0.4 t hour$^{-1}$. The total time taken to complete this work is $(x_1/8) + (x_2/0.4)$ hours, and this total cannot exceed 25 hours. The time constraint is therefore

$$\frac{1}{8}x_1 + \frac{1}{0.4}x_2 \leqslant 25$$

or

$$x_1 + 20x_2 \leqslant 200 \qquad (3.5)$$

In linear programming formulations, the constant terms appearing in the constraints are conventionally written on the right-hand side of the inequality or equation and hence are termed *right-hand side values* or simply *RHS values*. Furthermore, RHS values are restricted to non-negative numbers. Note that any negative RHS value can easily be changed to a positive value by multiplying both sides of the constraint by $-1$, and, in the case of an inequality constraint, reversing the inequality sign (i.e. $\geqslant$ becomes $\leqslant$, and vice versa). The RHS values in our example are 1500, 500, 700 and 200 (*see* equations (3.2)–(3.5)).

A constraint which is automatically satisfied by any values of the decision variables that satisfy one or more of the other constraints is termed *redundant*. For example, if we had the additional constraint $x_1 + 30x_2 \leqslant 300$ in our fertilizer example, then the values of $x_1$ and $x_2$ satisfying the time constraint $x_1 + 20x_2 \leqslant 200$ would automatically satisfy this additional constraint. This constraint is therefore redundant. Redundant constraints can be ignored when formulating a linear programme. It is generally quite easy to spot redundant constraints in problems with a small number of decision variables, but becomes increasingly difficult with larger problems. Fortunately, redundant constraints do not have to be eliminated from a problem that is to be solved on a computer.

## SPECIAL FORM

The objective function and constraints are in linear form as equations (3.1)–(3.5) contain no squared, cubic, or higher-order or mixed-product terms. Also they are deterministic as the coefficients of the decision variables in both the objective function and the constraints and the RHS values are constants. The decision variables $x_1$ and $x_2$ cannot assume negative values, as applying a negative amount of pig slurry or compound fertilizer has no physical meaning. So the *non-negativity conditions* are satisfied, namely

$$x_1 \geqslant 0 \tag{3.6}$$

$$x_2 \geqslant 0 \tag{3.7}$$

In addition, $x_1$ and $x_2$ can assume any integral or fractional values, provided they are non-negative and satisfy the constraints imposed upon them; therefore $x_1$ and $x_2$ are continuous. So the decision variables, objective function and constraints are of the special form necessary for solution by linear programming.

SUMMARY

The problem can be formulated as a linear programme and its complete statement (given by equations (3.1)–(3.7)) is

| | | | |
|---|---|---|---|
| minimize | $z = 2.5x_1 + 130x_2$ | [objective function] | (3.8a) |
| subject to | $6x_1 + 250x_2 \geqslant 1500$ | [nitrogen constraint] | (3.8b) |
| | $1.5x_1 + 100x_2 \geqslant 500$ | [phosphate constraint] | (3.8c) |
| | $4x_1 + 100x_2 \geqslant 700$ | [potash constraint] | (3.8d) |
| | $x_1 + 20x_2 \leqslant 200$ | [time constraint] | (3.8e) |
| | $x_1 \geqslant 0$ | [non-negativity condition] | (3.8f) |
| | $x_2 \geqslant 0$ | [non-negativity condition] | (3.8g) |

## Graphical solution

Our example is two dimensional as it contains only two decision variables, and can therefore be solved graphically. Each constraint and non-negativity condition are illustrated graphically in *Figure 3.1*. Let us first consider the nitrogen constraint, given by equation (3.8b). It can be seen from *Figure 3.1a* that this constraint is satisfied by values of $x_1$ and $x_2$ lying on or to the right of the continuous line $6x_1 + 250x_2 = 1500$, as shown by the shaded area. Similarly, *Figure 3.1b* shows that the phosphate constraint is satisfied by values of the decision variables lying on or to the right of the line $1.5x_1 + 100x_2 = 500$. Values of $x_1$ and $x_2$ satisfying each of the other constraints and the non-negativity conditions are shown in *Figures 3.1c,d,e*.

*Figure 3.2* shows all the constraints and non-negativity conditions illustrated in *Figure 3.1* combined on to one graph; the arrows indicate the direction of permissible values of $x_1$ and $x_2$ for each constraint. The constraints and non-negativity conditions taken together restrict the permissible values of $x_1$ and $x_2$ to points within or on the shaded area ABCDE. The area ABCDE is called the *feasible region*, as it contains all points that satisfy all the constraints and non-negativity conditions simultaneously. Each point within or on the feasible region is called a *feasible solution* to the linear programme. The feasible region therefore is the collection of the infinite number of feasible solutions which exist to the problem.

The objective function for various values of $z$ (i.e. lines of equal fertilizer costs) is shown in *Figure 3.3*. The solution to our problem lies at a point within or on the

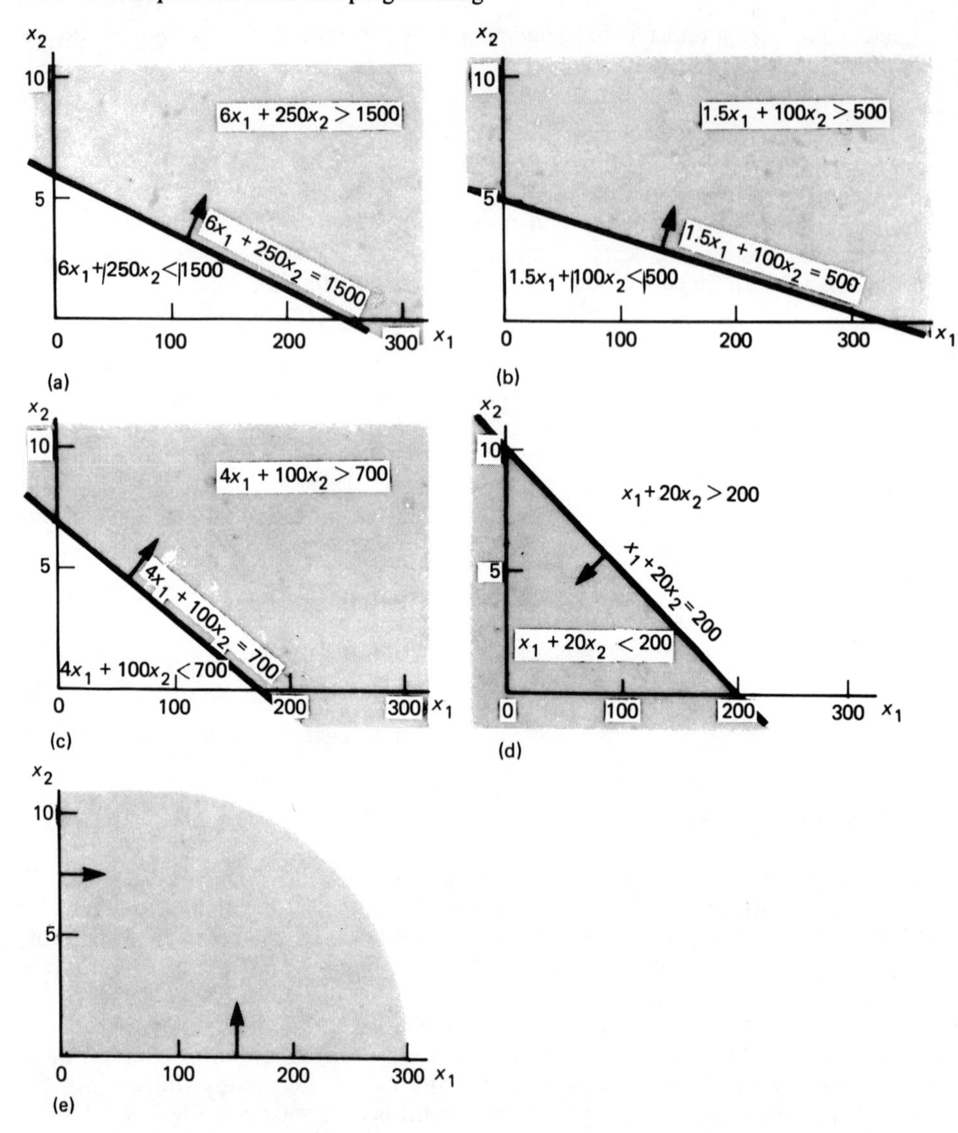


*Figure 3.1* Values of the decision variables satsifying each of the constraints and the non-negativity conditions

feasible region ABCDE, so we need only concern ourselves with lines of equal cost that either bisect or touch ABCDE. Thus, of the lines shown in *Figure 3.3*, the lines $z = 900$, $z = 800$ and $z = 711.1$ are of interest to us, whereas $z = 600$ is not. To find the optimal solution to our problem, we simply determine the lowest line of equal cost that has at least one point in or on the feasible region. The line satisfying this condition in our example is $z = 711.1$. This line has only one point in common with the feasible region, that is the corner (or *vertex*) labelled C, whose coordinates are $x_1 = 111.1$, $x_2 = 3.3$, formed by the intersection of the constraints $1.5x_1 + 100x_2 = 500$ and $6x_1 + 250x_2 = 1500$. Hence the optimal solution is $z = 711.1$, $x_1 = 111.1$,

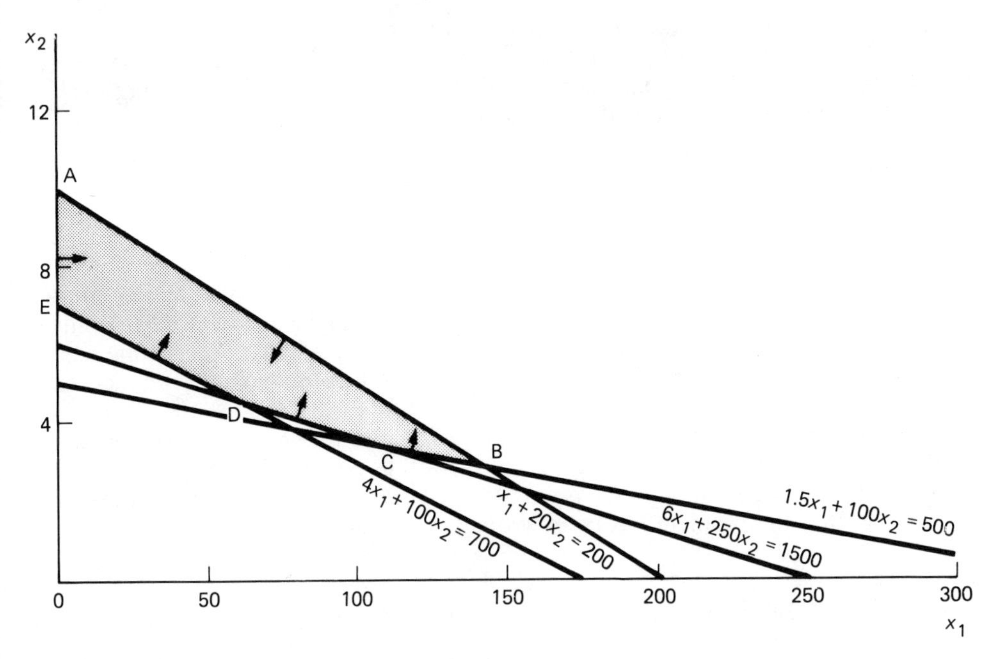


*Figure 3.2* The feasible region (shaded area ABCDE)

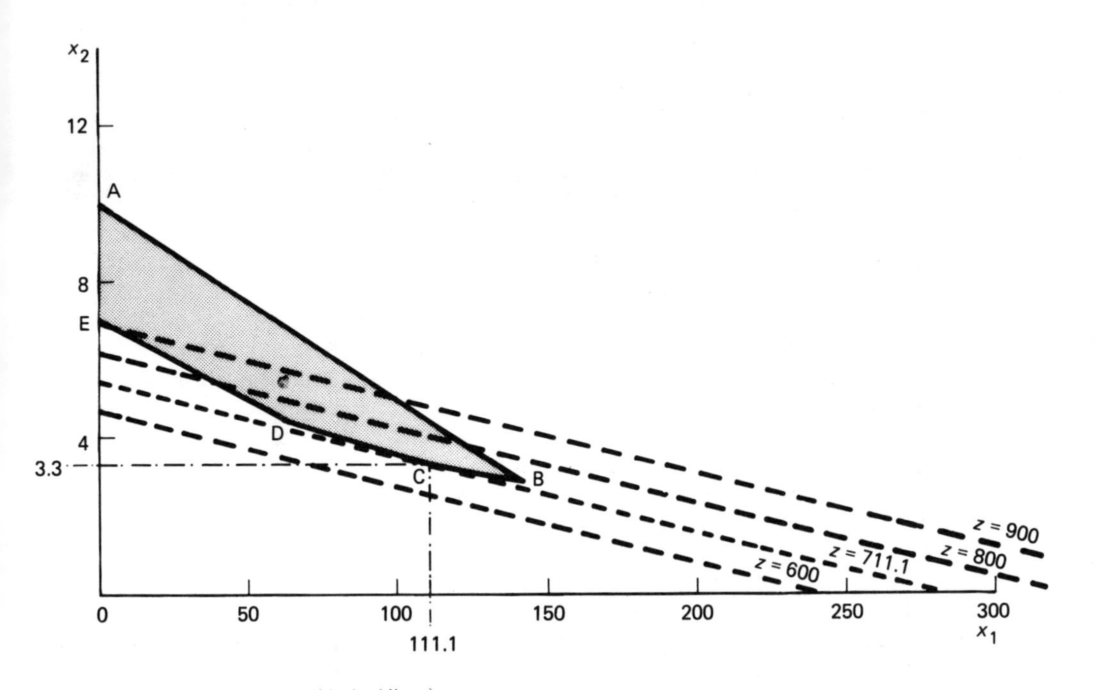


*Figure 3.3* Lines of equal cost (dashed lines)

and $x_2 = 3.3$ and the physical interpretation of this solution is that the farmer should spread 111.1 t of pig slurry and 3.3 t of compound fertilizer on his grassland in order to minimize his total fertilizer costs, the minimum total cost being £711.1. For this particular solution, the nitrogen constraint $6x_1 + 250x_2 \geqslant 1500$ and the phosphate constraint $1.5x_1 + 100x_2 \geqslant 500$ are satisfied as exact equalities (i.e. they are *binding*) but the potash constraint $4x_1 + 100x_2 \geqslant 700$ and the time constraint $x_1 + 20x_2 \leqslant 200$ are not (i.e. they have *slack*).

If one of the boundaries of the feasible region lies at infinity, then there is no upper limit to the value that the objective function can take. Thus, if the objective function were to be maximized, the linear programme would yield an *unbounded solution*. If the objective function were to be minimized however, the linear programme would yield a finite solution. In practice, if a linear programme has an unbounded solution, it almost certainly means that the problem has been incorrectly formulated.

With some linear programmes it is not possible to find non-negative values for the decision variables that satisfy all the constraints and non-negativity conditions. Such problems are said to have *no feasible solution*. Again, this often indicates in practice that the problem has been wrongly formulated.

## Computer solution

Practical problems cannot usually be solved graphically as they almost always involve more than two decision variables. Such problems are invariably solved on a computer using a purpose-built linear programming package. In this section, the mathematical concepts embodied in such packages are briefly explained and the use of a typical package is illustrated with reference to ICL LP400 (International Computers Limited, 1970).

Any linear programming problem with only two decision variables can be represented graphically in two dimensions, as is illustrated by our fertilizer example. Each constraint and non-negativity condition form a *half-space*, restricting values of the two decision variables to points in this half-space. The feasible region is that area of points common to all half-spaces and describes a *polygon*. The feasible region forms what is called a *convex set*. A convex set is a collection of points such that if $P_1$ and $P_2$ are *any* two points in the collection, then the line segment joining them is also in the collection. The boundary of the feasible region is formed by straight lines, each of which corresponds to a constraint or non-negativity condition satisfied as an equality. The points on the boundary where two of these straight lines intersect form the corners of the feasible region and are called *vertices* or *extreme points*. The feasible region has only a finite number of *vertices* because a linear programming problem has a finite number of constraints.

A linear programming problem with three decision variables can be represented in three-dimensional space. Again, each constraint and non-negativity condition forms a half-space, restricting values of the three decision variables to points in this half-space. The feasible region comprises those points common to all half-spaces, is convex, and forms a *polyhedron*. The surface of the polyhedron is defined by a series of planes, each plane corresponding to a constraint or non-negativity condition satisfied as an equality. The edges of the polyhedron are formed by the intersection to two of these planes, and the corners, or vertices, correspond to the points where three planes intersect. Again, there is only a finite number of vertices.

These concepts for two and three decision variables can be generalized to problems in $n$ decision variables. Each constraint and non-negativity condition restrict permissible values of the decision variables to points in a half-space whose boundary is a *hyperplane* in $n$-dimensional space. These hyperplanes taken together delineate the boundary of the feasible region and the feasible region describes a convex *polytope*. Any intersection of $n$ hyperplanes forms a vertex of the feasible region.

Let us return to our fertilizer example. It can be seen from *Figure 3.3* that the optimal solution to the problem lies at a vertex of the feasible region, namely, the point C of the polygon ABCDE. Furthermore, by considering lines of equal cost in the direction of increasing $z$, it is apparent from *Figure 3.3* that the objective function attains its maximum value at the vertex A of ABCDE. Thus, if the objective had been to maximize fertilizer costs, the optimal solution would lie at the point A. The objective function therefore attains both its maximum and minimum values at a vertex of the feasible region. This is true not only for this particular example, but for all linear programming problems. The optimal solution to a linear programme, provided a finite optimal solution exists, always lies at a vertex of the feasible region. Therefore, to solve any linear programme one needs only consider the vertices of the feasible region. A feasible solution that corresponds to a vertex is called a *basic feasible solution*.

The basic computational procedure for solving any linear programming problem is the *simplex method*, devised by Dantzig (1951). Most linear programming packages use a variant of the simplex method known as the *revised simplex* (or *inverse matrix*) *method*, which was developed by Dantzig and others at the Rand Corporation as a procedure for solving linear programmes on digital computers (Dantzig, 1953; Orchard-Hays, 1954).

All the inequality constraints appearing in a linear programming formulation must be converted to equalities before the simplex method (or any of its variants) can be applied. This is done by introducing an additional non-negative variable, called a *slack* variable, to each constraint so as to represent the difference between the two sides of the inequality. For example, the inequality constraints in our fertilizer example (equations (3.8b,c,d,e)) when converted to equalities become

$$6x_1 + 250x_2 - s_1 = 1500 \quad \text{[nitrogen constraint]} \tag{3.9a}$$

$$1.5x_1 + 100x_2 - s_2 = 500 \quad \text{[phosphate constraint]} \tag{3.9b}$$

$$4x_1 + 100x_2 - s_3 = 700 \quad \text{[potash constraint]} \tag{3.9c}$$

$$x_1 + 20x_2 + s_4 = 200 \quad \text{[time constraint]} \tag{3.9d}$$

The slack variables associated with $\geqslant$ inequalities ($s_1$, $s_2$ and $s_3$ in our example) are called negative slacks because they appear in the equalities with negative signs. Similarly, slack variables associated with $\leqslant$ inequalities ($s_4$ in our example) are called positive slacks because they appear in the equalities with positive signs. In the input to a linear programming package, the user specifies whether a constraint is of the form $\geqslant$, $\leqslant$ or $=$. The package then converts all the inequality constraints into equations using slack variables. Slack variables are often printed out with decision variables in an optimal solution.

The simplex method is essentially a two-phase procedure. The first phase involves finding an initial basic feasible solution or vertex. The second phase is an iterative procedure for moving from one vertex of the feasible region to another having a lower cost associated with it (or higher profit in the case of a maximization problem). Phase two continues until no further improvement in the objective function is possible (i.e. an optimal solution has been found). It is not essential for the linear programmer to know the exact computational details of the simplex method. He will rarely need to apply the simplex method manually as computer codes are readily available; anyway, real problems are almost invariably too large for manual solution. Thus, from a practical point of view, it is more important that the user familiarize himself with the various aspects of problem formulation, the essential features of linear programming packages commercially available, and the interpretation of the output they provide. For a full description of the simplex and revised simplex algorithms, the reader is referred to Beale (1968), and to Hadley (1962) and Dantzig (1963) for the underlying mathematical theory.

The use of a typical linear programming package is now illustrated by using ICL LP400 to solve our fertilizer example, as defined by equations (3.8a–g). In order to solve a linear programming problem using LP400, two files must first be created—a COMPILER file containing the control language program and an EXECUTOR file containing the data. *Figure 3.4* shows a simple control language

```
          PROGRAM
* INITIALIZE STANDARD DEMAND ROUTINES
          INITIALZ
* DEFINE PROBLEM NAME
          MOVE       ZPBNAM,'FERTLZR'
* INPUT PROBLEM
          INPUT
* SET PROBLEM UP ON THE MATRIX FILE
          SETUP
* CALCULATE THE INVERSE OF THE INITIAL BASIS
          INVERT
* OPTIMIZE THE PROBLEM AND OUTPUT THE SOLUTION
          NORMAL
          SOLUTION
* TERMINATE RUN OF EXECUTOR PROGRAM
          EXIT
          END
```

*Figure 3.4* A simple control language program

program suitable for solving routine linear programmes; lines starting with an asterisk denote comment statements. Note that LP400 always *minimizes* the objective function unless instructed to maximize. *Figure 3.5* gives the data file for the fertilizer example. The type and name of the objective function and constraints are defined in the ROWS section of the file (the non-negativity conditions are implicitly assumed in the package). *F* is used to define the objective function (a *F*ree row, one that is not a constraint), *N* defines a $\geqslant$ constraint (a row with *N*egative slack), *P* is used for a $\leqslant$ constraint (*P*ositive slack), and *Z* for an = constraint (*Z*ero slack). The COLUMNS section of the file is for specifying the decision variables and their coefficient in each of the rows (i.e. the objective function and the constraints); any zero coefficients can be omitted as unspecified coefficients are taken as zero. In the RHS section, a name is given to the vector of RHS values (called RHSIDE in our example) and each RHS value is specified. A

```
INPUT
ROWS
  F OBJECTIV
  N NITROGEN
  N PHOSPHAT
  N POTASH
  P TIME
COLUMNS
      X1        OBJECTIV   2.5       NITROGEN   6
      X1        PHOSPHAT   1.5       POTASH     4
      X1        TIME       1
      X2        OBJECTIV   130       NITROGEN   250
      X2        PHOSPHAT   100       POTASH     100
      X2        TIME       20
RHS
      RHSIDE    NITROGEN   1500      PHOSPHAT   500
      RHSIDE    POTASH     700       TIME       200
ENDATA
```

*Figure 3.5* The LP400 data file for the fertilizer example

full description of how to construct COMPILER and EXECUTOR files is found in the manual (International Computers Limited, 1970).

Having created the COMPILER and EXECUTOR files, LP400 can now be run. *Figure 3.6* is a reproduction of part of the computer solution for the fertilizer problem. Only the essential elements of this output are discussed. The optimal solution is shown as $711.\dot{1}$ and was reached after three iterations. An inspection of the ROWS section shows that the nitrogen and phosphate constraints are binding but the potash and time constraints have slack. The optimal values of the decision variables are given in the COLS section and are shown as $x_1 = 111.\dot{1}$ and $x_2 = 3.\dot{3}$. Linear programming packages always print out the optimal basic feasible solution. Remember that the optimal solution to a linear programming problem lies at a vertex of the feasible region and a feasible solution corresponding to a vertex is called a basic feasible solution. In a basic feasible solution, the variables—the decision variables and the slack variables—divide into two mutually exclusive groups. A number of them equal to the number of constraints (excluding the non-negativity conditions) are called *basic variables* (or *variables in the basis*), and the remainder *nonbasic variables* (or *variables out of the basis*). Therefore, in a basic feasible solution to a linear programme having $m$ constraints and $n$ variables, $m$ of the $n$ variables will be basic variables and the remaining $n - m$ variables will be nonbasic variables. The essential difference between nonbasic and basic variables is that nonbasic variables are *always* set equal to zero, whereas basic variables can assume any non-negative value, with most of them usually positive. A fuller understanding of the concepts of 'basic' and 'basis' requires a knowledge of advanced linear algebra; detailed explanations of the concepts are given by Hadley (1962) and Dantzig (1963). In the LP400 printout, decision variables in the optimal basis are indicated by a B under the STATUS heading in the COLS section. Slack variables in the optimal basis can be determined by an inspection of the ROWS section; a B preceding a constraint row name indicates that the slack variable associated with that constraint is a basic variable in the optimal solution and its

```
START OF PROCEDURE   SOLUTION
- - - - - - - - - - -


      PROBLEM NAME IN ZPBNAM IS 'FERTLZR '
      OBJECTIVE ROW NAME IS     'OBJECTIV'
      RIGHT HAND SIDE NAME IS   'RHSIDE  '

      MINIMISATION
SOLUTION

FUNCTION VALUE EQUALS      711.11111

TOTAL NUMBER OF ITERATIONS WAS      3

STATUS   SEQ.  TYPE   NAME       ROW          SLACK        LOWER         UPPER        COST          REDUCED
         NO.                     ACTIVITY     LEVEL        RHS LIMIT     RHS LIMIT                  COST

ROWS SECTION

   B       1   F    OBJECTIV   711.11111   -711.11111                                1.0000000     1.0000000
           2   N    NITROGEN   1500.0000                 1500.0000                                -0.24444444
           3   N    PHOSPHAT   500.00000                 500.00000                                -0.68888889
   B       4   N    POTASH     777.77778   -77.777778    700.00000
   B       5   P    TIME       177.77778    22.222222                 200.00000


SOLUTION

STATUS   SEQ.   TYPE     NAME        LEVEL        LOWER     UPPER      COST        REDUCED
         NO.                                      BOUND     BOUND                  COST

COLS SECTION

   B 65499    P    X1              111.11111                          2.5000000
   B 65498    P    X2              3.3333333                          130.00000

      ****SOLUTION COMPLETE

      ****DEMAND(S) SET :-   NONE
```

*Figure 3.6* Part of the LP400 output for the fertilizer example

value is the modulus of the number given under the SLACK LEVEL heading. Our fertilizer example has four constraints and six variables, two of which are decision variables and four slack variables (*see* equations (3.9a–d)). Thus a basic feasible solution to the problem will have four basic variables and two (i.e. 6 − 4) nonbasic variables. *Figure 3.6* shows that the basic variables and their values in the optimal basic feasible solution are $x_1 = 111.1^{\bullet}$, $x_2 = 3.3^{\bullet}$, the potash slack $s_3 = 77.7^{\bullet}$, and the time slack $s_4 = 22.2^{\bullet}$; and that the nonbasic variables are the nitrogen and phosphate slacks $s_1$ and $s_2$, which both have zero value.

The solution to a linear programme contains useful supplementary information as well as the values of the variables in the optimal solution, particularly the *reduced costs* of the nonbasic variables. The reduced costs of the slack variables are of major significance in linear programming and are known as the *shadow prices* on the rows of the problem. The shadow prices give the rate of change of the objective function with respect to the right-hand sides of the constraint equations (i.e. $\partial z/\partial b_i$, $i = 1, 2, \ldots, m$; as in equation (3.22b), p. 56). The reduced costs for our fertilizer problem are shown in the final column of *Figure 3.6*. The shadow price of the nitrogen slack $s_1$ is $-0.24^{\bullet}$. This means that if $s_1$ is increased by a small positive amount $\theta$ (i.e. our nitrogen constraint is decreased by $\theta$ whilst the other constraints are satisfied), the fertilizer costs would decrease by $0.24^{\bullet}\theta$ [£]. Similarly, the shadow price of the phosphate slack $s_2$ is $-0.68^{\bullet}$. This means reducing the phosphate requirements by $\theta$ would save $0.68^{\bullet}\theta$ [£] in fertilizer costs. However, the solution does not indicate the range of $\theta$ for which these reduced costs hold. This type of question is answered by using parametric programming, which is dealt with later in the chapter. If either of our decision variables $x_1$ and $x_2$ had finished up as nonbasic variables, the reduced costs would give how much the cost of $x_1$ or $x_2$ had to come down before it would enter the optimal basis.

## A worked example

A farmer wishes to formulate, as cheaply as possible, a rearing mix for feeding to his Hereford × Friesian steers which will satisfy certain nutritional requirements. The mix must have a metabolizable energy concentration of at least 13 MJ kg$^{-1}$ and a crude protein content of no less than 160 g kg$^{-1}$; the mix must also have a minimum mineral content of 7 g kg$^{-1}$ of calcium, 7 g kg$^{-1}$ of phosphorus, 3 g kg$^{-1}$ of sodium, and 2 g kg$^{-1}$ of magnesium. The farmer can obtain the following ingredients, at the costs shown in parentheses, with which to formulate the mix: barley

**TABLE 3.1. Nutritive value of the ingredients**

| | Nutritive value | | | | | |
|---|---|---|---|---|---|---|
| *Ingredient* | *Metabolizable energy* [MJ kg$^{-1}$] | *Crude protein* [%] | *Calcium* [%] | *Phosphorus* [%] | *Sodium* [%] | *Magnesium* [%] |
| Barley | 13.7 | 10.8 | 0.05 | 0.38 | 0.02 | 0.13 |
| Maize | 14.2 | 9.8 | 0.02 | 0.27 | 0.01 | 0.10 |
| White fish meal | 11.1 | 70.1 | 7.93 | 4.37 | 1.61 | 0.22 |
| Soyabean meal | 12.3 | 50.3 | 0.23 | 1.02 | 0.50 | 0.31 |
| Mineral–vitamin supplement | — | — | 12 | 6 | 6 | 3 |

(105.20 £t$^{-1}$), maize (142.75 £t$^{-1}$), white fish meal (245.00 £t$^{-1}$), soyabean meal (144.00 £t$^{-1}$), and a mineral–vitamin supplement (595.60 £t$^{-1}$). He wants the mix to contain at least 10% maize, so as to ensure the mix is sufficiently palatable and to avoid excessive acid production in the rumen of the young steers, and at least 1% mineral–vitamin supplement, so as to ensure that the cattle have a sufficient balance of important vitamins and trace elements in their diet. The nutritive value of each ingredient is shown in *Table 3.1*.

## Formulation

### DECISION VARIABLES

The farmer wishes to determine the amount of each ingredient to include in the mix. The decision variables, therefore, are the quantity of each ingredient in a unit quantity of mix. Let 1 kg of mix contain $x_1$ kg of barley, $x_2$ kg of maize, $x_3$ kg of white fish meal, $x_4$ kg of soyabean meal, and $x_5$ kg of mineral–vitamin supplement.

### OBJECTIVE FUNCTION

The farmer wants to formulate the mix as cheaply as possible. Let $z$ denote the cost (in pence) of a kilogram of mix. The objective function is therefore

$$\text{minimize} \quad z = 10.52x_1 + 14.275x_2 + 24.5x_3 + 14.4x_4 + 59.56x_5 \tag{3.10}$$

### CONSTRAINTS

The constraints on the decision variables are that the mix must have a metabolizable energy concentration of at least 13 MJ kg$^{-1}$, a crude protein content of at least 160 g kg$^{-1}$, a minimum mineral content of 7 g kg$^{-1}$ of calcium, 7 g kg$^{-1}$ of phosphorus, 3 g kg$^{-1}$ of sodium, and 2 g kg$^{-1}$ of magnesium. Furthermore, the mix must contain at least 10% maize and at least 1% mineral–vitamin supplement, and the sum of the ingredients in a unit quantity of mix must add up to one. First, let us consider the energy constraint. With reference to *Table 3.1*, the metabolizable energy concentration of the mix [MJ kg$^{-1}$] is given in terms of the decision variables by

$$13.7x_1 + 14.2x_2 + 11.1x_3 + 12.3x_4 + 0x_5$$

and this expression must not be less than 13 MJ kg$^{-1}$. The energy constraint is therefore given by

$$13.7x_1 + 14.2x_2 + 11.1x_3 + 12.3x_4 \geqslant 13 \tag{3.11}$$

Similar reasoning gives the protein constraint as

$$108x_1 + 98x_2 + 701x_3 + 503x_4 \geqslant 160 \tag{3.12}$$

the calcium constraint as

$$0.5x_1 + 0.2x_2 + 79.3x_3 + 2.3x_4 + 120x_5 \geqslant 7 \tag{3.13}$$

the phosphorus constraint as

$$3.8x_1 + 2.7x_2 + 43.7x_3 + 10.2x_4 + 60x_5 \geqslant 7 \tag{3.14}$$

the sodium constraint as

$$0.2x_1 + 0.1x_2 + 16.1x_3 + 5x_4 + 60x_5 \geqslant 3 \tag{3.15}$$

and the magnesium constraint as

$$1.3x_1 + x_2 + 2.2x_3 + 3.1x_4 + 30x_5 \geqslant 2 \tag{3.16}$$

The other constraints are as follows. The mix must contain at least 10% maize, in other words 1 kg of mix must contain at least 0.1 kg of maize. This lower limit on the amount of maize to include forms a lower bound on $x_2$:

$$x_2 \geqslant 0.1 \tag{3.17}$$

Similarly, the mix must contain at least 1% mineral–vitamin supplement, giving a lower bound on $x_5$:

$$x_5 \geqslant 0.01 \tag{3.18}$$

Finally, the sum of the ingredients in a unit quantity of mix must be one:

$$x_1 + x_2 + x_3 + x_4 + x_5 = 1 \tag{3.19}$$

SPECIAL FORM

Equations (3.10)–(3.19) contain no squared, higher-order, or mixed-product terms and therefore the objective function and constraints are in linear form. Also, the coefficients of the decision variables and the RHS values appearing in these equations are all constants, and so the objective function and constraints are deterministic. The decision variables are non-negative as including a negative quantity of a particular ingredient in the mix has no physical meaning. So the non-negativity conditions are satisfied

$$x_1, x_2, x_3, x_4, x_5 \geqslant 0 \tag{3.20}$$

In addition, the decision variables are continuous as $x_1$, $x_2$, $x_3$, $x_4$ and $x_5$ can assume any values, provided they do not violate the constraints and non-negativity conditions. So the decision variables, objective function and constraints are of the special form necessary for solution by linear programming.

SUMMARY

The rearing mix problem can be formulated as a linear programme and its complete statement (given by equations (3.10)–(3.20)) is

minimize $z = 10.52x_1 + 14.275x_2 + 24.5x_3 + 14.4x_4 + 59.56x_5$
[objective function] (3.21a)

subject to

$$13.7x_1 + 14.2x_2 + 11.1x_3 + 12.3x_4 \geqslant 13$$
[energy constraint] (3.21b)

$$108x_1 + 98x_2 + 701x_3 + 503x_4 \geqslant 160$$
[protein constraint] (3.21c)

$$0.5x_1 + 0.2x_2 + 79.3x_3 + 2.3x_4 + 120x_5 \geqslant 7$$
[calcium constraint] (3.21d)

$$3.8x_1 + 2.7x_2 + 43.7x_3 + 10.2x_4 + 60x_5 \geqslant 7 \quad \text{[phosphorus constraint]} \tag{3.21e}$$

$$0.2x_1 + 0.1x_2 + 16.1x_3 + 5x_4 + 60x_5 \geqslant 3 \quad \text{[sodium constraint]} \tag{3.21f}$$

$$1.3x_1 + x_2 + 2.2x_3 + 3.1x_4 + 30x_5 \geqslant 2 \quad \text{[magnesium constraint]} \tag{3.21g}$$

$$x_2 \geqslant 0.1 \quad \text{[lower bound]} \tag{3.21h}$$

$$x_5 \geqslant 0.01 \quad \text{[lower bound]} \tag{3.21i}$$

$$x_1 + x_2 + x_3 + x_4 + x_5 = 1 \quad \text{[total weight constraint]} \tag{3.21j}$$

$$x_1, x_2, x_3, x_4, x_5 \geqslant 0 \quad \text{[non-negativity conditions]} \tag{3.21k}$$

**Solution**

The rearing mix linear programme, as defined by equations (3.21a–k), is solved using ICL LP400. In this particular linear programme, the decision variables $x_2$ and $x_5$ have positive lower bounds associated with them (equations (3.21h,i). In many practical situations, a number of the decision variables appearing in a linear programme are similarly bounded from below or above or both. If there are many upper and lower bounds, and they are treated as ordinary constraints, they can rapidly build up the size of the basis. Most linear programming codes have a special facility for dealing with upper and lower bounds without increasing the size of the basis, based on modifications to the simplex algorithm first proposed by Dantzig (1954). Computationally, it is more efficient to use such a facility than to treat bounds as ordinary constraints.

*Figure 3.7* gives the EXECUTOR file for the rearing mix problem, and this data file is constructed in the same way as for the fertilizer example but for the added complication of bounds. All upper bounds can be entered as an *upper bound row* and all lower bounds as a *lower bound row*. The type and name of a bound row are defined in the ROWS section of the file and the bound row definitions must appear first in the ROWS section, before any other row definition. U is used to define an upper bound row type and L a lower bound row type. The names of upper and lower bounds rows must begin with U and L respectively and the second and subsequent letters of a name must be the same for any pair of corresponding upper and lower bound rows. For example, if the name UBOUND is chosen for an upper bound row, the name LBOUND must be used for the corresponding lower bound row and vice versa. The bounds are specified by entering their values as elements in the appropriate bound rows. In the rearing mix example, there are no upper bounds but only lower bounds on the decision variables $x_2$ and $x_5$, and these are specified as the coefficients of X2 and X5 in a lower bound row called LBOUND (*see Figure 3.7*). The COMPILER file for this example is the same as that used for the fertilizer example (*Figure 3.4*) with an additional line to initiate the bounds procedure:

```
_________ MOVE _____ ZBOUND, 'UBOUND'
```

```
INPUT
ROWS
  L LBOUND
  F OBJECTIV
  N ENERGY
  N PROTEIN
  N CALCIUM
  N PHOSPHRS
  N SODIUM
  N MAGNESUM
  Z TOTALWT
COLUMNS
    X1         OBJECTIV   10.52        ENERGY     13.7
    X1         PROTEIN    108          CALCIUM    0.5
    X1         PHOSPHRS   3.8          SODIUM     0.2
    X1         MAGNESUM   1.3          TOTALWT    1.0
    X2         OBJECTIV   14.275       ENERGY     14.2
    X2         PROTEIN    98           CALCIUM    0.2
    X2         PHOSPHRS   2.7          SODIUM     0.1
    X2         MAGNESUM   1.0          LBOUND     0.1
    X2         TOTALWT    1.0
    X3         OBJECTIV   24.5         ENERGY     11.1
    X3         PROTEIN    701          CALCIUM    79.3
    X3         PHOSPHRS   43.7         SODIUM     16.1
    X3         MAGNESUM   2.2          TOTALWT    1.0
    X4         OBJECTIV   14.4         ENERGY     12.3
    X4         PROTEIN    503          CALCIUM    2.3
    X4         PHOSPHRS   10.2         SODIUM     5.0
    X4         MAGNESUM   3.1          TOTALWT    1.0
    X5         OBJECTIV   59.56        CALCIUM    120
    X5         PHOSPHRS   60           SODIUM     60
    X5         MAGNESUM   30           LBOUND     0.01
    X5         TOTALWT    1.0
RHS
    RHSIDE     ENERGY     13           PROTEIN    160
    RHSIDE     CALCIUM    7            PHOSPHRS   7
    RHSIDE     SODIUM     3            MAGNESUM   2
    RHSIDE     TOTALWT    1
ENDATA
```

*Figure 3.7* The LP400 data file for the rearing mix example

inserted immediately before the INPUT statement. The _ symbol indicates a space. In this MOVE statement, the name appearing in quotes must always be the appropriate upper bound row name.

*Figure 3.8* gives the LP400 solution to the rearing mix example and shows the optimal composition of the mix to be 74.4% barley, 10% maize, 3.4% white fish meal, 9.1% soyabean meal and 3.1% mineral–vitamin supplement, and that this mix costs 13.2 pence (kg mix)$^{-1}$. Our linear programming formulation contains seven ordinary constraints (equations (3.21b–g) and (3.21j)), so there will be seven basic variables in the optimal basic feasible solution. An inspection of *Figure 3.8* shows that the seven variables in the optimal basis are the four decision variables $x_1$, $x_3$, $x_4$ and $x_5$, and the ENERGY, PHOSPHRS and MAGNESUM slack variables. If the two lower bounds in our formulation had been treated as ordinary constraints, there would be nine variables in the basis—an increase in basis size of 29%. Finally, note that the reduced cost of the nonbasic variable $x_2$ is 3.84, indicating that the price of maize would have to fall by 3.84 pence kg$^{-1}$ for $x_2$ to enter the optimal basis.

```
START OF PROCEDURE   SOLUTION
- - - - - - - - -


      PROBLEM NAME IN ZPBNAM IS 'CATTLMIX'
      BOUND ROW NAME IS         'UBOUND  '
      OBJECTIVE ROW NAME IS     'OBJECTIV'
      RIGHT HAND SIDE NAME IS   'RHSIDE  '

      MINIMISATION

SOLUTION

FUNCTION VALUE EQUALS      13.227227

TOTAL NUMBER OF ITERATIONS WAS        6

STATUS   SEQ. TYPE   NAME        ROW          SLACK       LOWER       UPPER         COST       REDUCED
         NO.                   ACTIVITY       LEVEL     RHS LIMIT   RHS LIMIT                    COST

ROWS SECTION

    B      1   F   OBJECTIV   13.227227   -13.227227                             1.0000000    1.0000000
    B      2   N   ENERGY     13.114940   -0.11493986  13.000000
           3   N   PROTEIN    160.00000                160.00000                             -0.00019654
           4   N   CALCIUM    7.0000000                7.0000000                             -0.01741019
    B      5   N   PHOSPHRS   7.3629738   -0.36297377  7.0000000
           6   N   SODIUM     3.0000000                3.0000000                             -0.78563058
    B      7   N   MAGNESUM   2.3407691   -0.34076917  2.0000000
           8   7   TOTALWT    1.0000000                1.0000000   1.0000000                 -10.332942


SOLUTION

STATUS   SEQ. TYPE   NAME       LEVEL        LOWER       UPPER       COST       REDUCED
         NO.                                 BOUND       BOUND                  COST

COLS SECTION

     B 65499   P   X1       0.74422535                           10.520000
       65498   P   X2       0.10000000   0.10000000              14.275000    3.8407516
     B 65497   P   X3       0.03448752                           24.500000
     B 65496   P   X4       0.09075130                           14.400000
     B 65495   P   X5       0.03053582   0.01000000              59.560000

      ****SOLUTION COMPLETE

      ****DEMAND(S) SET:-    NONE
```

*Figure 3.8* Part of the LP400 output for the rearing mix example

## Special topics

Some extensions of linear programming and other aspects of mathematical programming that have found application in agriculture are now outlined.

### Matrix generators and report writers

As our fertilizer example illustrates, linear programming packages require the data file to be of a particular form—usually of matrix form. Sometimes, this is not the most convenient form in which to input the data to the computer. In such situations, a separate computer program is usually written to generate the data file from the raw data.

The simplest matrix generators are essentially format converters. The user may wish to specify the coefficients and RHS values of a linear programme in one format, and the program then converts them to the standard format. For example, the linear programmer may wish to input his problem row by row instead of column by column as is required by most linear programming codes. The more sophisticated matrix generators are usually special purpose programs written for a particular application. These generators often carry out a number of calculations prior to producing the data file. Sometimes it is possible to calculate the coefficients and RHS values of a linear programme from a relatively small number of parameters; obviously, much time and effort can be saved if the rules for calculation are incorporated in a matrix generator. Sometimes, there is a need to change units; for example, an application rate might be given in $\text{t hour}^{-1}$, but the coefficient to go into the time constraint must be its reciprocal in $\text{hour t}^{-1}$. Sometimes, there is arithmetic to be performed on the cost coefficients; for example, adding up different components of cost, or in the case of multi-time period models, discounting the costs by a suitable factor to give their net present value.

Report writers, too, are essentially format converters. They are computer programs which translate the output from a linear programming package to a non-technical form readable to the layman.

### Parametric programming

The optimal solution to a linear program can depend critically on the accuracy of certain items of input data, and it is therefore important to explore an optimal solution once one has been obtained. Parametric programming is the technique used to examine the way the optimal solution changes as one or more of the coefficients varies, and most linear programming packages have ancillary procedures which enable these kinds of variations to be made speedily and readily.

There are two basic forms of parametric programming, namely, parametric programming on the objective function and parametric programming on the right-hand side. Now, the general linear programming problem in $m$ constraints and $n$ variables (decision variables and slacks) can be stated algebraically, having first transformed each inequality constraint to an equality by the suitable addition or subtraction of a non-negative slack variable, as

$$\text{minimize} \quad z = \sum_{j=1}^{n} c_j x_j \tag{3.22a}$$

$$\text{subject to} \quad \sum_{j=1}^{n} a_{ij}x_j = b_i \quad (i = 1, 2, \ldots, m) \tag{3.22b}$$

$$\text{and} \quad x_j \geq 0 \quad (j = 1, 2, \ldots, n) \tag{3.22c}$$

where $c_j$ is the cost associated with the $j$th variable, $a_{ij}$ is the coefficient of the $j$th variable in the $i$th constraint, and $b_i$ is the RHS value in the $i$th constraint. Parametric programming on the objective function involves solving the problem

$$\text{minimize} \quad z + \theta z^* = \sum_{j=1}^{n} c_j x_j + \theta \sum_{j=1}^{n} c_j^* x_j$$

subject to equations (3.22b and c) for $0 \leqslant \theta \leqslant \theta_{\max}$. In practice, $z^*$ is often a function of a single variable $x_k$ say, with $c_k^*$ set at $+1$ or $-1$, so as to explore the effect of changing a single cost coefficient $c_k$. Parametric programming on the right-hand side involves minimizing equation (3.22a) subject to

$$\sum_{j=1}^{n} a_{ij}x_j = b_i + \theta b_i^*$$

and equation (3.22c) for $0 \leqslant \theta \leqslant \theta_{\max}$. Solving this problem enables changes in the RHS values to be explored systematically.

### COST RANGING AND RIGHT-HAND-SIDE RANGING

Cost ranging and RHS ranging are particularly useful techniques related to parametric programming. Essentially, cost ranging is using parametric programming to vary each cost coefficient in the objective function in turn to determine how big a change can be made without altering the values of the variables in the optimal solution, as well as which variables would enter the optimal basis if these limits are slightly exceeded. RHS ranging is essentially using parametric programming to vary each RHS value in turn to determine what change can be made without altering the list of variables appearing in the optimal basis, as well as what variables would leave the basis if these limits are slightly exceeded. It should be noted that the actual numerical values of some of the variables in the basis will change with any changes in the RHS values.

An example of cost and RHS ranging, in relation to the rearing mix problem, is shown in *Figure 3.9*. This illustrates that in many situations fairly large inaccuracies in the costs can be tolerated before the values of the variables in the optimal solution change, though the value of the objective function will naturally change. To invoke the cost and RHS ranging procedures in LP400, the statements

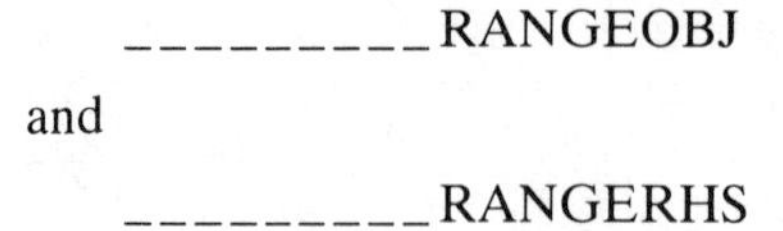

```
_________RANGEOBJ
```

and

```
_________RANGERHS
```

must be inserted immediately after the SOLUTION statement in the COMPILER file.

The reader is referred to Beale (1968) for a more detailed account of parametric programming.

RANGEOBJ

| STATUS | SEQ NO | TYPE | NAME | LEVEL | INPUT COST | LOWER COST | UPPER COST | LOWER VECTOR IN | SEQ NO | STATUS | UPPER VECTOR IN | SEQ NO | STATUS |
|---|---|---|---|---|---|---|---|---|---|---|---|---|---|
| COLS SECTION ************ | | | | | | | | | | | | | |
| B | 65499 | P | X1 | .74422535 | 10.520000 | 9.0140734 | 10.615629 | CALCIUM | 4 | | PROTEIN | 3 | |
| | 65498 | P | X2 | .10000000 | 14.275000 | 10.434248 | ** | | | | | | |
| B | 65497 | P | X3 | .03448753 | 24.500000 | 23.898976 | 48.118334 | PROTEIN | 3 | | SODIUM | 6 | |
| B | 65496 | P | X4 | .09075130 | 14.400000 | 14.300415 | 15.067254 | PROTEIN | 3 | | CALCIUM | 4 | |
| B | 65495 | P | X5 | .03053582 | 59.560000 | 22.318218 | 60.361022 | SODIUM | 6 | | PROTEIN | 3 | |

****DEMAND(S) SET : NONE

RANGERHS

| STATUS | SEQ NO | TYPE | NAME | RHS | UNIT COST | LOWER RHS | UPPER RHS | LOWER VECTOR OUT | SEQ NO | STATUS | UPPER VECTOR OUT | SEQ NO | STATUS |
|---|---|---|---|---|---|---|---|---|---|---|---|---|---|
| B | 2 | N | ENERGY | 13.000000 | .00000000 | ** | 13.114940 | | | | | | |
| | 3 | N | PROTEIN | 160.00000 | .00019654 | 129.47100 | 243.69444 | PHOSPHRS | 5 | L | X5 | 65495 | L |
| | 4 | N | CALCIUM | 7.0000000 | .01741019 | 6.0282316 | 10.377685 | PHOSPHRS | 5 | L | MAGNESUM | 7 | L |
| B | 5 | N | PHOSPHRS | 7.0000000 | .00000000 | ** | 7.3629738 | | | | | | |
| | 6 | N | SODIUM | 3.0000000 | .78563058 | 2.4957325 | 3.4097543 | MAGNESUM | 7 | L | ENERGY | 2 | L |
| B | 7 | N | MAGNESUM | 2.0000000 | .00000000 | ** | 2.3407692 | | | | | | |
| | 8 | Z | TOTALWT | 1.0000000 | 10.332942 | .99165321 | 1.4296000 | ENERGY | 2 | L | X4 | 65496 | L |

****DEMAND(S) SET : NONE

*Figure 3.9* Part of the LP400 output for cost and RHS ranging on the rearing mix example

## Integer programming

Integer programming is the name given to solving linear programming problems where decision variables have to take integer values. The field is divided into *pure* integer programming, in which all the decision variables are required to be integers, and *mixed* integer programming, in which some of the variables have to be integers. Methods of integer programming can be grouped into three categories: cutting planes, enumeration and heuristics.

Cutting plane methods owe their origin to Gomory (1958). Essentially, these methods add new constraints (or cutting planes) in a systematic fashion to the original set of constraints. These cutting planes change the convex set of solutions so that it will contain as an optimum vertex a point which has the necessary coordinates as integers.

Enumeration methods are normally used to solve integer programming problems in practice and the most widely used enumeration method is branch and bound, the original work on which is due to Land and Doig (1960). In branch and bound, a sequence of ordinary linear programmes is solved in which bounds are imposed on the integer variables. The method begins by optimizing the problem as a continuous problem (i.e. ignoring the integer requirements). This solution becomes the first node. Branch and bound uses a tree enumeration procedure, with each branch generated from a node. If the optimal solution to the continuous problem contains integer variables at non-integer values, then one of these variables is selected to form two new branches. These branches define two new problems. In the first, the integer variable being branched on is bounded so as to take only values less than or equal to the integer part of its present level. In the second, the integer variable being branched on is bounded so as to assume only values greater than or equal to the smallest integer greater than its present level. One of these problems is put on a backing list of problems to be solved, whilst the other is optimized. The optimal solution thus obtained is then examined to see whether it is unfeasible, feasible and integer, or feasible and non-integer. If it is unfeasible, the problem last added to the backing list is recovered and optimized. If it is feasible and integer, the solution becomes the 'best' integer solution so far found. If it is feasible and non-integer, this solution becomes a node for branching and two more branches are formed, one of which is put on the backing list whilst the other is optimized. The procedure is repeated until all the problems on the backing list with better costs than that of the best integer solution yet found have been explored. The best integer solution found is then the global integer solution. The method depends on a careful choice of branching and cut-off rules if excessive enumeration is to be avoided.

Heuristic (or approximate) methods are designed to provide 'good' solutions quickly and efficiently, but do not guarantee optimality. Methods of this kind are often useful in practice, as computational experience with more exact algorithms has not been entirely satisfactory. A variety of heuristic methods have been proposed, many of which involve finding local optima by direct search methods.

Integer programming in LP400 is performed using branch and bound. The integer programming routine is invoked by immediately following the SOLUTION statement by the statement

```
_________INTEGER
```

in the COMPILER file. General integer variables are defined in the COLUMNS section of the EXECUTOR file, by writing GX in columns two and three of the lines specifying the names and the coefficients of those variables which are required to take integer values; similarly, 0–1 integer variables (i.e. variables which can only assume the values 0 or 1) are defined by writing IX in columns two and three.

A more comprehensive account of integer programming can be found in Hadley (1964) and in Beale (1968).

## Separable programming

Separable programming, due to Miller (1963), is a nonlinear programming technique which can be used to find a global or local optimum to a large number of nonlinear problems. The technique allows nonlinear functions of *single* variables (e.g. $x_1^3$, $\ln x_2$, $\exp x_3$) to be used in either the constraints or the objective function of an otherwise linear programming problem. The limitation that nonlinear terms must be functions of a single variable is not particularly restrictive, as product terms

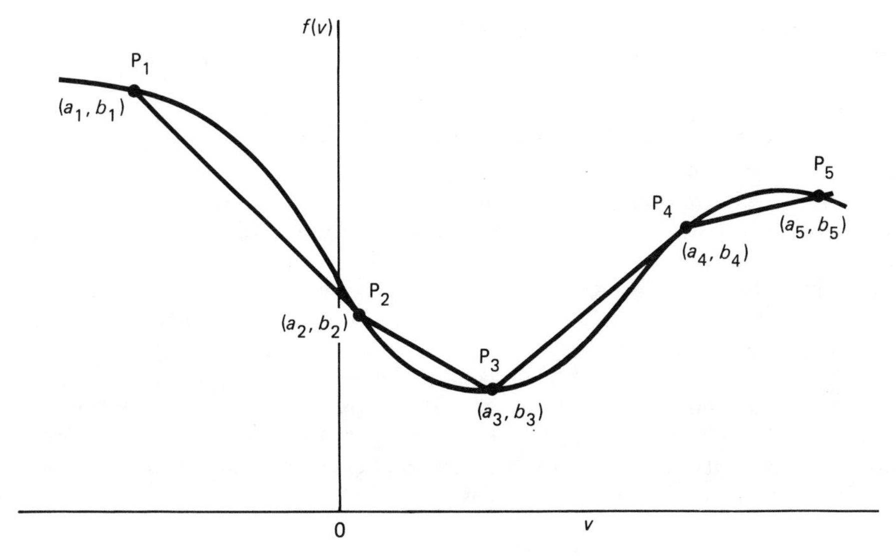


*Figure 3.10* A piecewise linear approximation to a nonlinear function of a single variable

can be separated into sums and differences of functions of a single variable using simple transformations. For example, the term $x_1x_2$ can be replaced by $u_1^2 - u_2^2$ where

$$u_1 = (x_1 + x_2)/2 \quad \text{and} \quad u_2 = (x_1 - x_2)/2$$

and the term $x_1^{\alpha_1} x_2^{\alpha_2} x_3^{\alpha_3} \ldots x_n^{\alpha_n}$ can be replaced by $e^u$ where

$$u = \alpha_1 \ln x_1 + \alpha_2 \ln x_3 + \alpha_3 \ln x_3 + \ldots + \alpha_n \ln x_n$$

Once all nonlinearities are in the form of functions of a single variable, each nonlinear function can be replaced by a piecewise linear approximation. Consider the nonlinear function $f(v)$ of a single variable $v$ shown in *Figure 3.10*. A piecewise linear approximation to $f(v)$ is shown on the figure as $P_1 P_2 P_3 P_4 P_5$, where the coordinates of the points $P_i$, $i = 1, 2, \ldots, 5$, are $(a_i, b_i)$. Now consider the equations

$$\sum_{i=1}^{5} \lambda_i = 1 \tag{3.23a}$$

$$\sum_{i=1}^{5} a_i \lambda_i = v \tag{3.23b}$$

$$\sum_{i=1}^{5} b_i \lambda_i = f(v) \tag{3.23c}$$

and

$$\lambda_i \geqslant 0 \quad \text{for } i = 1, 2, \ldots, 5 \tag{3.23d}$$

The vertices of the piecewise linear approximation to $f(v)$ are obtained by putting all but one of the $\lambda_i$ equal to zero in equations (3.23a–c). Any other point on the approximation is obtained by allowing two adjacent $\lambda$'s to take positive values whilst setting the others to zero. Incidentally, $v$ and $f(v)$ need not necessarily be non-negative, the points $P_i$ do not have to be equally spaced, and the interval $(a_1, a_5)$ may be larger than the interval of interest.

In separable programming, the $\lambda_i$ are called a *set of special variables* and equations (3.23a) and (3.23b) are known respectively as the *convexity row* and *reference row* for this set of special variables. The nonlinear problem can be linearized by substituting a piecewise linear approximation for each nonlinearity $f(v)$, and by adding a set of special variables to the list of variables and a convexity row and reference row to the constraints of the problem for each nonlinearity. The linearized problem can now be solved as an ordinary linear programme provided we restrict the special variables to be considered as candidates for entering the basis. The restrictions are that no more than two special variables from any one set can appear in the current basis, and that the members of a pair must be adjacent.

We illustrate the technique by solving the problem

$$\text{minimize} \quad z = x_1 + 2x_2 \tag{3.24a}$$

$$\text{subject to} \quad -2x_1 + x_2 \leqslant 6 \tag{3.24b}$$

$$x_1 + 5x_2 \leqslant 65 \tag{3.24c}$$

$$x_1 \leqslant 10 \tag{3.24d}$$

$$x_1 \geqslant 2 \tag{3.24e}$$

$$1.2x_1 + x_2 - 0.2x_1^2 \geqslant 1.8 \tag{3.24f}$$

$$x_1, x_2 \geqslant 0 \tag{3.24g}$$

The graphical solution to this two-dimensional problem is shown in *Figure 3.11*, the global optimum being $z = 2.4$, $x_1 = 2$, $x_2 = 0.2$. Before solving by separable programming, we first check that all nonlinearities are in the form of functions of a single variable. The only nonlinearity is the $x_1^2$ term in equation (3.24f), so this requirement is satisfied. To apply the technique, we arbitrarily choose $(0, 10)$ as

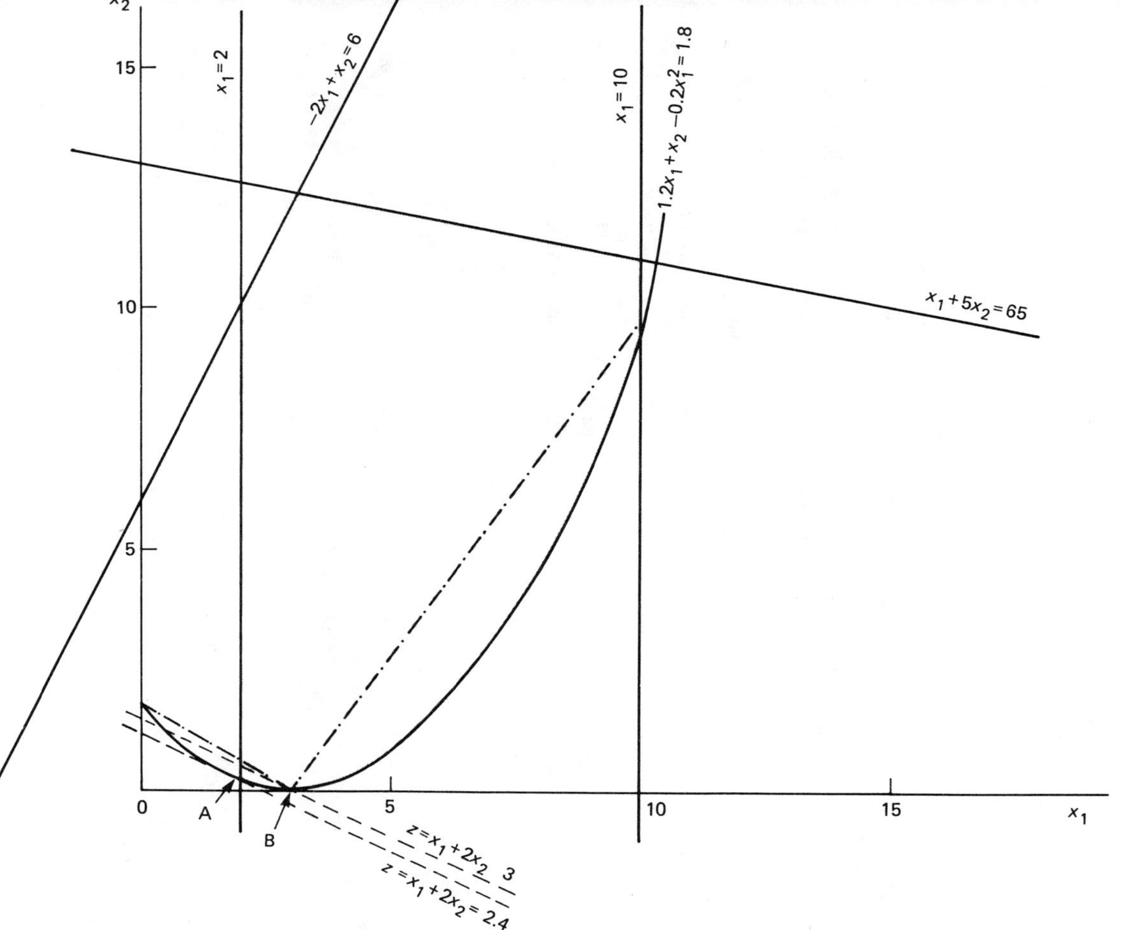


*Figure 3.11* The graphical solution to the problem defined by equations (3.24a–g). The global optimum is $z = 2.4$, $x_1 = 2$, $x_2 = 0.2$, which occurs at the point A. The separable programming solution, using the linear approximation to the curve $1.2x_1 + x_2 - 0.2x_1^2 = 1.8$ defined by equations (3.25a–e) and shown by the dotted lines, gives the local optimum $z = 3$, $x_1 = 3$, $x_2 = 0$. This occurs at point B

our interval of interest for $x_1$ and the points $P_1(0,0)$, $P_2(3,9)$, $P_3(10,100)$ as the vertices of our piecewise linear approximation to $f(x_1) = x_1^2$. Therefore, a linear approximation to $x_1^2$ is given by

$$\lambda_1 + \lambda_2 + \lambda_3 = 1 \quad (3.25a)$$

$$0\lambda_1 + 3\lambda_2 + 10\lambda_3 = x_1 \quad (3.25b)$$

$$0\lambda_1 + 9\lambda_2 + 100\lambda_3 = x_1^2 \quad (3.25c)$$

$$\lambda_1, \lambda_2, \lambda_3 \geqslant 0 \quad (3.25d)$$

$$\lambda_1, \lambda_2, \lambda_3 = \text{special variables} \quad (3.25e)$$

The problem is linearized by using this piecewise linear approximation for $x_1^2$; the resultant linear approximation to equation (3.24f) is shown by the dotted lines in *Figure 3.11*. The graphical solution to the linearized problem is also shown in *Figure 3.11*, giving the local optimum $z = 3$, $x_1 = 3$, $x_2 = 0$ to the original problem.

To implement a computer solution using LP400, we rewrite equations (3.24a–g) and (3.25a–e) as follows:

| | | | |
|---|---|---|---|
| minimize | $z = x_1 + 2x_2$ | [objective function] | (3.26a) |
| subject to | $-2x_1 + x_2 \leqslant 6$ | [constraint 1] | (3.26b) |
| | $x_1 + 5x_2 \leqslant 65$ | [constraint 2] | (3.26c) |
| | $x_1 \leqslant 10$ | [constraint 3] | (3.26d) |
| | $x_1 \geqslant 2$ | [constraint 4] | (3.26e) |
| | $1.2x_1 + x_2 - 0.2x_1^2 \geqslant 1.8$ | [constraint 5] | (3.26f) |
| | $\lambda_1 + \lambda_2 + \lambda_3 = 1$ | [convexity row] | (3.26g) |
| | $3\lambda_2 + 10\lambda_3 - x_1 = 0$ | [reference row] | (3.26h) |
| | $9\lambda_2 + 100\lambda_3 - x_1^2 = 0$ | [function row] | (3.26i) |
| | $x_1, x_2, \lambda_1, \lambda_2, \lambda_3 \geqslant 0$ | [non-negativity conditions] | (3.26j) |
| | $\lambda_1, \lambda_2, \lambda_3 = \text{special variables}$ | | (3.26k) |

Note that in LP400 terminology, the equations of the type $f(v) = \Sigma b_i \lambda_i$ are called *function rows*. The EXECUTOR file for the problem, as defined by equations (3.26a–k), is given in *Figure 3.12*. The separable programming routine is invoked by inserting the statements

```
_________STARTSEP
_________INTERP
_________ENDSEP
```

before the EXIT statement in the COMPILER file. A powerful facility in LP400 separable programming is that the piecewise linear approximations to the nonlinear functions can be refined automatically. When a local optimum has been obtained, the grid is refined in the region of this solution using the procedure INTERP. This enables global or near-global optima to be obtained in most situations. A control language program to do this is given with the solution to Exercise 3.5 on p. 291.

```
INPUT
TITLE       SEPARABLE PROGRAMMING EXAMPLE
ROWS
  F  OBJECTIV
  P  CNSTRT1
  P  CNSTRT2
  P  CNSTRT3
  N  CNSTRT4
  N  CNSTRT5
 CZ  CONX1
  Z  REFX1
  Z  FUNX1
COLUMNS
    X1          OBJECTIV   1                CNSTRT1    -2
    X1          CNSTRT2    1                CNSTRT3    1
    X1          CNSTRT4    1                CNSTRT5    1.2
    X1          REFX1      -1
    X2          OBJECTIV   2                CNSTRT1    1
    X2          CNSTRT2    5                CNSTRT5    1
    X1SQ        CNSTRT5    -0.2             FUNX1      -1
 H1 X1000HHH    CONX1                       REFX1
    X1000HHH    FUNX1
 HF X100000F    FUNX1                       'X2'       1
  W X1000000    REFX1      0
  W X1000001    REFX1      3
  W X1000002    REFX1      10
RHS
    RHSIDE      CNSTRT1    6                CNSTRT2    65
    RHSIDE      CNSTRT3    10               CNSTRT4    2
    RHSIDE      CNSTRT5    1.8              CONX1      1
ENDATA
```

*Figure 3.12* The EXECUTOR file for the separable programming example (equations (3.26a–k))

Using this (having first deleted the MOVE statement as the objective function in this example is to be minimized not maximized), in conjunction with the EXECUTOR file shown in *Figure 3.12*, produces the global optimum to our problem: $z = 2.4$, $x_1 = 2$, $x_2 = 0.2$.

The reader is referred to Beale (1968) and International Computers Limited (1970) for a fuller account of the practicalities of separable programming, and to Hadley (1964) for a more mathematical treatment.

## Dynamic programming

Dynamic programming is a method for solving optimization problems which can be formulated as a sequence of decisions. Unlike most other mathematical programming techniques, it is not a variant of linear programming. The method was developed by Bellman and colleagues at the Rand Corporation (Bellman, 1957; Bellman and Dreyfus, 1962), and is based on Bellman's *principle of optimality*, which states:

> *An optimal policy has the property that, whatever the initial state and initial decision are, the remaining decisions must constitute an optimal policy with respect to the state resulting from the first decision.*

The technique utilizes the mathematical notion of recursion and is best introduced by way of an example. Consider the shortest route problem given in *Figure 3.13*, which shows a network of towns represented by circles, and roads represented by lines. The lengths of the roads (in miles) are indicated. A traveller wishes to go

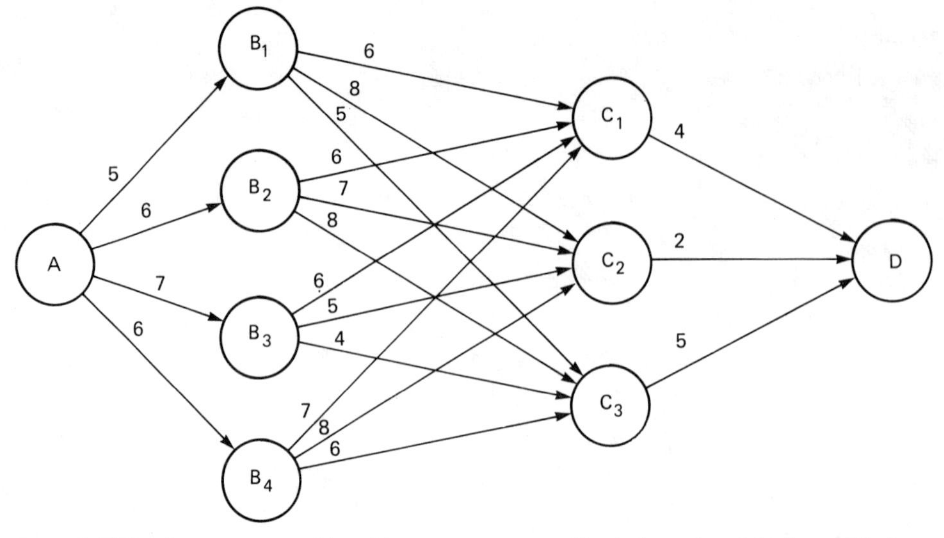


*Figure 3.13* The shortest route problem. Towns are represented by circles and roads by lines, with the lengths of the roads indicated

from town A to D by the shortest possible route, moving in the direction indicated by the arrows. What is his optimal path?

Before tackling this problem, we define the terms: state, stage and policy, as used in dynamic programming. A *state* is a configuration of a system, and being at a particular town corresponds to a state in the shortest route problem. A *stage* is a transition from one state to an adjacent state. In *Figure 3.13*, travelling from one town to the next comprises a stage and every route from A to D has three stages:

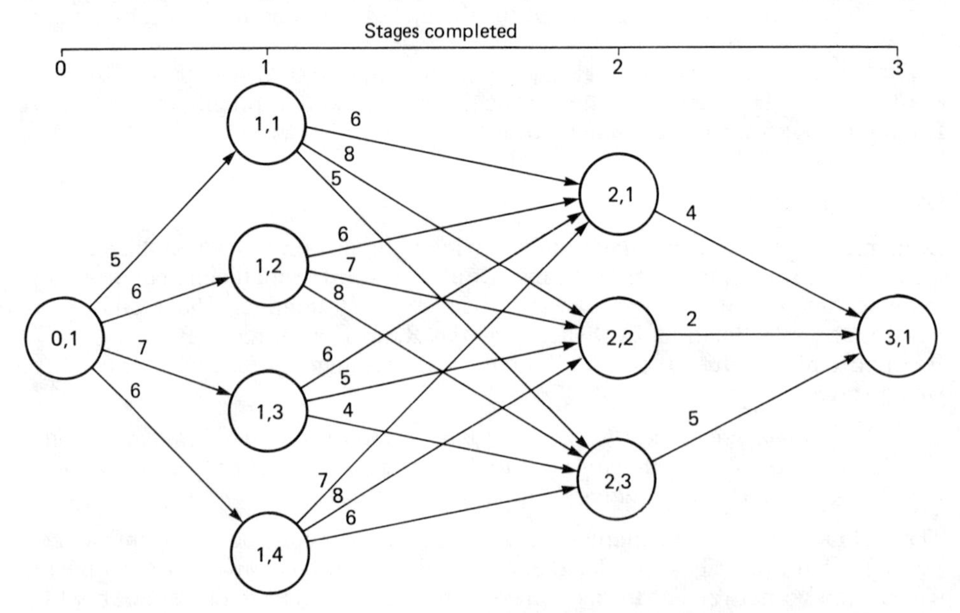


*Figure 3.14* The shortest route problem with towns denoted by stage and state variables

A → B, B → C, C → D. A *policy* is a set of actions in accordance with a particular objective. In our problem, the actions $A \rightarrow B_3 \rightarrow C_2 \rightarrow D$ and $A \rightarrow B_4 \rightarrow C_1$ are both examples of a policy.

The solution of a dynamic programming problem involves two steps. The first step is the construction of a recurrence relation relating the various stages, and the second is using the recurrence relation to compute an optimal policy. For computational purposes, stage and state variables are defined. Let $n$ be the number of stages completed and $i$ the individual state at each stage; state $(n,i)$ is thus the $i$th state at the $n$th stage. The shortest route problem with towns denoted by stage and state variables is shown in *Figure 3.14*.

To construct a recurrence relation for the shortest route problem, we define $f(n, i)$ as the minimum distance from Town (0,1), i.e. A, to Town $(n, i)$, and $r(n - 1, j : n, i)$ as the distance from Town $(n - 1, j)$ to Town $(n, i)$, e.g. $r(1,1 : 2,2) = 8$ miles in our problem. By the principle of optimality, we have the recurrence relation:

$$f(n, i) = \underset{j}{\text{minimum}}\, [f(n - 1, j) + r(n - 1, j : n, i)] \tag{3.27}$$

where, by definition, $f(0,1) = 0$. Equation (3.27) is typical of many recurrence relations arising in dynamic programming.

Repeated application of equation (3.27) yields the following.

Stage 1:

$$\begin{aligned} f(1,1) &= f(0,1) + r(0,1 : 1,1) = 5 \\ f(1,2) &= f(0,1) + r(0,1 : 1,2) = 6 \\ f(1,3) &= f(0,1) + r(0,1 : 1,3) = 7 \\ f(1,4) &= f(0,1) + r(0,1 : 1,4) = 6 \end{aligned}$$

Stage 2:

$$f(2,1) = \text{minimum} \begin{Bmatrix} f(1,1) + r(1,1 : 2,1) = 5 + 6 = 11 \\ f(1,2) + r(1,2 : 2,1) = 6 + 6 = 12 \\ f(1,3) + r(1,3 : 2,1) = 7 + 6 = 13 \\ f(1,4) + r(1,4 : 2,1) = 6 + 7 = 13 \end{Bmatrix}$$

therefore

$$f(2,1) = f(1,1) + r(1,1 : 2,1) = 11$$

Similarly,

$$f(2,2) = f(1,3) + r(1,3 : 2,2) = 12$$

and

$$f(2,3) = f(1,1) + r(1,1 : 2,3) = 10$$

Stage 3:

$$f(3,1) = \text{minimum} \begin{Bmatrix} f(2,1) + r(2,1 : 3,1) = 11 + 4 = 15 \\ f(2,2) + r(2,2 : 3,1) = 12 + 2 = 14 \\ f(2,3) + r(2,3 : 3,1) = 10 + 5 = 15 \end{Bmatrix}$$

therefore

$$f(3,1) = f(2,2) + r(2,2 : 3,1) = 14$$

The minimum distance from Town (0,1) to Town (3,1) is 14 miles, and the optimal route is found by back substitution, that is

$$\begin{aligned} f(3,1) &= f(2,2) + r(2,2:3,1) \\ &= f(1,3) + r(1,3:2,2) + r(2,2:3,1) \\ &= f(0,1) + r(0,1:1,3) + r(1,3:2,2) + r(2,2:3,1) \end{aligned}$$

The optimal policy is thus the route Town (0,1) → Town (1,3) → Town (2,2) → Town (3,1), i.e. A → $B_3$ → $C_2$ → D.

In solving the shortest route problem, we started from the initial state and worked *forwards* through the problem. This is known as forward recurrence, or dynamic programming of the forward type. The problem, however, could equally well have been solved by a system of backward recurrence (*see* Solution 3.6, p. 292). Indeed, a backward recurrence formulation is generally adopted in practice as the backward approach is usually computationally more efficient than the forward one.

The reader is referred to Hastings (1973) for a comprehensive and readable introduction to both deterministic and stochastic dynamic programming, and to Bellman and Dreyfus (1962) for a more advanced treatment.

## Exercises

### Exercise 3.1

Solve graphically the linear programme:

$$\begin{aligned} \text{maximize} \quad & z = x_1 + 2x_2 \\ \text{subject to} \quad & -2x_1 + x_2 \leqslant 6 \\ & x_1 + 5x_2 \leqslant 65 \\ & x_1 \leqslant 10 \\ & x_1, x_2 \geqslant 0 \end{aligned}$$

and shade in the region representing the feasible solutions. Check your graphical solution by resolving the problem using a linear programming package.

### Exercise 3.2

A farmer wishes to determine a daily ration for the coming fortnight for an 80 kg ewe carrying twin lambs. The ewe, which is six weeks from lambing, has a metabolizable energy (ME) requirement of 16 MJ day$^{-1}$ and a crude protein (CP) requirement of 160 g day$^{-1}$. The feeds available are hay (ME value 7.74 MJ per kg dry matter (DM), CP value 116 g per kg DM) and a concentrate mix (ME value 12.78, CP value 159). The farmer wishes to minimize the use of concentrates. The ewe's appetite limit for hay is 12.79 g DM per kg liveweight per day. The consumption of concentrates depresses the ewe's appetite for hay by 0.63 g DM hay per g DM concentrates consumed. Formulate the problem as a linear programme and solve it graphically. Use a linear programming package to check your graphical solution.

### Exercise 3.3

A feed compounder wishes to formulate a least-cost finishing mix for pigs which will satisfy certain nutritional requirements. The mix must have a digestible energy (DE) concentration of at least 13 MJ $kg^{-1}$, a minimum digestible crude protein (DCP) content of 12%, a minimum lysine content of 0.6% and a maximum fibre content of 5%. In addition, the mix must contain at least two parts per hundred of a mineral–vitamin supplement. The available ingredients, together with their nutritive value, fibre content and cost are shown below.

| *Ingredient* | *DE* [MJ $kg^{-1}$] | *DCP* [g $kg^{-1}$] | *Lysine* [g $kg^{-1}$] | *Fibre* [%] | *Cost* [£ $t^{-1}$] |
|---|---|---|---|---|---|
| Soyabean meal | 15 | 410 | 28.0 | 5.2 | 145 |
| Weatings | 11.9 | 99 | 6.4 | 7.5 | 110 |
| Maize meal | 14.5 | 73 | 2.6 | 2 | 160 |
| Barley meal | 12.7 | 77 | 3.2 | 4.6 | 130 |
| Mineral–vitamin supplement | — | — | — | — | 400 |

Calculate the optimal mix. By how much does the cost of maize meal have to fall before it is used?

### Exercise 3.4

A company manufacturing tractor components intends to build a new factory. Three possible sites have been identified for the development. The cost of building a new factory at a particular site involves a fixed purchase cost for the land and a variable building cost depending on the size of the factory. The company intends that the factory be at least 2 ha in size, and the information about each site is given below.

| *Site* | *Area of site* [ha] | *Purchase cost* [£ × $10^3$] | *Building cost* [£ × $10^3$ $ha^{-1}$] |
|---|---|---|---|
| 1 | 2.2 | 9.9 | 300 |
| 2 | 2.8 | 16.8 | 295 |
| 3 | 3.2 | 16.0 | 296 |

Use integer programming to determine which site the company should select, so as to minimize the cost of the development.

### Exercise 3.5

Use separable programming to

maximize $z = x + y$
subject to $x^2 + y^2 \leqslant 4$ and $x, y \geqslant 0$

**Exercise 3.6**

Re-solve the shortest route problem given in *Figure 3.13* (p. 64) using a backward recurrence dynamic programming formulation.

Chapter 4

# Testing and evaluation of models

## Introduction

The testing and evaluation of a model is a continuous process that should be carried out right from the beginning of a modelling project. Any other approach is usually ineffective. While much of the material in this chapter is more relevant to dynamic deterministic models of the mechanistic or semi-mechanistic type, the most empirical models are also based on agricultural or biological assumptions, which should be explicitly stated and considered. A useful account of this topic is given by Penning de Vries (1977), who emphasizes the lack of an agreed terminology and the need for clear definitions when these matters are discussed. Some modellers speak at length of validation and verification, although we feel that the two terms, testing and evaluation, suit present purposes adequately.

The term *testing* is taken here to mean checking for methodological correctness. That is: the mathematical equations must correctly represent the stated agricultural and biological assumptions; the equations must be self-consistent and dimensionally homogeneous; any algebra and analysis must be free from error; any computer coding must be correct and achieve the stated and intended aims; and so on. In some areas of agriculture and biology, the modelling literature is full of such failing. In fact, in large and complex models, it can be so difficult not to make errors, that the preferred approach is to assume that errors will be made, and to adopt a self-checking method of working that catches the errors as or shortly after they are made. To find them later can be nearly impossible. In the Popperian sense, testing is an objective process which has the result positive or negative (Popper, 1958).

*Evaluation*, on the other hand, is not a wholly objective process and it is less easy to define. Evaluation is concerned with aspects such as appropriateness (to objectives), plausibility, goodness-of-fit, elegance, economy, simplicity, and utility; it is rare indeed for a model to possess all these attributes, and people will always vary in the weights they attach to the individual items. Final evaluation of a model can only properly take place after testing has been carried out, and one is sure of the model's methodological correctness.

Assuming that objectives have been adequately defined elsewhere, most modelling projects can be divided into four parts:

1. the structure of the model, including its agricultural and biological assumptions;
2. the mathematical expression of 1, and any further mathematical analysis;

3. the solution of the equations derived in 2, usually by means of a computer program;
4. examining and interpreting the predictions of the model, especially in relation to 1, and fitting the predictions to experimental data where available.

Testing and evaluation should be carried out at each step, and satisfactorily completed before pressing on to the next part. Often these steps do overlap and interact, and one may sometimes be compelled to go back over ground again that one thought had been satisfactorily won. For example, the discipline imposed by mathematical description may force new and unexpected elements into the underlying biology (1).

## Model structure

A mathematical model only represents a set of agricultural or biological assumptions. These assumptions are always a simplification of reality. There are always colleagues, critics and referees who are unable to accept the degree of simplification chosen—the model is either 'too complex' or 'over-simplified'. As discussed in Chapter 1, the modeller may wish to develop ideas, or describe and summarize, or produce a practical tool for management. The degree of simplification, which is often connected to the empirical/mechanistic content, should be appropriate to the objectives.

Much of the skill of modelling lies in assembling the assumptions and deciding on approximations. These can be Procrustean in scope, and colleagues who see their own particular hobby-horse put aside may be deeply offended. One tries to make 'reasonable' assumptions, that are biologically plausible. The modeller should not deceive himself by thinking that there are objective methods for this—guesswork is used; hopefully it is informed guesswork, but it is inevitably based on one's own experience, technical skills and research attitudes.

Since the assumptions are so important, it is essential that the modeller familiarizes himself thoroughly with the biology of the problem. Sometimes one sees mathematicians or statisticians trying to do biological modelling, but keeping the biology at arm's length. It rarely works. The collaborating biologist usually fails to grasp what is being attempted in the mathematical representation, and the mathematician may be unable to assess and rank the biological possibilities. The result is likely to be work of little value, and frustration for the concerned parties.

The structure of a model can only be evaluated and not tested (other than for self-consistency). The evaluation depends upon assessing and relating current knowledge to the modelling objectives, and deciding the type of model that is most likely to achieve them. How this is done is, at present, a matter of judgement, rather than a definable objective procedure.

## Mathematical equations

The accurate translation of agricultural and biological ideas into mathematics is essential. This requires mathematical fluency, and also a sound understanding of the biological ideas being translated. To accomplish this without error, there are some simple rules that can be followed, which reduce the possibility of error and help in detecting errors. As with computer programming, one must expect to make errors, and use an approach that makes it easy to pick them up.

The first step is to define the symbols. It is worth giving this careful thought, since equations are much easier to read, understand and check if similar symbols are used for similar quantities, and if the similar symbols have the same units. For example, rate constants with dimensions time$^{-1}$ may be denoted by $k_1, k_2, \ldots$. In a plant model, the components of dry weight may be shown by $W_\ell$, $W_s$, $W_r$ for the leaf, stem and root. Where there is a consensus in the literature about the use of certain symbols for certain quantities, the traditional symbols should be used unless there are good reasons for doing otherwise. The use of computer language notation in mathematical analysis (or in scientific papers) is, in our view, mistaken. It is less efficient, less readable, and less easily checked than the more conventional mathematical notation that has evolved over many centuries using the Latin and Greek alphabets, with upper or lower case, and subscripts and superscripts as needed. Computer notation, while still quite primitive, is rapidly evolving. Work presented in such notation may quickly become inaccessible. Indeed, many journals do not allow computer notation to be used within the body of a scientific communication.

The second step is to check dimensions. In an equation, each term must have the same dimensions as all the others. For this purpose, a symbol table should be constructed, which has a verbal definition of each symbol used, and also its dimensions. Sometimes, it is helpful to work out also the dimensions of groups of symbols that often occur together. A single system of units (preferably SI) should be used throughout the model, even when these are not the customary units. This avoids troublesome conversion factors for quantities like g to kg, or m$^3$ to litres, which can very easily give rise to errors.

A third step is to check for mathematical consistency and completeness. There must be enough equations to define the problem, but the problem must not be over-defined. For instance, for a dynamic model with three state variables, three difference equations or first-order differential equations are required. For a simple static problem with five variables, five equations are needed, although this is not generally true for a linear-programming problem (p. 37).

A fourth useful check is for biological consistency and completeness at the whole-system level. In both animal and plant models, carbon (C) and nitrogen (N) accounting can be carried out. One can write

$$\frac{\mathrm{d}}{\mathrm{d}t}(\text{total C or N in system}) = \text{system inputs} - \text{system outputs} \qquad (4.1)$$

Internal transfers, say from pool $i$ to pool $j$, $T_{ij}$, occur twice in the mathematical equations of the model: positively in the differential equation for pool $j$, and negatively in that for pool $i$. Summing the equations should give cancellation of all internal transfers. For instance, however complicated a plant model is, some of its equations should sum to

$$\text{gross photosynthetic rate} = \text{growth rate} + \text{respiration rate} \qquad (4.2)$$

where these quantities are expressed in the same units. An analogous equation for an animal model relates dietary inputs to growth and outputs.

## Solving the equations of the model

This process is usually carried out by computer, using a language such as BASIC, PASCAL, ALGOL or FORTRAN, or possibly a special purpose language such as

CSMP or LP400. The general principle to be followed is that mistakes will be made, so that programs should be written such that the mistakes are easily located and corrected. The advice in programming manuals on writing clear self-documenting modular programs should be taken very seriously; more rapid progress is often made if programs are always written as though they were to be understood and used by others. All the familiar points about careful definition of program symbols, modular program construction and checking, and frequent comment statements, are relevant.

Where possible, self-consistency checks of the type in equations (4.1) and (4.2) should be written into the program; these may pin-point programming errors or mathematical errors in model formulation. In the early runs of a program, it is often worthwhile to print out every left-hand-side quantity in the program, and sometimes errors can be located by doing a detailed check on a hand calculator working direct from the mathematical equations (not from the programmed version of these equations).

Checks should also be applied against the possibility of integration errors, due to an inappropriate integration method or to too large an integration interval. The results of running the program should be reasonably stable against variation in integration method and interval. Some machines have rather a short word-length, and rounding errors can be soon encountered if very short integration intervals are used (p. 26).

## Prediction and comparison with experiment

When the model has been carefully tested, and is free from mathematical, computational, and numerical errors, its predictions then truly reflect the consequences of the assumptions (biological, agricultural, empirical) on which it is based. The model can now be used for whatever purposes were intended.

It is usual to first examine the qualitative behaviour of the model, and if this is satisfactory and if appropriate data are available (which is often not the case), to proceed to a quantitative fitting as described on pp. 27–31. This latter process, of estimating parameters by fitting to a set of data, is sometimes referred to as calibrating the model.

## Sensitivity analysis, parameter ranking, and model simplification

Consider a model with a single adjustable parameter $P$, which is fitted to a data set by minimizing a residual sum of squares $R$ with $v$ degrees of freedom (pp. 27–31). The variance of $P$, $V(P)$, is given by

$$V(P) = \frac{R}{v} \frac{1}{\partial^2 R / \partial P^2} \tag{4.3}$$

For comparing the effects of different parameters on model performance, a dimensionless quantity that is independent of the absolute value of the parameter is required. The coefficient of variation of $P$, $CV(P)$, is such a quantity:

$$CV(P) = \frac{[V(P)]^{1/2}}{P} \tag{4.4}$$

Thus, referring to *Figure 2.3* (p. 29), curve S has a low value of $CV(P)$, whereas curve I has a high value of $CV(P)$.

For a model with several adjustable parameters, equation (2.23) is applied to calculate the variance of the $i$th parameter $P_i$, and equation (4.4) becomes

$$CV(P_i) = \frac{[V(P_i)]^{1/2}}{P_i} \tag{4.5}$$

The coefficients of variation can now be used to rank the parameters, a low value of $CV(P_i)$ indicating that the parameter has a considerable effect on the fit of the model to that particular data set, and vice versa. Different data sets may give different results. Experience with crop and plant models, using high quality data from controlled environments, suggests that values for $CV(P_i)$ in the range 0.05 to 0.2 are reasonable, whereas values much above 0.2 may indicate that the part of the model to which $P_i$ belongs should be scrutinized.

A sensitivity analysis with a parameter ranking may be used to indicate ways in which a model can be simplified. It might be possible to remove a parameter with a very large coefficient of variation entirely from a model. However, there may be good biological (or other) reasons for retaining a parameter in a model, even where it is having little effect on the model's predictions. Since this analysis depends on the data sets considered, and the method chosen for calculating a residual, its application requires caution.

Consider next a model that predicts (amongst other things) a quantity $Y$ at a single particular time point—$Y$ might be the total lactation yield of a dairy cow, or the yield of a crop at maturity. It is supposed that the model has been satisfactorily tested and evaluated, including giving a reasonable fit to experimental data. Some of the parameters of the model will be physiological, and some will be environmental; some of the environmental parameters will be controlled by management, e.g. feeding regimes of an animal, or fertilizer inputs to a crop. Objectives for programmes of animal or plant breeding, or priorities for management, may be formulated more objectively if it is possible to rank the parameters $P_i$ with respect to their relative effects on yield $Y$. A suitable dimensionless quantity for measuring the sensitivity of $Y$ to parameter $P_i$ is $S(Y, P_i)$, defined by

$$S(Y, P_i) = \frac{\partial Y}{\partial P_i}\frac{P_i}{Y} \approx \frac{\delta Y}{Y}\frac{P_i}{\delta P_i} \tag{4.6}$$

$\delta P_i$ denotes a small finite change in parameter $P_i$, and $\delta Y$ is the small change this causes in $Y$. $S(Y, P_i)$ denotes the sensitivity of the quantity $Y$ to the parameter $P_i$. For computing the $S(Y, P_i)$, a 5% parameter increment is usually sufficient ($\delta P_i/P_i = 0.05$). If $S(Y, P_i) = 1$, a given fractional change in the parameter value produces the same fractional change in yield $Y$. Parameters with $S(Y, P_i) > 1$ have larger effects on yield, and vice versa.

An example of use of a model to rank parameters is given by Thornley, Hurd and Pooley (1981, Table 1), who rank the parameters of a leaf growth model in terms of the marginal contribution of the parameters to the plant carbon budget.

## Exercises

### Exercise 4.1

Using SI units (kg, m, day), what are the units of the parameters $a$, $n$, $b$ and $k$ in the equation

$$\frac{dw}{dt} = at^n e^{-bt} - kt$$

where $w$ [kg] and $t$ [day] are mass and time variables?

### Exercise 4.2

If the fractional nitrogen content of tissue, $f_N$, is defined by the equation $W_N = f_N W$, where $W_N$ [kg nitrogen] and $W$ [kg total matter] are the respective masses of nitrogen and total matter, derive the units of $f_N$. Can these units be simplified? What are the units of leaf area index (LAI)?

### Exercise 4.3

Economists often talk about demand elasticities and cost elasticities. Find out what these are, and compare the ideas with the analysis of model sensitivity and parameter ranking in equations (4.3) to (4.6).

Chapter 5

# Growth functions

## Introduction

In both the plant and animal sciences, growth functions have been used for many years, usually to provide a mathematical summary of time-course data on the growth of an organism or part of an organism. The term growth function is generally used to denote an analytical function which can be written down in a single equation. Thus, a general growth function connecting dry weight $W$ to time $t$ is

$$W = f(t) \tag{5.1}$$

where $f$ denotes some functional relationship.

The development of this area has occurred mostly in the plant sciences, and the early papers of Richards (1959, 1969) are well worth reading. More recently, textbooks on the topic have been published: Hunt (1982) gives a general and readable account of the subject; Causton and Venus (1981) have written a more specialized text concentrating on the growth function known as the Richards function (p. 85).

The use of growth functions is largely empirical: the form of the function $f$ will sometimes be chosen by simply looking at the data and making an informed guess; however, it is preferable to try to select or construct a function that has some biological plausibility, and whose parameters may be meaningful—that is, they may characterize some underlying physiological or biochemical mechanism or constraint. For example, autocatalysis produces exponential growth; limited substrate or nutrient gives rise to an asymptote; and senescence or differentiation may cause diminishing growth rates and asymptotic behaviour, as in the Gompertz equation. Scientists no longer have the naive expectation of finding that chimera: an analytical function which will describe animal or plant growth under a wide range of conditions and whose parameters have biological significance. This is reflected in the growth of animal and plant modelling, and its emergence as a discipline in its own right. For it is recognized that it is most unlikely that one would ever be able to describe a complicated system such as an animal or plant in a simple manner. The use of two or more equations, many parameters, numerical methods and simulation, is now relatively widespread. Admittedly, the application of such procedures will only define numerically the dry weight–time relationship of equation (5.1), although now, this will be one of several predicted relationships which are based on the theoretical structure or model, which itself incorporates various physiological or biochemical assumptions about mechanism. Although all models are empirical at some level, growth functions may be of value at the organ, tissue and cell levels, as well as at the whole-organism level.

It is convenient, and also traditional, to approach the topic of growth functions by first discussing the rate of growth, or $\mathrm{d}W/\mathrm{d}t$ in the case of dry weight $W$. Differentiation of equation (5.1) with respect to time gives

$$\frac{\mathrm{d}W}{\mathrm{d}t} = g(t) \quad \text{where} \quad g(t) = \frac{\mathrm{d}f(t)}{\mathrm{d}t} \tag{5.2}$$

Elimination of the time variable $t$ between equations (5.1) and (5.2) then gives

$$\frac{\mathrm{d}W}{\mathrm{d}t} = h(W) \tag{5.3}$$

where $h$ is a function. Note that equation (5.3) is now in the familiar 'rate = function of state' form (p. 16) with $W$ as a state variable. An equation in the form of equation (5.3) can often be interpreted biologically, usually in biochemical terms; however, for some growth functions (certain exponential polynomials, for example), it may not be practical to derive an analytical equation (5.3) from equations (5.1) and (5.2). Many growth functions are obtained by first writing down a relation of the equation (5.3) type, although occasionally, as with the Gompertz equation (5.29), the form

$$\frac{\mathrm{d}W}{\mathrm{d}t} = u(W, t) \tag{5.4}$$

where $u$ denotes some function of $W$ and $t$, may be preferred.

Since growth in dry weight is of most interest, our discussion will be in these terms, although clearly the results may be equally applicable to other variables such as liveweight, leaf area, fresh weight, and so on. Also, although units should be chosen to suit the occasion, it is convenient here to make a definite choice: day as the unit of time, and kg as the unit of weight.

The quantity

$$\frac{1}{W}\frac{\mathrm{d}W}{\mathrm{d}t}$$

which has dimensions of time$^{-1}$, is frequently encountered in discussions of growth functions. Unfortunately, various terms are used to describe this growth rate derivative. Plant scientists talk about the *relative* growth rate, although this is hardly a satisfactory use of the word *relative*. Most biologists refer to the *specific* growth rate; the recommended scientific meaning of *specific* is 'divided by mass' (Royal Society, 1975, p. 10), which then strictly precludes its use for

$$\frac{1}{X}\frac{\mathrm{d}X}{\mathrm{d}t}$$

where $X$ is a volume, area or some other quantity. In other areas (for example, economics) the term *proportional* rate of growth is widely used for all quantities of this type; this seems to be an explicit and pertinent use of *proportional*, although it is unfamiliar to biologists. We will use any one of the three words, depending upon context, and hope this will not cause confusion.

## Growth equations

In this section, a survey is given of the more commonly encountered growth equations. Where possible, these are derived from a simple model, usually by integration of a differential equation of the type of equation (5.3) or equation (5.4). This allows some meaning to be associated with the parameters of the W : *t* equation. Although the common practice of plotting growth data on a semi-logarithmic scale is sound, it has dangers for the unwary: for example, a sigmoid curve might appear non-sigmoid. For this reason, a typical member of each type of equation is shown on both linear and semi-logarithmic scales.

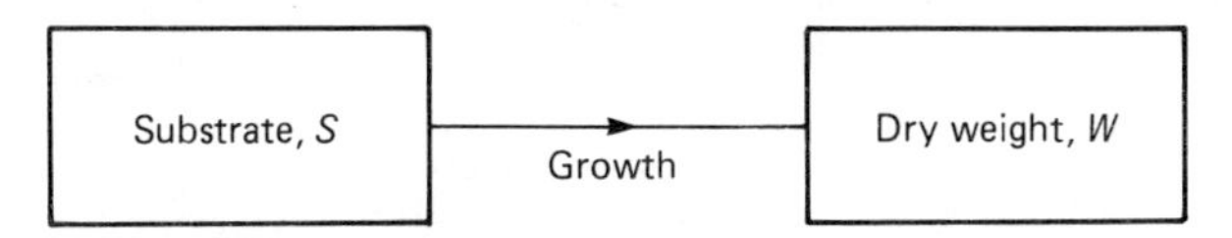


*Figure 5.1* A closed two-compartment model for considering growth. The system is closed with no inputs or outputs, and the growth process transfers material from the substrate compartment to the second compartment without loss. Various assumptions concerning how the rate of the growth process depends on $W$ and $S$ enable different growth equations to be derived

In *Figure 5.1*, a simple two-compartment model is defined. With the assumption that there is no gain or loss of material from the system, therefore

$$\frac{dW}{dt} = -\frac{dS}{dt} \tag{5.5}$$

$$\frac{dW}{dt} + \frac{dS}{dt} = \frac{d}{dt}(W + S) = 0$$

so that

$$W + S = \text{a constant} = W_0 + S_0 = W_f + S_f = C \tag{5.6}$$

$W_0$ and $S_0$ are the initial values of $W$ and $S$ at time $t = 0$; $W_f$ and $S_f$ are the final values of $W$ and $S$ approached as $t \rightarrow \infty$ (assuming that a steady state is eventually reached; $C$ is a constant. The next step is to write the growth rate as a function $v$ of $W$ and $S$, so that

$$\frac{dW}{dt} = v(W, S) \tag{5.7}$$

Since from equation (5.6), $S = C - W$, and by substituting for $S$ in equation (5.7), therefore

$$\frac{dW}{dt} = v(W, C - W) = h(W) \tag{5.8}$$

where $h$ is a function of the single variable $W$ alone, as in equation (5.3). The problem has become one of the single state variable, $W$. The crucial step is now what function $v$ one assumes for equation (5.7); this determines the form of equation (5.8), and after integration, the growth equation itself, equation (5.1). We now proceed to derive, by making different assumptions for $v$ of equation (5.7), several distinct growth equations, all of which may be interpreted in terms of the model of *Figure 5.1*.

**Simple exponential growth with an abrupt cut-off**

The assumptions are: the quantity of growth machinery is proportional to dry weight $W$; the growth machinery works at maximal rate so long as there is any substrate available at all; growth is irreversible and it stops once the substrate is exhausted. Equation (5.7) becomes

$$\frac{dW}{dt} = \mu W \tag{5.9}$$

where $\mu$ is a parameter known as the specific or relative growth rate. $\mu$ depends on the proportion of $W$ which constitutes growth machinery, but also on the efficiency or speed with which this machinery can operate. Integration of equation (5.9) gives

$$W = W_0 e^{\mu t} \quad \text{for} \quad 0 \leqslant t < t_f \tag{5.10a}$$

and

$$W = W_f \quad \text{for} \quad t \geqslant t_f \tag{5.10b}$$

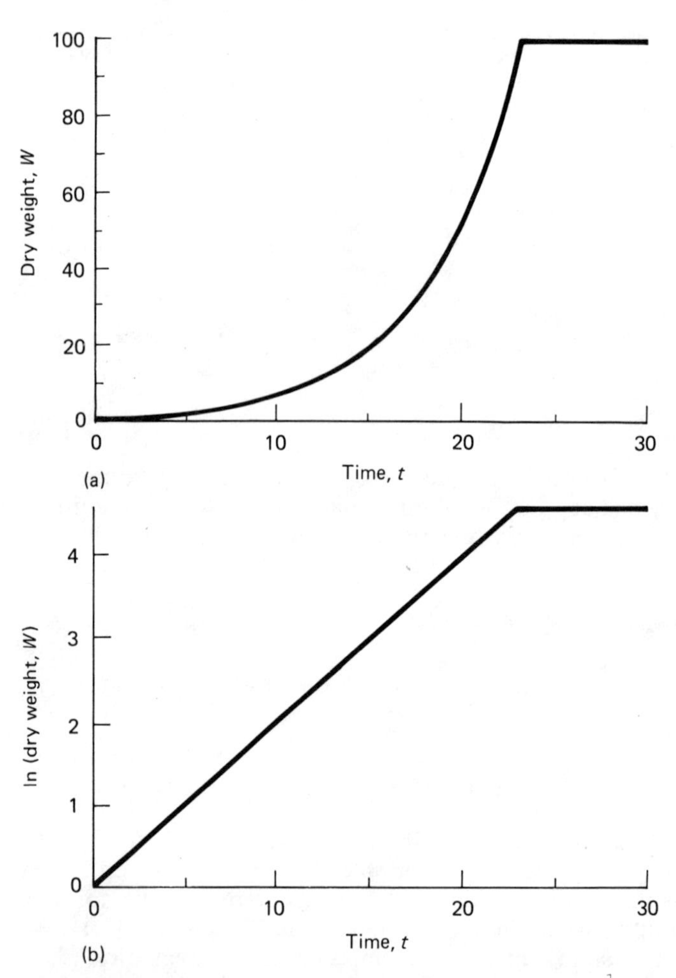


*Figure 5.2* Simple exponential growth with an abrupt cut-off. $W$ is the dry weight, $t$ is time, both in arbitrary units. The curves describe equations (5.10) with $W_0 = 1$, $W_f = 100$, $\mu = 0.2$, and $t_f = 23.0$

When $W = W_f$, $S = 0$, so from equation (5.6)

$$W_f = W_0 + S_0 \tag{5.11}$$

and growth stops abruptly when (putting $t = t_f$ and $W = W_0 + S_0$ in equation (5.10a))

$$t_f = \{\ln[(W_0 + S_0)/W_0]\}/\mu \tag{5.12}$$

Simple exponential growth limited by the amount of substrate available is illustrated in *Figure 5.2*. On the semi-logarithmic plot the growth curve is linear, since from equation (5.10)

$$\ln W = \ln W_0 + \mu t \tag{5.13}$$

### The monomolecular equation

This equation describes the progress of a simple irreversible first-order chemical reaction. The assumptions are: the quantity of growth machinery is constant and independent of dry weight $W$; this machinery works at a rate proportional to the substrate level $S$; growth is irreversible. Instead of equation (5.9), we now have

$$\frac{dW}{dt} = kS \tag{5.14}$$

where $k$ is a constant. From equation (5.6) with $S_f = 0$, $S = W_f - W$ may be substituted in equation (5.14), giving

$$\frac{dW}{dt} = k(W_f - W) \tag{5.15}$$

This becomes

$$\int_{W_0}^{W} \frac{dW}{W_f - W} = \int_0^t k\,dt$$

and therefore

$$\ln\left(\frac{W_f - W_0}{W_f - W}\right) = kt$$

This can be written as

$$W = W_f - (W_f - W_0)e^{-kt} \tag{5.16}$$

or, if one takes the initial dry weight $W_0 = 0$, equation (5.16) takes the simpler form

$$W = W_f(1 - e^{-kt})$$

Equation (5.16) is illustrated in *Figure 5.3*. The growth rate decreases continually and there is no point of inflexion. This can be seen from the second differential:

$$\frac{d^2W}{dt^2} = -k^2(W_f - W) = -(W_f - W_0)k^2 e^{-kt}$$

which is only zero for $t \to \infty$ or $W \to W_f$.

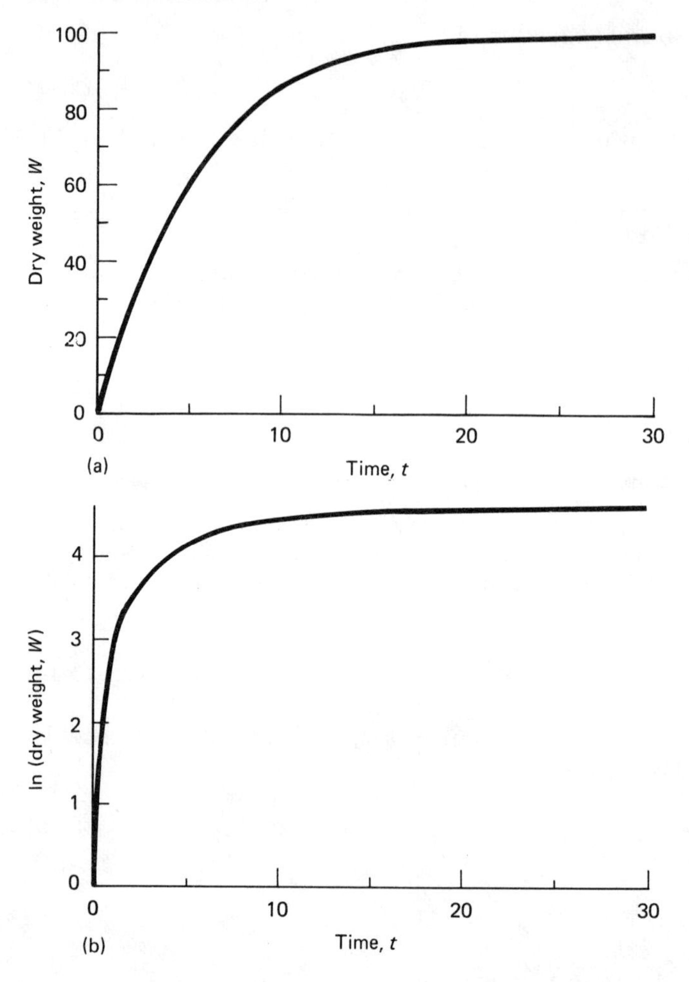


*Figure 5.3* Monomolecular growth (equation (5.16)). $W$ is the dry weight, $t$ is time, both in arbitrary units. The curves describe equation (5.16) with $W_0 = 1$, $W_f = 100$, and $k = 0.2$

## The logistic growth equation

In the last two sections, two extreme situations were analysed: in equation (5.9) growth rate depends only on the quantity of growth machinery (proportional to dry weight $W$) and not at all on the availability of substrates; in equation (5.14) growth rate depends only on the substrate level and not on the dry weight. To derive the logistic growth equation, a composite assumption is made: the quantity of growth machinery is proportional to dry weight $W$; this growth machinery works at a rate proportional to the amount of substrate $S$; growth is irreversible. Corresponding to equations (5.9) and (5.14), we now have

$$\frac{dW}{dt} = k'WS \qquad (5.17)$$

where $k'$ is a constant. Again, $S = W_f - W$ may be substituted (from equation (5.6) with $S_f = 0$), giving

$$\frac{dW}{dt} = k'W(W_f - W) \tag{5.18}$$

It is convenient to work with a constant $\mu$, defined in terms of $k'$ by

$$k' = \mu/W_f \tag{5.19}$$

so that

$$\frac{dW}{dt} = \mu W\left(1 - \frac{W}{W_f}\right) \tag{5.20}$$

This can be written (using the method of partial fractions, p. 280)

$$\int_{W_0}^{W} \left(\frac{1}{W_f - W} + \frac{1}{W}\right) dW = \int_0^t \mu \, dt$$

which after integration and rearrangement gives

$$W = \frac{W_0 W_f e^{\mu t}}{W_f - W_0 + W_0 e^{\mu t}} \tag{5.21}$$

More commonly, the logistic equation is written in the form

$$W = \frac{W_0 W_f}{W_0 + (W_f - W_0)e^{-\mu t}} \tag{5.22}$$

Consideration of either of these last two equations shows that, for $W_0 << W_f$, for low values of $t$ (putting $W_0 = 0$ in the denominator)

$$W \approx W_0 e^{\mu t} \tag{5.23}$$

giving exponential growth initially with relative growth rate $\mu$. As $t \to \infty$, $W \to W_f$, giving limited growth with an asymptote.

Differentiation of equation (5.20) gives

$$\frac{1}{\mu}\frac{d^2W}{dt^2} = \frac{dW}{dt}\left(1 - \frac{2W}{W_f}\right)$$

so there is a point of inflexion (equating the above to zero) at

$$W = \tfrac{1}{2}W_f \tag{5.24}$$

Substituting equation (5.24) into equation (5.22), this occurs at a time $t = t^*$, where

$$t^* = \frac{1}{\mu}\ln\left(\frac{W_f - W_0}{W_0}\right) \tag{5.25}$$

The growth curve is illustrated in *Figure 5.4*. The logistic equation is the first of the growth curves considered here to give smooth sigmoid behaviour.

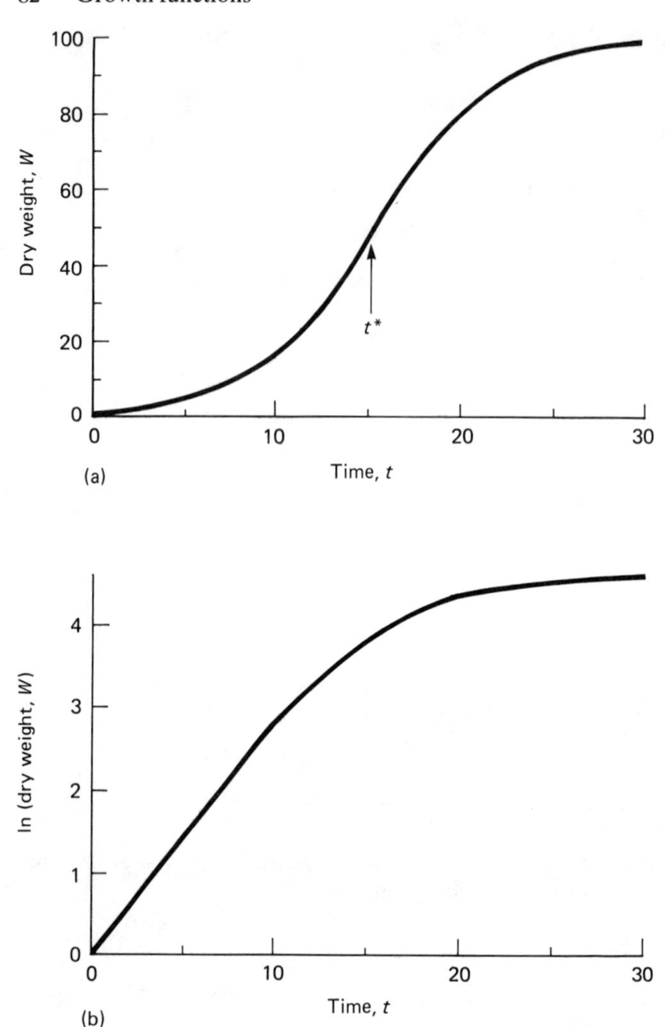


*Figure 5.4* Logistic growth (equation (5.22)). $W$ is the dry weight, $t$ is time, both in arbitrary units. The curves describe equation (5.22) with $W_0 = 1$, $W_f = 100$, and $\mu = 0.3$. $t^*$ denotes the point of inflexion: by equation (5.25), $t^* = 15.32$

## The Gompertz growth equation

In deriving the logistic equation, it is assumed that the rate of the autocatalytic growth is modified by substrate availability, according to equation (5.17). The Gompertz equation may be derived by assuming: the substrate is non-limiting, so that the growth machinery is always saturated with substrate; the quantity of growth machinery is proportional to the dry weight $W$, with a constant of proportionality $\mu$; the effectiveness of the growth machinery decays with time, according to first-order kinetics giving exponential decay. This decay may be viewed as due to degradation (possibly of enzymes), or senescence, or development and differentiation. There are alternative sets of assumptions that will give rise to the Gompertz equation. Formalizing the above, therefore

$$\frac{dW}{dt} = \mu W \tag{5.26}$$

as in equation (5.9), but now the specific growth rate parameter $\mu$ is no longer constant, but is governed by

$$\frac{d\mu}{dt} = -D\mu \tag{5.27}$$

$D$ is an additional parameter, describing the decay in the specific growth rate. Integration of equation (5.27) gives

$$\mu = \mu_0 e^{-Dt} \tag{5.28}$$

where $\mu_0$ is the value of $\mu$ at time $t = 0$. Substituting equation (5.28) into equation (5.26), therefore

$$\frac{dW}{dt} = \mu_0 W e^{-Dt} \tag{5.29}$$

This equation is of the type of equation (5.4), and is easily integrated:

$$\int_{W_0}^{W} \frac{dW}{W} = \mu_0 \int_0^t e^{-Dt}\, dt \tag{5.30}$$

giving

$$\ln\left(\frac{W}{W_0}\right) = \frac{\mu_0}{D}(1 - e^{-Dt}) \tag{5.31}$$

This can be written

$$W = W_0 \exp[\mu_0 (1 - e^{-Dt})/D] \tag{5.32}$$

For small values of $t$, $e^{-Dt} \approx 1 - Dt$, giving $1 - e^{-Dt} \approx Dt$, and therefore growth is exponential, with

$$W \approx W_0\, e^{\mu_0 t} \tag{5.33}$$

As $t \rightarrow \infty$, an asymptote at $W = W_f$ is approached, where

$$W_f = W_0\, e^{\mu_0/D} \tag{5.34}$$

Differentiation of equation (5.29) gives

$$\frac{1}{\mu_0}\frac{d^2W}{dt^2} = \frac{dW}{dt} e^{-Dt} - DWe^{-Dt}$$

Equating the above to zero, and substituting for $dW/dt$ with equation (5.29), there is a point of inflexion at time $t = t^*$, where

$$t^* = \frac{1}{D}\ln\left(\frac{\mu_0}{D}\right) \quad \text{and} \quad W(t = t^*) = \frac{W_f}{e} \tag{5.35}$$

By substituting for $\exp(-Dt)$ from equation (5.31), equation (5.29) can be put into the 'rate is a function of state' form (equation (5.3)), giving

$$\frac{dW}{dt} = \mu_0 W\left[1 - \frac{D}{\mu_0}\ln\left(\frac{W}{W_0}\right)\right] \tag{5.36}$$

Using equation (5.34), $D$ or $\mu_0$ may be eliminated in favour of $W_f$, giving two alternative and frequently encountered forms of the Gompertz equation:

$$\frac{dW}{dt} = \mu_0 W\left[\frac{\ln(W_f/W)}{\ln(W_f/W_0)}\right] = DW\ln\left(\frac{W_f}{W}\right) \tag{5.37}$$

The Gompertz equation is sketched out in *Figure 5.5*. Compared with the logistic equation, it has the same number of parameters (three), but the inflexion point no

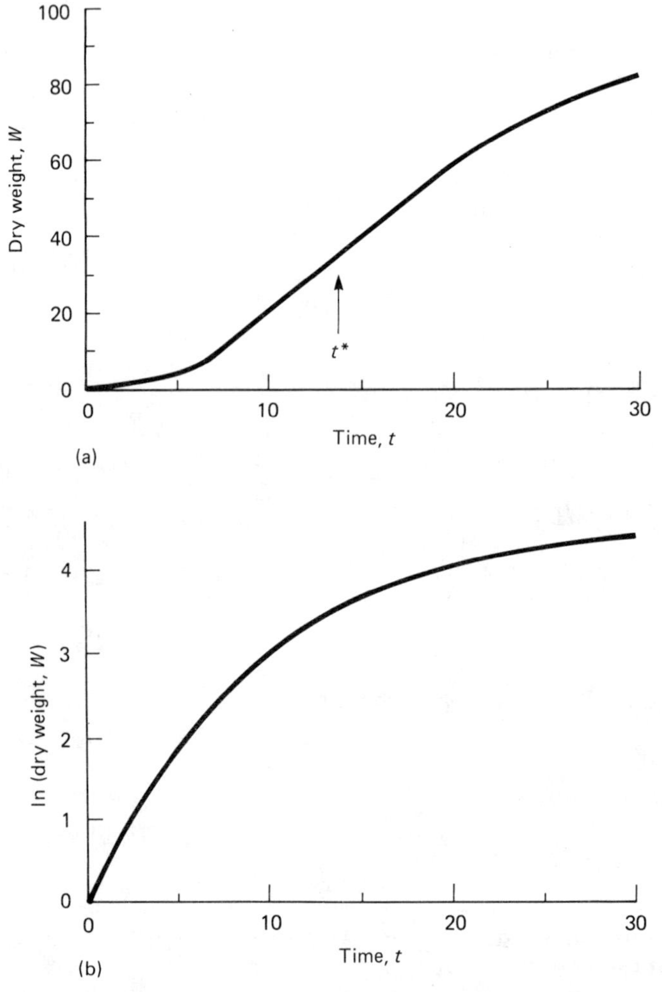


*Figure 5.5* Gompertz growth function (equation (5.32)). $W$ is the dry weight, $t$ is the time, both in arbitrary units. The curves describe equation (5.32) with $W_0 = 1$, $\mu_0 = 0.5$, and $D = 0.1086$. By equation (5.34), $W_f = 100$. $t^*$ denotes the point of inflexion: by equation (5.35), $t^* = 14.07$

longer occurs at half the final weight (equation (5.24)), but at 1/e times the final weight (equation (5.35)). Comparison of *Figures 5.4* and *5.5* shows that, for the same initial and final weights and a similar time of inflexion, the Gompertz shows faster early growth, but a slower approach to the asymptote, with a longer linear period about the inflexion point.

### The Richards growth equation

In the plant sciences, Richards (1959, 1969) was the first to apply a growth equation developed by von Bertalanffy (1957) to describe the growth of animals. While the Richards function should be viewed as an empirical construct, it has a generality that may sometimes be an advantage: for particular values of an additional parameter $n$, it encompasses the three previously discussed growth equations—the monomolecular ($n = -1$), the logistic ($n = 1$), and the Gompertz ($n = 0$). The differential equation for the Richards equation may be written (for example, see Causton, Elias and Hadley, 1978)

$$\frac{dW}{dt} = \frac{kW(W_f^n - W^n)}{nW_f^n} \tag{5.38}$$

where $k$, $n$ and $W_f$ are constants. $k$ and $W_f$ are positive. $n \geqslant -1$, since $n < -1$ is non-physiological, giving infinite growth rates as $W \to 0$.

With $n = -1$, this becomes

$$\frac{dW}{dt} = k(W_f - W)$$

identical with equation (5.15), the monomolecular equation. With $n = 1$, it becomes

$$\frac{dW}{dt} = kW\left(1 - \frac{W}{W_f}\right)$$

which, replacing $k$ by $\mu$, is identical to equation (5.20), the logistic equation. With $n = 0$, one has to exercise a little care in taking the limit $n \to 0$, and use (for $W_f^n$ and $W^n$) the expansion

$$x^n = e^{n \ln x} = 1 + n \ln x + \frac{1}{2!}(n \ln x)^2 + \ldots$$

It is then easily shown that

$$\frac{dW}{dt} = kW \ln\left(\frac{W_f}{W}\right)$$

is identical with equation (5.37) from which the Gompertz equation is derivable, if $k$ is replaced by $D$.

Equation (5.38) can be integrated by first writing it as

$$n \int_{W_0}^{W} \left(\frac{1}{W} + \frac{W^{n-1}}{W_f^n - W^n}\right) dW = \int_0^t k \, dt \tag{5.39}$$

giving

$$W_0^n W_f^n = W^n[W_0^n + (W_f^n - W_0^n)e^{-kt}]$$

which can be written

$$W = \frac{W_0 W_f}{[W_0^n + (W_f^n - W_0^n)e^{-kt}]^{1/n}} \tag{5.40}$$

At $t = 0$, $W = W_0$; for $t \to \infty$, $W \to W_f$. To find the inflexion point (if present), differentiation of equation (5.38) gives

$$\frac{d^2W}{dt^2} = \frac{k}{nW_f^n} \frac{dW}{dt} [W_f^n - (n + 1)W^n]$$

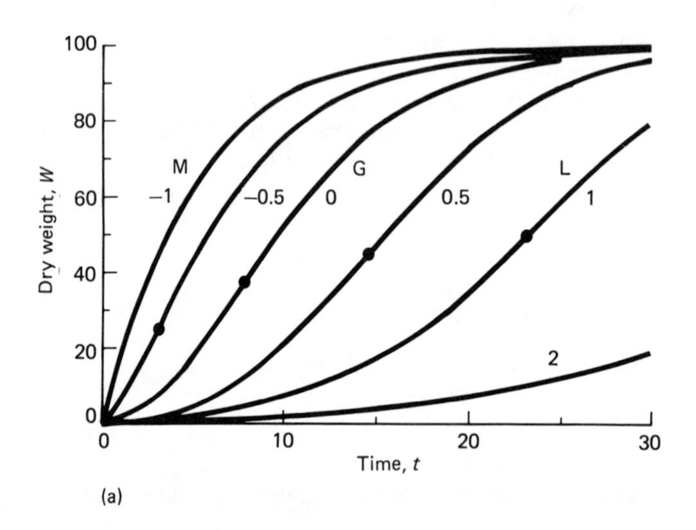


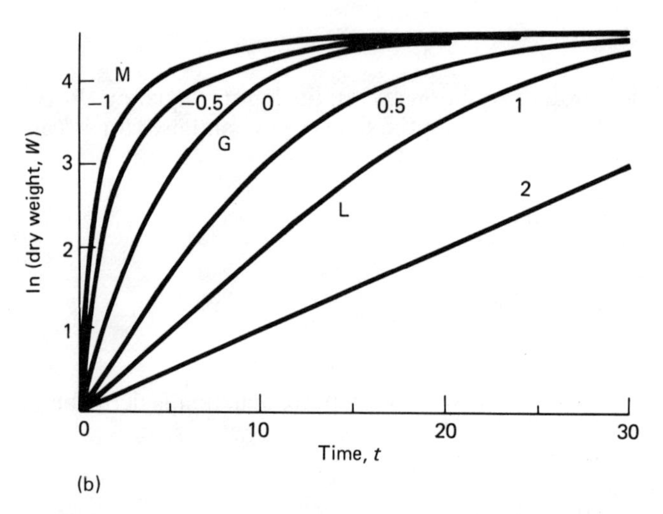


*Figure 5.6* Richards growth function (equation (5.40)). $W$ is the dry weight, $t$ is the time, both in arbitrary units. The curves describe equation (5.40) with $W_0 = 1$, $W_f = 100$, $k = 0.2$, and for six $n$-values as given. M denotes the monomolecular ($n = -1$, *Figure 5.3*); G denotes the Gompertz ($n = 0$, *Figure 5.5*); L denotes the logistic ($n = 1$, *Figure 5.4*). Points of inflexion in (a) are shown by solid circles (equations (5.41) and (5.42))

Equating this to zero at time $t = t^*$ (the inflexion point), therefore either $\mathrm{d}W/\mathrm{d}t = 0$, or

$$W(t = t^*) = W_\mathrm{f}\left(\frac{1}{n+1}\right)^{1/n} \tag{5.41}$$

By substituting $W = W(t = t^*)$ from equation (5.41) and $t = t^*$ into equation (5.40), the time of inflexion $t^*$ is

$$t^* = \frac{1}{k}\ln\left(\frac{W_\mathrm{f}^n - W_0^n}{nW_0^n}\right) \tag{5.42}$$

The family of curves obtained by varying the parameter $n$ in equation (5.40) is sketched in *Figure 5.6*. Compared with the Gompertz and the logistic, the point of inflexion is able to occur at any fraction of the final dry weight (equation (5.41)), as $n$ varies over the range $-1 < n < \infty$.

**The Chanter growth equation**

Chanter (1976, Appendix 2) devised a function which is a hybrid of the logistic and Gompertz equations, and whose parameters may be similarly interpreted. By analogy with equations (5.20) and (5.29), an equation

$$\frac{\mathrm{d}W}{\mathrm{d}t} = \mu W\left(1 - \frac{W}{B}\right)\mathrm{e}^{-Dt} \tag{5.43}$$

may be constructed, where $\mu$, $B$ and $D$ are constants. Thus, in equation (5.43) the specific growth rate $(1/W)(\mathrm{d}W/\mathrm{d}t)$ is modified by two factors: the first, $(1 - W/B)$, depends linearly on substrate level as in the logistic equation (5.17); the second, $\exp(-Dt)$, depends on the passage of time, and may be interpreted as differentiation, development or senescence, as in the Gompertz equation (5.29). Integration of equation (5.43) gives

$$W = \frac{W_0 B}{W_0 + (B - W_0)\exp\{-[\mu(1 - \mathrm{e}^{-Dt})/D]\}} \tag{5.44}$$

This is similar but not identical to the Richards function (equation (5.40)). The four parameters $W_0$, $B$, $\mu$ and $D$ can in this case all be assigned some biological meaning.

At time $t = 0$, the initial weight is given by $W = W_0$. However, the final weight, obtained as $t \to \infty$, is given by a more complicated quantity

$$W_\mathrm{f} = \frac{W_0 B}{W_0 + (B - W_0)\mathrm{e}^{-\mu/D}} \tag{5.45}$$

Note that the final weight, $W_\mathrm{f}$, may be determined by substrate availability, $B$, or by differentiation, $D$, or by a combination of both. For example, using

$$W_\mathrm{L} = B \quad \text{and} \quad W_\mathrm{G} = W_0\,\mathrm{e}^{\mu/D} \tag{5.46}$$

for the limiting final weights that would be obtained with the logistic or Gompertz factors acting alone (cf. equation (5.34)), then equation (5.45) can be written

$$W_\mathrm{f} = \frac{W_\mathrm{L} W_\mathrm{G}}{W_\mathrm{L} + W_\mathrm{G} - W_0} \tag{5.47}$$

so that the final weight always lies below both $W_L$ and $W_G$, but may approach the lower of $W_L$ and $W_G$ very closely.

From equation (5.45), one can derive

$$B = \frac{W_f W_0(e^{\mu/D} - 1)}{W_0 e^{\mu/D} - W_f} \tag{5.48}$$

which can be used to substitute for $B$ in equation (5.44), giving a growth equation in terms of $W_0$, $W_f$, $\mu$ and $D$.

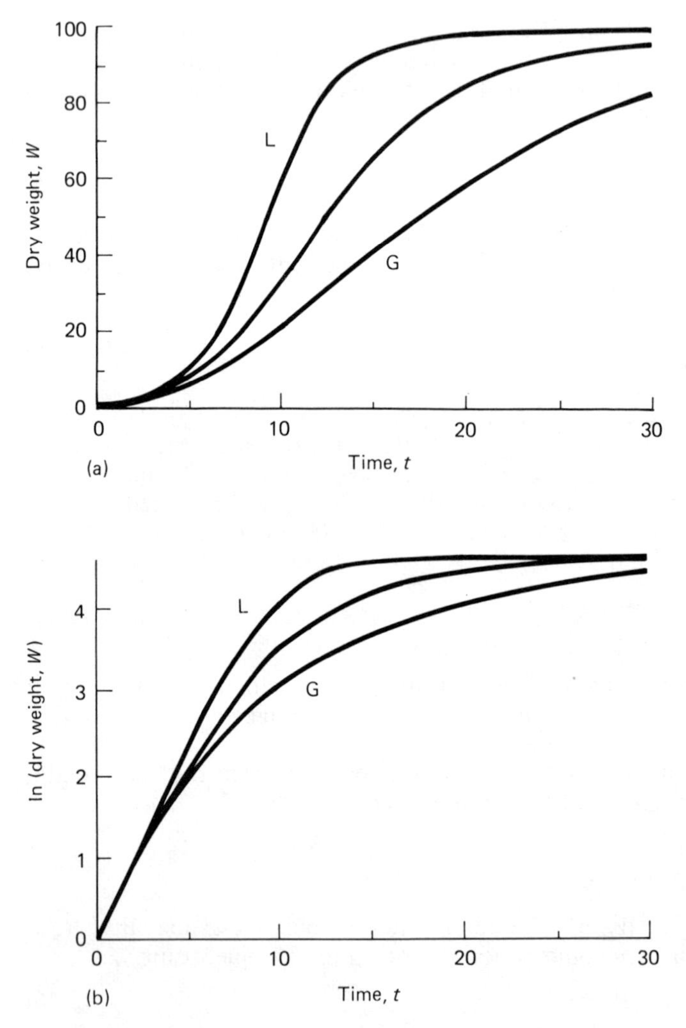


*Figure 5.7* Chanter growth function (equation (5.44)). $W$ is the plant dry weight, $t$ is the time, both in arbitrary units. The curves describe equation (5.44) with $W_0 = 1$, $W_f = 100$, $\mu = 0.5$ and for three values of $D$, to give the logistic (denoted by L, $D \to 0$), an intermediate case ($D = 0.05$), and the Gompertz (denoted by G, $D \to \mu/[\ln(W_f/W_0)] = 0.1086$). In equation (5.44), the value of $B$ is obtained from equation (5.48). All the curves shown above have the same initial and final dry weights, and the same initial specific growth rate

By solving equation (5.44) for $\exp(-Dt)$ and substituting into equation (5.43), equation (5.43) can be put into a proper rate:state form without the time variable $t$ on the right side (cf. equations (5.3) and (5.4)), giving

$$\frac{dW}{dt} = \mu W\left(\frac{B-W}{B}\right)\left\{1 - \frac{D}{\mu}\ln\left[\left(\frac{W}{B-W}\right)\left(\frac{B-W_0}{W_0}\right)\right]\right\} \tag{5.49}$$

By taking the limit $D \to 0$ (no senescence), equations (5.44) and (5.49) become identical to the logistic equations (5.22) and (5.20); in the limit $B \to \infty$ (no substrate limitation), the equations become identical to the Gompertz equations (5.32) and (5.36).

Differentiation of equations (5.43) or (5.49) to find the inflexion point gives an equation with only numerical solutions for the time of inflexion, $t = t^*$, or the dry weight at the inflexion point $W(t = t^*)$.

The family of curves described by equation (5.44) is sketched out in *Figure 5.7*. To obtain these, the values of the initial and final dry weights, $W_0$ and $W_f$, are fixed, and also the value of the specific growth rate parameter, $\mu$. For each assumed value of $D$, $B$ is calculated from equation (5.48), and this is used in equation (5.44) to give the $W(t)$ response. For given values of $W_0$, $W_f$ and $\mu$, $D$ may only lie in the range

$$0 \leqslant D \leqslant \mu/[\ln(W_f/W_0)] \tag{5.50}$$

the upper limit being set by the vanishing denominator of equation (5.48) which makes $B$ infinite. By varying $D$ over this range, a family of curves lying between the corresponding logistic and Gompertz curves is generated, all having the same initial and final dry weights, and the same initial specific growth rate. The point of inflexion is shifted progressively to later times as $D$ increases.

### Exponential polynomials

In contrast with the preceding six growth functions, which can all be derived from relatively simple rate : state relations giving them some mechanistic basis, the exponential polynomial class of functions is empirical, and in general has no physiological interpretation. These equations may be written

$$W = \exp(a_0 + a_1 t + a_2 t^2 + a_3 t^3 + \ldots) \tag{5.51}$$

where $a_0$, $a_1$, ... are constants. Taking logarithms, therefore

$$\ln W = a_0 + a_1 t + a_2 t^2 + a_3 t^3 + \ldots \tag{5.52}$$

Differentiating equations (5.51) or (5.52), therefore

$$\frac{dW}{dt} = (a_1 + 2a_2 t + 3a_3 t^2 + \ldots)\exp(a_0 + a_1 t + a_2 t^2 + \ldots) \tag{5.53}$$

which may be written

$$\frac{1}{W}\frac{dW}{dt} = (a_1 + 2a_2 t + 3a_3 t^2 + \ldots) \tag{5.54}$$

also

$$\frac{d^2W}{dt^2} = W[(2a_2 + 6a_3 t + \ldots) + (a_1 + 2a_2 t + 3a_3 t^2 + \ldots)^2] \tag{5.55}$$

The initial dry weight is given by

$$W_0 = W(t = 0) = \exp a_0 \tag{5.56}$$

The final dry weight $W(t \rightarrow \infty)$ is infinite or zero. The time of inflexion, $t^*$, may be obtained by solving (from equation (5.55))

$$0 = 2a_2 + 6a_3 t + \ldots + (a_1 + 2a_2 t + \ldots)^2 \tag{5.57}$$

The time at which the maximum dry weight is attained may be found by solving (from equation (5.53))

$$0 = a_1 + 2a_2 t + 3a_3 t^2 + \ldots \tag{5.58}$$

and the maximum weight itself by substituting back into equation (5.51).

Equation (5.51) may be easily fitted to data by statistical methods, for example, see Causton and Venus (1981) (in contrast to some of the growth functions discussed earlier where fitting may be more difficult), and a number of computer programs have been written to carry out this task (Hughes and Freeman, 1967; Nicholls and Calder, 1973; Hunt and Parsons, 1974). Hurd (1977) and others have argued convincingly that one should not go beyond a quadratic (that is, put $a_3 = a_4 = \ldots = 0$), due to dangers of over-fitting and spurious responses.

For the exponential *quadratic*, a simple and quick method of obtaining rough estimates of the three parameters $a_0$, $a_1$, and $a_2$ is as follows. From equation (5.52) $a_0$ is given by

$$a_0 = \ln W_0 \tag{5.59}$$

From equation (5.54)

$$a_1 = \frac{1}{W}\frac{\mathrm{d}W}{\mathrm{d}t}(t = 0) = \text{the initial specific growth rate} \tag{5.60}$$

From equation (5.52) this is the initial slope of the data when plotted on a semi-log plot. From equation (5.58), the maximum dry weight $W_m$ is attained at time $t_m$, where

$$t_m = -\frac{a_1}{2a_2} \tag{5.61}$$

and putting this into equation (5.51), therefore

$$W_m = W_0 \exp(-a_1^2/4a_2) \tag{5.62}$$

so that

$$a_2 = -\left(\frac{a_1^2}{4}\right) \Bigg/ \left[\ln\left(\frac{W_m}{W_0}\right)\right] \tag{5.63}$$

$a_0$, $a_1$ and $a_2$ have now been obtained. The times of inflexion $t^*$ are given by equation (5.57)

$$t^* = -\frac{a_1}{2a_2} \pm \frac{1}{(-2a_2)^{1/2}} \tag{5.64}$$

being spaced equally from the time of maximum dry weight.

A typical exponential quadratic curve is sketched out in *Figure 5.8*. Taking the curve as a whole, it is clear that this function is not suitable for organs or organisms

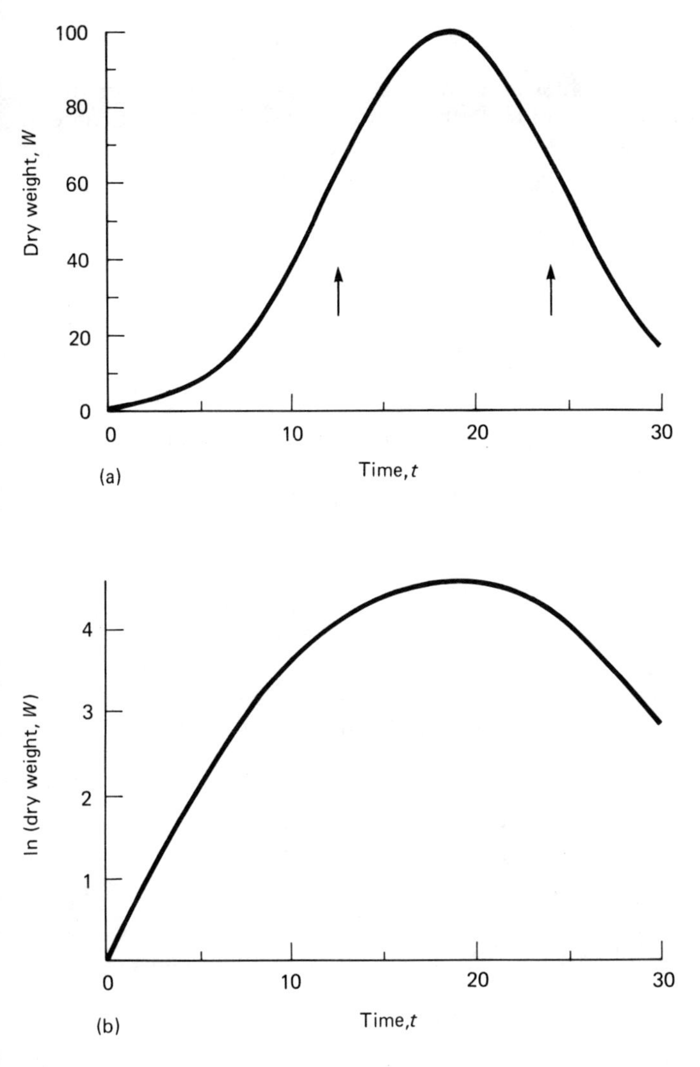


*Figure 5.8* Exponential quadratic growth function. $W$ is the dry weight, $t$ is the time, both in arbitrary units. The curves represent equations (5.51) and (5.52) with $a_0 = 0$, $a_1 = 0.5$, $a_2 = -0.0136$, $a_3 = a_4 = \ldots = 0$, so that $W_0 = 1$ and the maximum dry weight, $W_m$ (equation (5.62)) = 100. The arrows in (a) denote the times of the inflexion points (equation (5.64))

which show asymptotic dry weight : time behaviour, with a non-zero asymptote. More commonly, the physiologist would use only a part of this response, stopping short somewhere in the region of the maximum.

Although it was stated at the beginning of this section that exponential polynomials are completely empirical, this needs qualification. It can be shown (Thornley, unpublished) that an exponential quadratic describes the cell numbers in meristematic tissue where the progeny of cell division have an exponentially diminishing probability of remaining meristematic.

## Allometry

The allometric relationship was first used by Huxley (1924) with reference to the growth of animals and their parts, and later by Pearsall (1927) in the plant context. Suppose that $P$ and $Q$ are attributes (observable quantities) of an organism: for instance, $P$ and $Q$ may be the weights of different limbs of an animal, or $P$ might be the dry weight of a plant and $Q$ its leaf area. As the organism grows and develops, both $P$ and $Q$ will change with time, so that

$$P \equiv P(t) \quad \text{and} \quad Q \equiv Q(t) \tag{5.65}$$

indicating the time dependence of $P$ and $Q$. $P$ and $Q$ are said to be allometrically related if they obey the allometric equation, namely

$$P = aQ^b \tag{5.66}$$

where $a$ and $b$ are constants. $P$ and $Q$ vary with time in such a way that equation (5.66) is satisfied at all times.

It may be noted that equation (5.66) is symmetrical between $P$ and $Q$, and can just as well be written

$$Q = cP^d$$

where

$$c = \left(\frac{1}{a}\right)^{1/b} \quad \text{and} \quad d = \frac{1}{b} \tag{5.67}$$

Taking logarithms of equation (5.66) gives a linear equation

$$\ln P = \ln a + b \ln Q \tag{5.68}$$

Differentiating equation (5.68) with respect to time $t$ gives

$$\frac{1}{P}\frac{dP}{dt} = b\frac{1}{Q}\frac{dQ}{dt} \tag{5.69}$$

Thus the proportional rates of growth of $P$ and $Q$ are related by the constant factor $b$.

The allometric equation has been used in many different ways. For example

$$W_1 = aW_2^b \tag{5.70}$$

has been used to relate the weight of a limb of an animal to the weight of the whole animal. It has also been used to relate root dry weight to shoot dry weight in a growing plant. Plant scientists sometimes use allometry to relate leaf area to plant dry weight, $W$, by means of

$$A = aW^b \tag{5.71}$$

Further applications of the allometric equation to animal production are given in Chapter 11.

In summary, the use of the allometric equation is declining; the method is sometimes empirically convenient, but it has little or no explanatory power, and there is usually no reason for preferring the allometric equation to any other empirically selected equation.

## Growth functions in animal and plant modelling

The use of growth functions represents one of the simplest approaches to the problem of quantitatively describing animal or plant growth. However, difficulties can arise even when the method is only used on historic data. These may be summarized as follows:

1. The choice of function is usually fairly arbitrary.
2. Possibly no appropriate analytic function exists.
3. The parameters have little or no meaning.
4. There may be fitting difficulties.
5. Observed growth data often contain discontinuities which cannot be accommodated (due, for instance, to sudden changes in nutrition, in environment, or disease).
6. The lack of a mechanistic basis can make it difficult to build in biological parameters in a meaningful way.

A major objective of much agricultural research is concerned with the prediction of performance and behaviour of animals, plants and crops, but it is wished to accomplish this effectively and economically—that is, with minimum but sufficient inputs. It is doubtful if the larger, detailed and all-embracing animal and plant models will be able to fill this role adequately, although these models retain many other virtues. While one approach is to take a complex model and progressively simplify it, an alternative is to start with the simplicity contained in the growth-function method, and progressively to make matters more elaborate. How this might be achieved in relation to crop growth is discussed in Chapter 8.

# Exercises

### Exercise 5.1

Derive the bimolecular growth equation, by replacing equation (5.14) by $\mathrm{d}W/\mathrm{d}t = kS^2/W_\mathrm{f}$, and integrating. Sketch it when $W_0 = 1$, $W_\mathrm{f} = 100$, and $k = 0.2$ for $0 \leqslant t \leqslant 30$, and briefly discuss its properties.

### Exercise 5.2

Derive a growth equation for simultaneous monomolecular and bimolecular growth, from the growth rate equation $\mathrm{d}W/\mathrm{d}t = k_1 S + k_2 S^2/W_\mathrm{f}$. Show that the resulting equation reduces to equation (5.16) if $k_2 = 0$, and to the solution of Exercise 5.1 if $k_1 = 0$. Find the initial growth rate.

### Exercise 5.3

Derive an equation whose solution will give the inflexion point of the Chanter growth function (equation (5.43) *et seq.*).

**Exercise 5.4**

The logistic equation is based on a linear response to substrate (equation (5.17)), whereas the diminishing-returns type of response of a Michaelis–Menten equation is more commonly encountered in biology. Derive and discuss such a modified logistic. [Hint: divide the right side of equation (5.17) by $(K + S)$.]

**Exercise 5.5**

The logistic equation (equation (5.17) *et seq.*) or the modification of the previous example both describe growth that can continue indefinitely as long as substrate is available. For animals and many determinate plants, and particularly with organs, growth ceases before the substrate is exhausted, due to differentiation, which may be expressed as in the Gompertz equation (5.29). Combine this latter effect with that of Exercise 5.4 to give a new growth function. [Hint: multiply the right side of equation (5.17) by $\exp(-Dt)/(K + S)$.] Note that this may be viewed as a Michaelis–Menten modification of the Chanter growth equation (5.43).

Chapter 6

# Weather

## Introduction

Agricultural performance is a result of genetically determined characteristics in combination with the environment, of which the weather is the most important component for crops and pastures. It is said that climate, or average weather, determines what crops are grown in a region, and the yields achieved depend on the actual weather during the season. The quantitative relationships between weather and performance are a principal concern of the agricultural modeller, who is aiming at prediction, understanding and control.

Many models of growth, or of particular facets of animal or plant performance, take the general form

$$\frac{dW}{dt} = g(W, P, E_M, E_W) \qquad (6.1)$$

where, for instance, $W$ is weight, $t$ is time, $P$ denotes a set of parameters or constants that refer to the organism, $E_M$ denotes the parameters or variables that define the component of the environment under man's control (e.g. nutritional inputs to an animal; herbicide or fertilizer inputs to a growing crop), and $E_W$ denotes the quantities referred to as weather variables (e.g. temperature, radiation, rainfall), which usually lie outside the control of the experimenter; the growth function $g$ indicates some function of $W$, $P$, $E_M$ and $E_W$. To test equation (6.1), it is usual to do an experiment in which the environment is monitored throughout the course of the experiment, which might last days, weeks, or many months. Insertion of these measured weather data into equation (6.1), together with an assumed set of parameter values $P$, allows equation (6.1) to be integrated numerically, to give predicted values of weight $W$ over the same time period. Model evaluation then proceeds in the normal manner by comparison of predicted and measured values. However, for some purposes (e.g. sensitivity analysis, model development and exploration), one may not wish to use actual weather data, which either may not exist, or be rather atypical. It may be both preferable and convenient to use expressions (either deterministic or stochastic), suitable for computation, which give an approximate description of the average weather at the site in question (Larsen and Penge, 1981; Richardson, 1981). This chapter is concerned with these matters.

## Time

The day is generally the most convenient and appropriate unit of time for describing growth and development of animals and plants, both at the organ and organism levels. Throughout this chapter, the unit day and a decimal day for within-day behaviour are used.

Let $t$ be the time variable. Next, define $t_i$ and $t_d$ by

$$t_i = \text{integer part } (t) \tag{6.2a}$$

and

$$t_d = \text{decimal part } (t) \tag{6.2b}$$

Thus, if $t = 16.243$ day, then $t_i = 16$ day and $t_d = 0.243$ day; $t_i$ is the day number minus one ($t = 16.243$ occurs on the 17th day), and $t_d$ gives us the time, as a decimal part of a day, on the $(t_i + 1)$th day (Exercise 6.1). Frequently, it is convenient to use the integer $t_i + 1$ to index daily temperature (or other) means, giving a series $T_1, T_2, \ldots$, where $T$ is temperature; $T_1$ is the mean temperature on the first day, between $t = 0$ and $t = 1$. Note also that for much simulation work, it can be practical and appropriate to use a time interval of $\Delta t = 0.01$ day, equivalent to 14.4 minutes.

### Calendar conversion

Agronomists and climatologists sometimes number the days of the year starting from 1 March (the Julian day number); this avoids the problem of leap years. Let

$$N = \text{climatological day number } (1 \leqslant N \leqslant 366) \tag{6.3}$$

If $t = 0$ is at the beginning of the first day of March, then (equation (6.2a))

$$N = \text{integer part } (t) + 1 = t_i + 1 \tag{6.4}$$

Stuff and Dale (1973) describe an easy method of converting day number $N$ to month and day of month. Let

$$D = \text{day of month } (1 \leqslant D \leqslant 31) \quad \text{and} \quad M = \text{month } (1 \leqslant M \leqslant 12) \tag{6.5}$$

to convert $D$ and $M$ to $N$:

$$\text{if } M \leqslant 2, \text{ then } M := M + 12$$

where := has the meaning 'is assigned the value of'. $N$ is then obtained by

$$N = \text{integer part } (30.6M + D - 91.3) \tag{6.6}$$

To convert the other way from day number $N$ to day of month $D$ and month $M$, then

$$M = \text{integer part } [(N + 91.3)/30.6]$$

$$D = \text{integer part } (N - 30.6M + 92.3)$$

and

$$\text{if } M \geqslant 13, \text{ then } M := M - 12 \tag{6.7}$$

(*See* Exercise 6.2)

### Equation of time

Apparent local solar time is the time shown by a sundial at that location. It differs from mean solar time by a quantity that varies over the year; this quantity is given by an equation known as the equation of time. This is due to the orbit of the earth round the sun being an ellipse rather than a circle, and also to the obliquity of the ecliptic: the plane of the equator is at an angle to the plane containing the sun and the orbit of the earth. The correction given by the equation of time can be calculated from an empirical formula (Usher, 1970), namely

$$\begin{aligned}\text{equation of time} = \\ &- 0.001\,98 - 7.129\,65 \cos y - 1.840\,02 \sin y \\ &- 0.688\,41 \cos 2y + 9.922\,99 \sin 2y \\ &+ 0.302\,60 \cos 3y + 0.106\,35 \sin 3y \\ &+ 0.035\,08 \cos 4y - 0.212\,11 \sin 4y \\ &- 0.008\,95 \cos 5y - 0.007\,73 \sin 5y + 0.000\,61 \cos 6y\end{aligned}$$

the result being in minutes and decimal minutes. $y$ is the year angle, and is defined in equation (6.8). The correction reaches its maximum values of about $-14$ minutes in February and $+16$ minutes in early November, and is zero on 15 April, 14 June, 1 September and 25 December. It is not usually important in the agricultural context, although it could be a factor in some horticultural applications involving the timed control of environments in relation to day light.

## Radiation

### Zenith and azimuth solar angles; daylength

Let $y$ denote the *year angle*, so that a period of one year corresponds to 360°. Assume that the year angle $y$ is zero at the vernal equinox on 21 March, and also that $t = 0$ falls on 1 March. Climatological day number $N$ is obtained from $t$ by equations (6.2a) and (6.4). Year angle $y$ is then given by

$$y = \left(\frac{N - 21}{365}\right) 360° \tag{6.8}$$

Thus at the vernal equinox on 21 March, $N = 21$ and $y = 0$.

The solar declination, $\delta$, is the angle in degrees between the line joining the sun and the earth, and the equatorial plane. $\delta$ depends only on time of year, being 0° at the equinoxes ($y = 0°$ and $y = 180°$), $+23.45°$ on 21 June ($y = 90°$), and $-23.45°$ on 21 December ($y = 270°$). Usher (1970) has given an empirical equation by which $\delta$ can be calculated:

$$\begin{aligned}\delta = {} & 0.380\,92 - 0.769\,96 \cos y + 23.265\,00 \sin y \\ & + 0.369\,58 \cos 2y + 0.108\,68 \sin 2y + 0.018\,34 \cos 3y \\ & - 0.166\,50 \sin 3y - 0.003\,92 \cos 4y + 0.000\,72 \sin 4y \\ & - 0.000\,51 \cos 5y + 0.002\,50 \sin 5y + 0.004\,42 \cos 6y\end{aligned} \tag{6.9}$$

Let $h$ denote the *hour angle* of the sun, this being the difference between the actual time of day, $t_d$, from equation (6.2b), and the time of apparent noon, $t_n$ (i.e. when the sun is highest in the sky).

$$h = (t_d - t_n)360° \qquad (6.10)$$

$h$ spans 360° (−180° to 180°). Since the sun is highest at midday, then $t_n = 0.5$, and at midday $t_d = 0.5$ and $h = 0°$.

Let $z$, the zenith angle, be the angular distance of the sun from the local vertical. $z$ is given by (Sellers, 1965)

$$\cos z = \sin \phi \sin \delta + \cos \phi \cos \delta \cos h \qquad (6.11)$$

where $\phi$ is the latitude. The sun is on the horizon when $z = 90°$, which, from equation (6.11), occurs at $h = h_o$ with

$$\cos h_o = -\tan \phi \tan \delta \qquad (6.12)$$

Taking the positive solution for $h_o$ lying in the range 0° to 180°, the daylength $g_N$ for the $N$th day is given by

$$g_N = \frac{2 \cos^{-1}(-\tan \phi \tan \delta)}{360} \qquad (6.13)$$

in units of decimal parts of a day. For example, at the equinoxes, $\delta = 0°$, $g_N = 2 \times 90°/360° = 0.50$ day.

In equation (6.13), the daylength is calculated assuming the zenith angle $z$ is 90°, that is, when the centre of the sun's disc is on the horizon, neglecting refraction in the atmosphere. More precisely, the times of sunrise and sunset are defined as when the upper part of the sun's disc just touches the horizon of a sea-level observer. Assuming a mean refraction of 0.57°, and the sun's angular semi-diameter to be 0.27°, this occurs when the zenith angle $z$ is equal to 90.83°. The daylength calculation can be reworked for this and other definitions. For instance, civil twilight begins and ends when the centre of the sun is 6° below the horizon. Nautical twilight uses a figure of 12° below the horizon; and astronomical twilight one of 18°. In all these cases, to find daylength $g_N$, one simply takes a zenith angle $z$ = 90.83°, 96°, 102°, or 108° in the equation

$$g_N = \frac{2}{360} \cos^{-1}\left(\frac{\cos z}{\cos \phi \cos \delta} - \tan \phi \tan \delta\right) \qquad (6.14)$$

which is derived from equation (6.11). It is uncertain which of these alternative daylength definitions is biologically most appropriate, and in any event, the daylength on a particular day may be much affected by the weather conditions on that day. Examples of the different daylengths resulting from equation (6.14) with various $z$ values are given in the solution to Exercise 6.3. A simple (first harmonic) sine wave approximation to the daylengths calculated by equation (6.14) is calculated in the solution to Exercise 6.5, and is shown in *Figure 6.1*.

Let $a$ be the azimuth angle of the sun, that is, the angle between geographical south and the sun, measured on a horizontal plane. The sine of $a$ is given by

$$\sin a = \frac{\cos \delta \sin h}{\sin z} \qquad (6.15a)$$

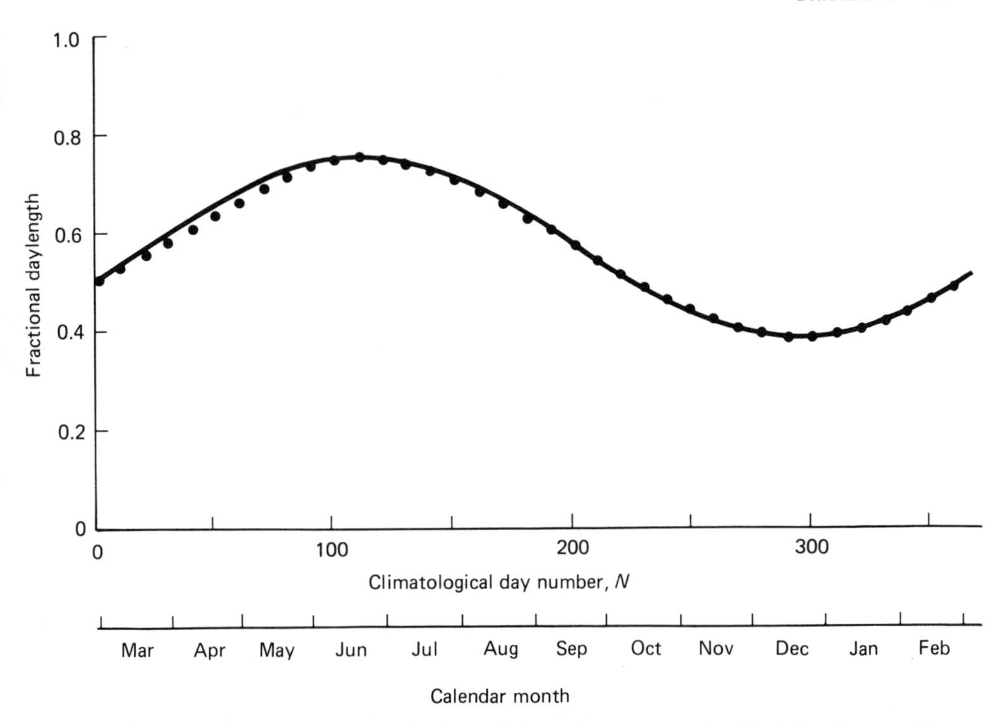


*Figure 6.1* Simple sine wave approximation to the daylength, $g_N$, for latitude 50.75°, defined by a zenith angle of 96° (the centre of the sun's disc is geometrically 6° below the horizon). The accurate calculations of Solution 6.3 (p. 297) are shown by ●; the solid line represents equation (S6.5.2)

If $a_o$ is the azimuth angle at sunrise or sunset, these occur at $\sin z = 1$, so that

$$\sin a_o = \cos \delta \sin h_o \qquad (6.15b)$$

where $h_o$ is obtained from equation (6.12). An alternative to equation (6.15b) is derived in Exercise 6.4.

## Seasonal variation in daily radiation receipt

### MEAN VALUES

Let $\bar{J}_N$ be the average over many years of the daily receipt of photosynthetically active radiation (PAR) on a horizontal plane for the $N$th day of the year, $J_N$. It has been observed (for example, by Rees and Thornley, 1973) that the change in $\bar{J}_N$ over the year may reasonably be represented by a sinusoid:

$$\bar{J}_N = (\bar{J}_{113} + \bar{J}_{296}) + (\bar{J}_{113} - \bar{J}_{296})\sin\left[\left(\frac{N-21}{365}\right)360\right] \qquad (6.16)$$

where $N = 1$ falls on 1 March, $N = 113$ is 21 June, and $N = 296$ is 21 December. For many purposes, equation (6.16) will be sufficiently accurate, although it only makes use of a small part of the available data. Greater precision can be obtained using Fourier analysis (*see* Glossary), to calculate the coefficients from all the data, and by using additional terms in the Fourier series, as in equation (6.9). These

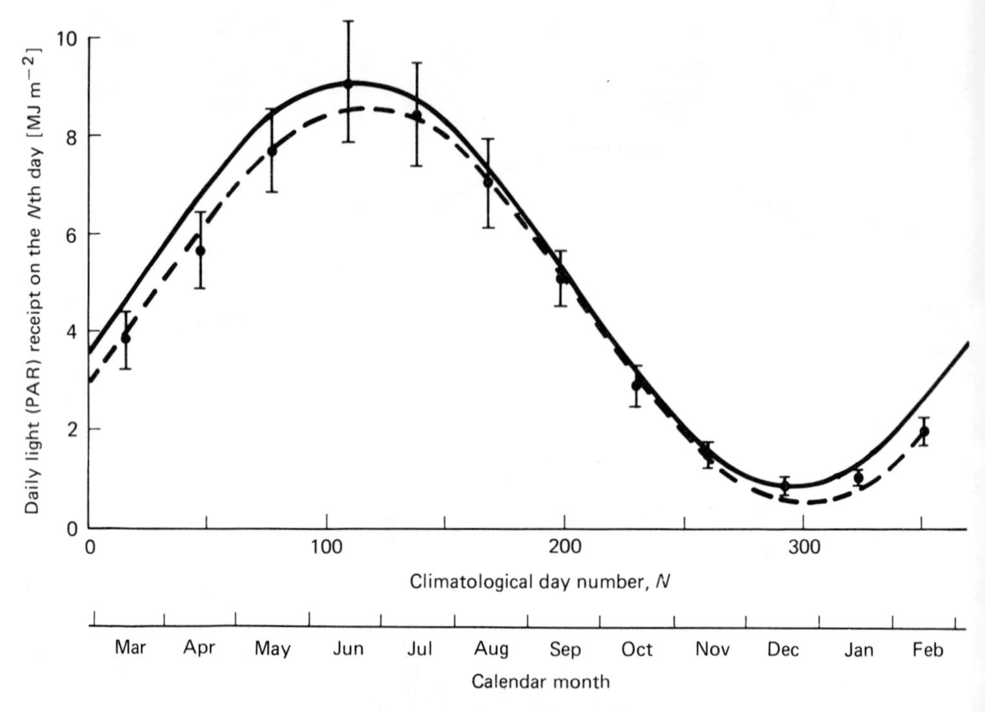


*Figure 6.2* Sine wave approximations to 10-year averages of monthly means of daily light receipts at Kew, London. Measured values are represented by ● and the bars show two standard deviations. The solid line represents the sine wave of equation (S6.6.3), which is calculated just from the maximum and minimum of the data points (p. 299). The dashed line represents the more accurately calculated sine wave approximation of equation (S6.6.5)

calculations are part of Exercise 6.6, and the results are shown in *Figure 6.2*. The standard deviations of the radiation data are also shown in *Figure 6.2*, illustrating that approximately it is the logarithm of the radiation that is normally distributed.

## VARIABILITY

The variability in $J_N$ is a topic that has not been extensively studied. A straightforward way of attempting to account for the variability in $J_N$ is to assume that $\ln J_N$ is normally distributed, so that

$$\ln J_N - \overline{\ln J_N} = \epsilon_N \tag{6.17}$$

where $\epsilon_N$ is a random variable with zero mean and standard deviation $\sigma_1$ obeying the normal distribution. Thus,

$$P(\epsilon) = \frac{1}{\sigma_1 \surd(2\pi)} \exp(-\epsilon^2/2\sigma_1^2) \tag{6.18}$$

is the probability density function for $\epsilon$.

However, since good and bad periods of weather tend to last for durations of several days, it may be more realistic to assume that today's radiation receipt is influenced by the outcome for the previous day. This can be represented by a simple Markov process (*see* Glossary) in which equation (6.17) is replaced by

$$\ln J_N - \overline{\ln J_N} = \epsilon_N + \rho_1 \epsilon_{N-1} \tag{6.19}$$

where $\rho_1$ is an autocorrelation coefficient, and $\epsilon_N$ is as before.

However, although this approach seems intuitively reasonable, the analysis of 22 years data, as in Exercise 6.7, covering both summer and winter, shows no significant correlation coefficient—a surprising result.

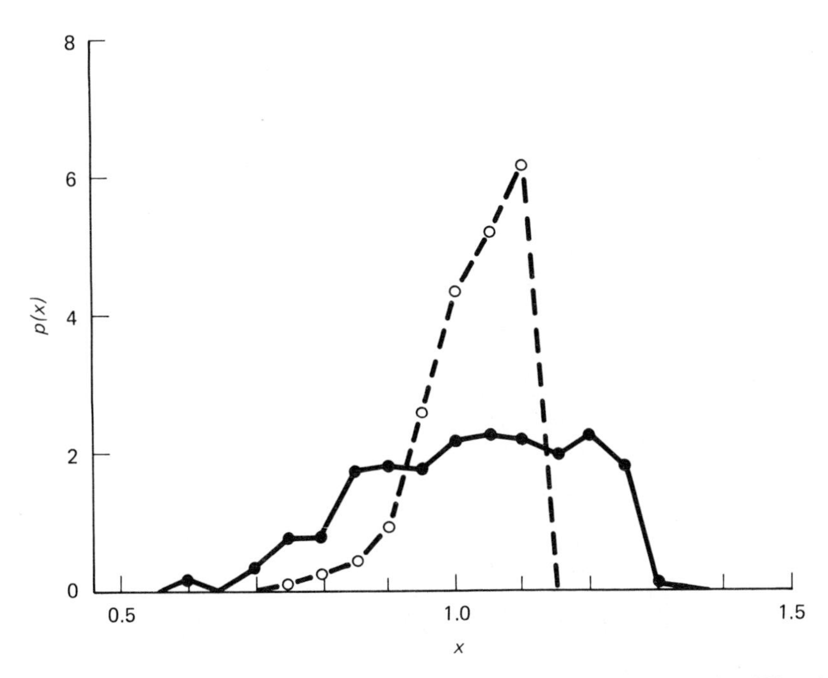


*Figure 6.3* Distribution of the logarithm of the daily light receipt, $J$, in December (●) and June (○), relative to the mean ($\overline{\ln J}$), and in the absence of seasonal effects. $p(x)\Delta x$ is the probability of $x$ falling in the range $x - \Delta x$ to $x$. The mean value of $x$, $\bar{x} = 1$. $x = (\ln J)/(\overline{\ln J})$

The distribution of daily light receipt, $J_N$, in the absence of seasonality has been examined using 231 measurements (11 years × 21) about midsummer (11 June to 1 July) and midwinter (11 December to 31 December), with the results shown in *Figure 6.3*. These are both, for winter as well as summer, suggestive of half-normal distributions, i.e. the normal distribution of equation (6.18) defined only for $\epsilon \leqslant 0$ and renormalized. A possible mechanism for this is to assume that 'cloudiness' (i.e. all factors that are attenuating light receipt) is half-normally distributed about the origin.

**Diurnal variation in radiation**

Following Charles-Edwards and Acock (1977, equation (7)), a full sine wave is used to describe diurnal response. If $J_N$ is the receipt of photosynthetically active

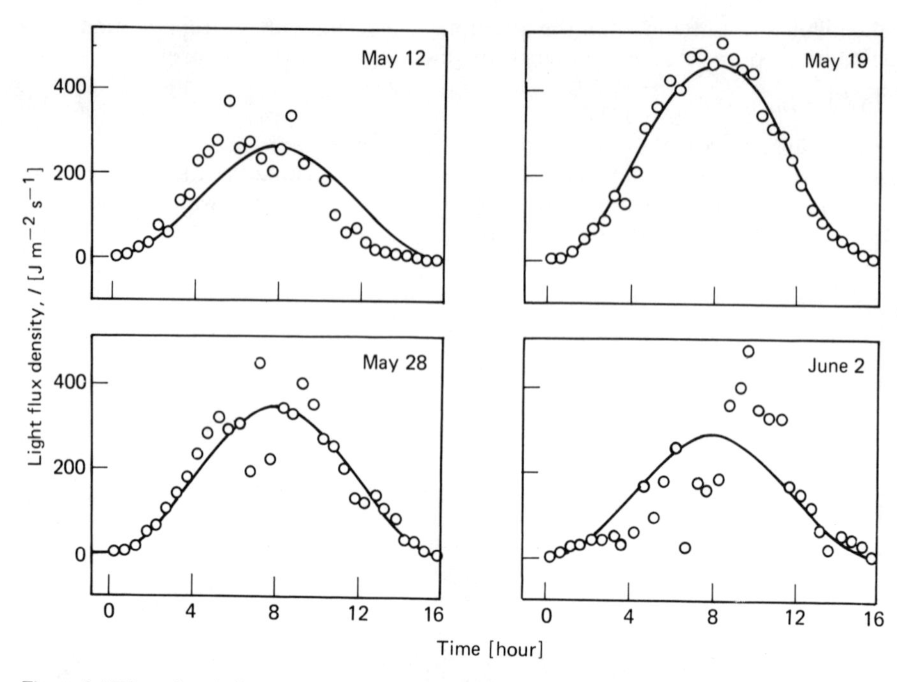


*Figure 6.4* Diurnal variation in radiation: measurements are compared with the simple sinusoid of equation (6.20). (After Charles-Edwards and Acock, 1977)

radiation on the $N$th day when the daylength is $g_N$, then the light flux density at time $t_d$, $I(t_d)$, is given by (*see Figure 6.4*)

$$I(t_d) = \frac{J_N}{g_N}\{1 + \cos[(t_d - 0.5)\frac{360}{g_N}]\} \quad \text{for} \quad 0.5 - \tfrac{1}{2}g_N \leqslant t_d \leqslant 0.5 + \tfrac{1}{2}g_N$$

$$\text{otherwise} \quad I(t_d) = 0 \tag{6.20}$$

$t_d$ is the decimal part of the day (equation (6.2)). This expression is constructed so that the term in the curly brackets varies from 0 to 2 and back to 0, and

$$\int_{0.5 - g_N/2}^{0.5 + g_N/2} I(t_d)\, dt_d = J_N$$

Greater precision may be obtained by including additional terms in the Fourier series.

Some authors (cf. Monteith, 1973, p. 31) have used a half sine wave for diurnal response, replacing equation (6.20) with

$$I(t_d) = \frac{\pi}{2g_N} J_N\{\sin[(t_d - 0.5 + \tfrac{1}{2}g_N)\frac{360}{2g_N}]\}$$

$$\text{for} \quad 0.5 - \tfrac{1}{2}g_N \leqslant t_d \leqslant 0.5 + \tfrac{1}{2}g_N$$

$$\text{otherwise} \quad I(t_d) = 0 \tag{6.21}$$

Equations (6.20) and (6.21) may be used in several ways. For example, $J_N$ may denote the *actual* light receipt on the $N$th day, and the terms in braces in these equations give an approximate diurnal variation. Alternatively, one might put $J_N = \bar{J}_N$ of equation (6.16), the mean daily light receipt for day $N$ over many years. In this case, there is an additional approximation and possible error, since $J_N$ (actual) may depart substantially from $\bar{J}_N$. A further alternative is to take a distribution of $J_N$ from equations (6.17) and (6.18), to account for the variability in $J_N$.

Another approach to describing light flux density during the day is using the methods of time-series analysis (*see* Glossary). However, very many measurements are needed in order to define the problem adequately, and for many purposes, it will be more efficient to make directly the appropriate measurements.

It is possible to use the method that was used for seasonal variability in equations (6.16) to (6.19) to describe diurnal variability, although this needs some modification. How this might be accomplished is now described.

Suppose that the experimental system being studied has a response time of $\Delta t$. This means that any light fluctuations that are faster than this are averaged. For photosynthesis, this time is of order of 1 minute or rather less. There is nothing to be gained by making radiation measurements with a higher time resolution than the photosynthetic response time (Thornley, 1974). Assume that radiation measurements are made at intervals of time $\Delta t$, and each measurement is an average over the preceding interval $\Delta t$. For the $N$th day with daylength $g_N$, there will be approximately

$$n = \text{integer part } (g_N/\Delta t + 1) \tag{6.22}$$

measurements, which may be written as

$$I_1, I_2, \ldots, I_n \tag{6.23}$$

To relate the continuous variables, time $t_d$ of equation (6.2b) and $I(t_d)$ to the index $i$ ($i = 1, 2, 3, \ldots, n$) and $I_i$, approximately

$$I(t_d) \approx I_i \tag{6.24}$$

where the subscript $i$ is given by

$$i = \text{integer part}\left(\frac{t_d - 0.5 + \frac{1}{2}g_N}{\Delta t} + 1\right) \tag{6.25}$$

during the interval $0.5 - \frac{1}{2}g_N \leqslant t_d \leqslant 0.5 + \frac{1}{2}g_N$.

Let $\bar{I}_i$ be the mean light flux density over a time interval $\Delta t$ at time $t_d$ (where $t_d$ and $i$ are related by equation (6.25)) for the $N$th day averaged over many years, then analogously to equations (6.17) to (6.19), a possible assumption is that

$$\ln(I_i/\bar{I}_i) = \epsilon_i + \rho_2 \epsilon_{i-1} \tag{6.26}$$

and

$$P(\epsilon) = \frac{1}{\sigma_2 \surd(2\pi)} \exp(-\epsilon^2/2\sigma_2^2) \tag{6.27}$$

where $\sigma_2$ is the standard deviation and $\rho_2$ is an autocorrelation coefficient. $\bar{I}_i$ is obtained from equation (6.20) or (6.21) with $J_N = \bar{J}_N$.

## Angular distribution of radiation

In the last two sections, the discussion of daily and within-day radiation receipts has been concerned with the radiation received by an unobstructed horizontal surface.

For some purposes, for example, light interception by foliage, building design, glasshouse construction, the size and nature of shade trees, the angular distribution of the radiation falling on a given surface is important. It is the intention to outline here the principal features of this topic.

### LENGTH OF MEASUREMENT PERIOD

The angular distribution perceived will clearly depend greatly upon whether the measurements are made over a period of a minute or two, an hour or so, or the whole day. Over a whole day, the sun will make a complete transit; over an hour most cloud patterns will change substantially; whereas over a minute there will usually be little change unless the cloud formations are particularly fast-moving. The remarks below must be interpreted with these considerations in mind.

### DIRECT RADIATION FROM THE SUN

At a particular time on the $N$th day, the zenith angle $z$ and the azimuth angle $a$ are obtained by equations (6.11) and (6.15a). For daily integrals, it is necessary to integrate over the track of the sun, allowing for its changing brightness over the day, which is at least partly due to the changing path-length through the atmosphere. There is no simple treatment of this problem.

### DIFFUSE RADIATION FROM A CLEAR SKY

Even in a cloudless sky, there can be considerable scattering from the molecules in the air, and also from dust particles or aerosol. The latter can be very variable, although both contributions depend greatly on the zenith angle of the sun. In general, the higher the clear sky diffuse radiation, the less is the direct solar radiation. Also, the diffuse radiation increases as one approaches the sun. The ratio of direct solar radiation to diffuse clear sky radiation is about one for high zenith angles (70° to 90°), but may rise to three at lower angles (Monteith, 1973, p. 27).

### OVERCAST SKIES

If skies are partly overcast, the angular distribution of radiation can be complex and rapidly changing. Only empirical treatments seem possible. For completely overcast skies, the situation simplifies. Two formulae are sometimes applied. The first describes the so-called uniform overcast sky (UOC), where the sky is assumed to be equally bright in all directions. Let $B(z,a)$ be the brightness of the sky in the direction specified by zenith angle $z$ and azimuth angle $a$, then in a uniform overcast sky

$$B(z, a) = I_o/\pi \qquad (6.28)$$

where $I_o$ is a constant. For a horizontal sensor, we must evaluate the integral

$$\int_0^{2\pi} \int_0^{\pi/2} (I_o/\pi)\cos z\,(\sin z\,\mathrm{d}z\,\mathrm{d}a) \qquad (6.29)$$

where the cos $z$ is to resolve on to the horizontal surface, and the second bracketted expression is the element of solid angle. Evaluation of the integral gives the result $I_o$ for the radiation flux density falling on the horizontal sensor.

However, since an overcast sky is usually brighter overhead than near the horizon, some workers prefer to use an almost equally arbitrary expression for a standard overcast sky (SOC), namely

$$B(z, a) = (9I_o/7\pi)\left(\frac{1 + 2\cos z}{3}\right) \tag{6.30}$$

With this expression, the sky overhead is three times as bright as on the horizon. Again, integration over the sky gives $I_o$ for the radiation falling on a horizontal sensor.

SUMMARY

The topic of angular distribution of radiation is not in a very satisfactory state. It is highly empirical: there has not been much experimental work, and there is no simple approximate theory. Fortunately for the plant scientist, the subject is of marginal importance for plant and crop growth. The radiation received by a horizontal plane is the main factor, and, as outlined earlier, this can be described semi-empirically to an acceptable degree of accuracy.

## Temperature

Both air and soil temperatures in temperate regions such as southern Britain vary sinusoidally over the year to a good approximation, and the mean temperature $\overline{T}_N$ on day $N$ may often be represented by

$$\overline{T}_N = a_y + b_y \sin\left[\left(\frac{N - N_o}{365}\right) 360\right] \tag{6.31}$$

where $a_y$ is the mean over the year, $b_y$ is the amplitude of the sine wave modulation, and $N_o$ gives the phase of the sine wave. Exercise 6.8 is concerned with the calculation of these parameters from air and soil temperature data from Littlehampton, Sussex, the results of which are shown in *Figures 6.5* and *6.6*.

A similar approach may be applied to diurnal temperature variation, but since the decimal time of day, $t_d$, varies between 0 and 1, the equation is (for the first harmonic of a Fourier series)

$$T(t_d) = a_d + b_d \sin[(t_d - t_o)360] \tag{6.32}$$

with parameters $a_d$, $b_d$ and $t_o$. The mean over the day is $a_d$, which has the average value of $\overline{T}_N$ of equation (6.31) for the $N$th climatological day.

Variability in the diurnal or season temperatures may be approached empirically much as for radiation (cf. equations (6.17) and (6.18)), but now equation (6.17) should be replaced by

$$T_N = \overline{T}_N + \epsilon_N \tag{6.33}$$

where $\epsilon_N$ is a random variable with zero mean and standard deviation $\sigma_N$. Temperature records at Littlehampton (Exercise 6.8) show that $\sigma_N$ does not vary significantly over the seasons, either for soil or air temperatures.

An excellent analysis of soil temperatures is given by West (1952).

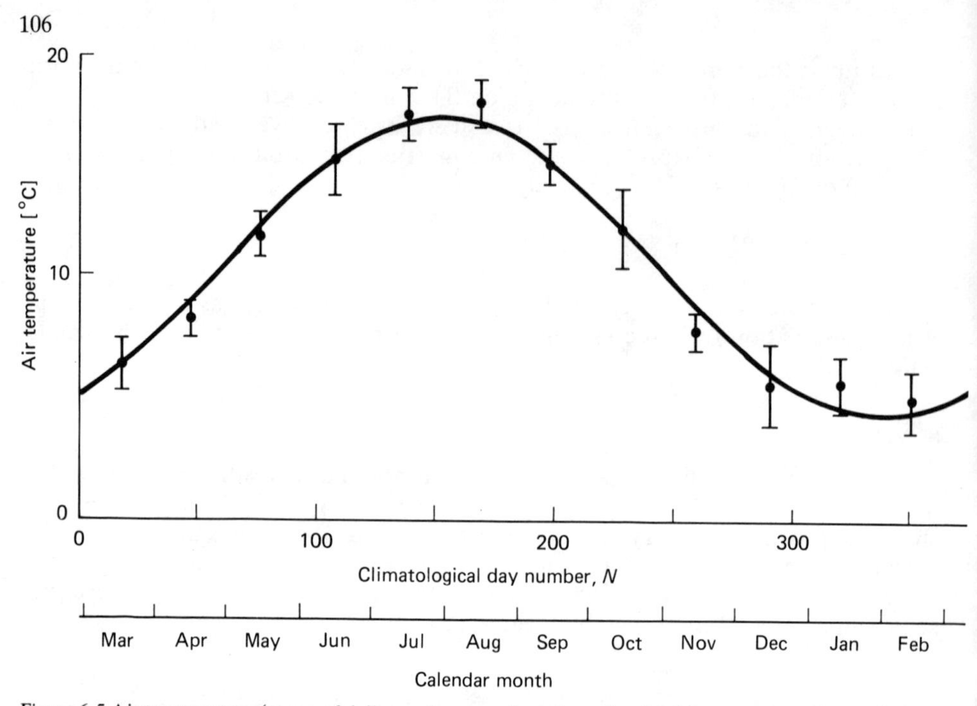


*Figure 6.5* Air temperature (mean of daily maximum and minimum) at Littlehampton, Sussex. The points are 10-year averages of monthly means, and the bars are two standard deviations. The solid line is the fitted sine wave of Exercise 6.8 and equation (S6.8.1) (p. 300)

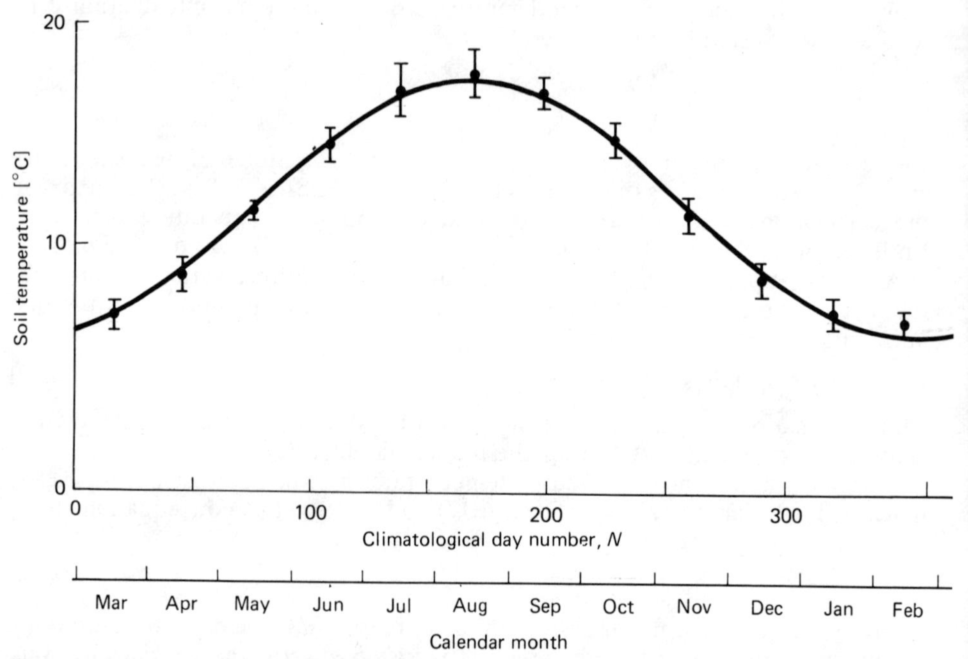


*Figure 6.6* Soil temperature at 100 cm at Littlehampton, Sussex. The points are 10-year averages of monthly means, and the bars are two standard deviations. The solid line is the fitted sine wave of Exercise 6.8 and equation (S6.8.2) (p. 300)

## Rainfall

Rainfall is almost invariably a major determinant of yield for most field crops, and it is only in small sectors of the industry, such as horticulture and protected crops, that rainfall can be neglected. The growth responses of plants are often a compromise between photosynthesis and transpiration, with an optimization of water use efficiency being a prime aim.

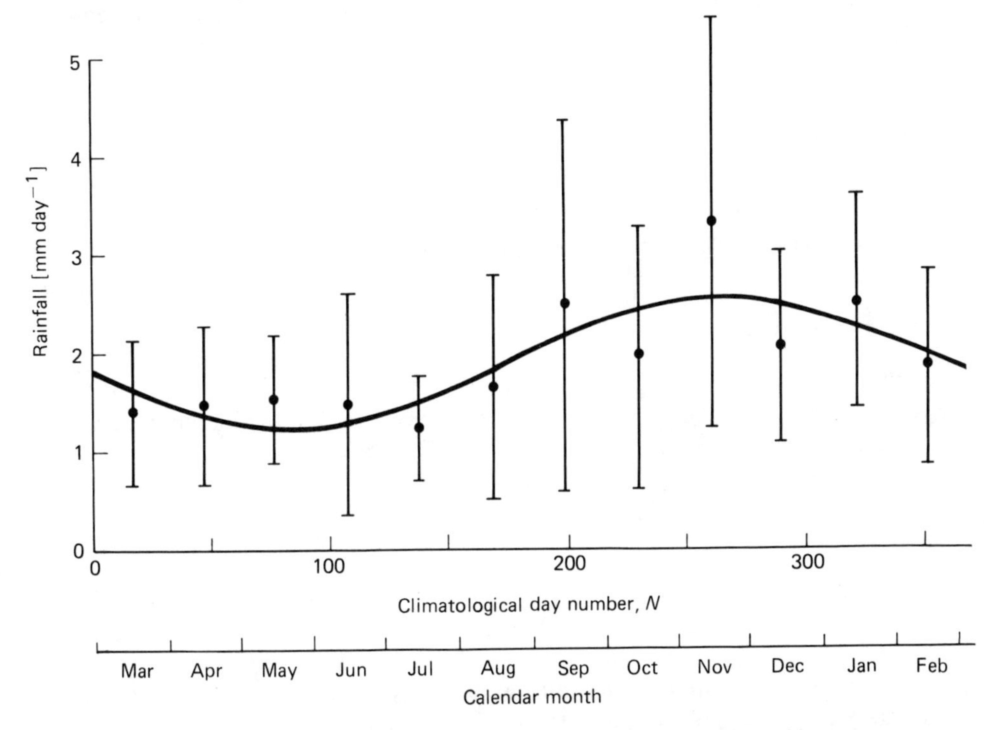


*Figure 6.7* Rainfall at Littlehampton, Sussex. The points are 10-year averages of monthly means, expressed on a daily basis, and the bars are two standard deviations. The solid line is a fitted sine wave, and represents equation (6.34) with $a = 1.91$ mm day$^{-1}$, $b = 0.66$ mm day$^{-1}$, and $N_o = 173$ day

In *Figure 6.7* are shown 10-year averages of monthly rainfall at Littlehampton, Sussex, expressed on a daily basis so that the trend is not obscured by the different lengths of the calendar months. There is a seasonal variation, and the procedure of fitting a simple sine wave to such data gives a very approximate description of rainfall. Thus, if $\bar{r}_N$ is the mean of the daily rainfall on day $N$ then

$$\bar{r}_N = a + b\sin\left[\left(\frac{N - N_o}{365}\right)360\right] \tag{6.34}$$

where $a$, $b$ and $N_o$ are the usual parameters.

The variability of rainfall is of crucial agronomic importance, and is an essential component of quantitative studies of irrigation and water use. Data from Littlehampton (*Figure 6.7*) show that the coefficient of variation (= standard deviation/mean) of the monthly rainfall totals is around 0.6 and is much the same for all calendar months. Meteorological data also show that roughly a third of the days in any month are rainy. Our aim in the rest of this section is to construct, as an example, a simple model of rainfall that will account for these two principal features, and may be suitable for use in simulation studies. An authoritative account of the problem of analysing and simulating daily rainfall is given by Stern and Coe (1982) and Stern, Dennett and Dale (1982a, b), who combine Fourier analysis, Markov chains and distribution functions of daily rainfall amount in their approaches to the topic.

Consider the rainfall over a period of $n$ days, and let us ignore any seasonal changes in rainfall expectation over this period. Assume that daily rainfall is an all-or-nothing event, so that

$$p = \text{probability of no rain} \tag{6.35a}$$

and

$$q = \text{probability of } h \text{ mm rain} \tag{6.35b}$$

falling on a given day. The mean daily rainfall $\bar{r}$ is

$$\bar{r} = qh \tag{6.36}$$

and $p$ and $q$ satisfy

$$p + q = 1 \tag{6.37}$$

Assuming that each day is independent of every other day, then the rainfall distribution is given by the binomial expansion (*see* Glossary):

$$(p+q)^n = p^n + np^{n-1}q + \frac{n(n-1)}{(1)(2)}p^{n-2}q^2 + \ldots + \binom{n}{i}p^{n-i}q^i + \ldots + q^n \tag{6.38}$$

where

$$\binom{n}{i} = \frac{n!}{r!(n-r)!}$$

The probability of $ih$ mm of rain during the $n$-day period is $\binom{n}{i}p^{n-i}q^i$. The mean rainfall over the month of $n$ days is

$$\text{h}\sum_{i=0}^{i=n} i\binom{n}{i}p^{n-i}q^i = nqh \tag{6.39}$$

in agreement with equation (6.36). The standard deviation ($SD$) may be obtained by

$$(SD)^2 = h^2 \sum_{i=0}^{i=n} (i - nq)^2\binom{n}{i} \tag{6.40}$$

giving

$$SD = h(npq) = h[nq(1-q)] \tag{6.41}$$

The coefficient of variation ($CV$) is therefore given by

$$CV = \frac{SD}{\text{mean}} = \left(\frac{1-q}{nq}\right) \tag{6.42}$$

The mean number of rainy days, $n_r$, is

$$n_r = nq \tag{6.43}$$

Inversion of equation (6.42) gives

$$q = \frac{1}{n(CV)^2 + 1} \tag{6.44}$$

Applying this simple theory to rainfall from Littlehampton for a 30-day month, $n = 30$, and taking $CV = 0.6$, equation (6.44) gives $q = 0.085$. However, application of equation (6.43) gives $n_r = 3$, only three rainy days a month, which conflicts with the ten rainy days a month that are observed. The problem arises at least partly because the approach assumes that each day is independent of every other day, and does not allow for the fact that good and bad weather tends to occur over periods of several days. Such correlations tend to increase variability, so it might then be possible to reconcile a coefficient of variation of 0.6 for the monthly means with about ten rainy days a month. A more rigorous approach would be to use conditional probabilities, possibly as with the Markov process of equation (6.19) for radiation. However, an approximate intuitive method is as follows.

Suppose that each period of good or bad weather lasts for $n_p$ days (e.g. five days), and during these periods, it either rains every day, or it does not rain at all. There are $n_{eff}$ periods of $n_p$ days in a month (for 5-day periods and a 30-day month, $n_{eff} = 6$), so equation (6.44) must be rewritten

$$q = \frac{1}{n_{eff}(CV)^2 + 1} \tag{6.45}$$

which gives

$$n_{eff} = \left(\frac{1-q}{q}\right) \Big/ (CV)^2 \tag{6.46}$$

together with the relationship

$$\text{number of days in month} = n_{eff}\, n_p = n \tag{6.47}$$

If there are ten rainy days (on average) in a month of 30 days, and $CV = 0.6$, then from equation (6.43) $q = 10/30 = 0.33$, which with equation (6.46) gives $n_{eff} = 5.6$, and therefore the length of the period of good or bad weather $n_p = 30/5.6 = 5.3$ days. The rainfall distribution is given, as before, by the binomial distribution of equation (6.38), but substituting $n_{eff}$ for $n$. The mean daily rainfall $\bar{r}$ is still given by equation (6.36), and enables $h$ to be obtained. Equations (6.35) are replaced by

$$p = \text{probability of no rain falling during } n_p \text{ days} \tag{6.48a}$$

and

$$q = \text{probability of } n_p h \text{ mm rain falling during } n_p \text{ days} \tag{6.48b}$$

these being exclusive events (*see* Exercise 6.9).

The use of this method in crop growth simulation is as follows. From local climate data, extract the coefficient of variation ($CV$) of the monthly means, and also the mean number of rainy days in a month, $n_r$. These two numbers may be the same for all calendar months, or may be different for each month. Use equation (6.43) to find $q$, and then equation (6.45) to find $n_{eff}$, the effective length of the periods of good or bad weather. In the computer model, $n_{eff}$ and $q$ are provided as parameters. Rainfall is calculated for periods of $n$ days at a time. At the beginning of a block of $n$ days, generate a random number between 0 and 1. If the random number is greater than $(1 - q)$, then it rains for the whole of the period of $n$ days, at the rate of $h$ mm per day.

## Windrun

The wind may be important in estimating heat losses from farm animals and buildings, but generally has only a small effect on the photosynthetic rate and growth of field crops. Its major influence on crop growth and yields is usually via the disease vectors, as it is one of the more important factors in the spread of many diseases. Strong winds are sometimes responsible for direct mechanical damage, causing loss of yield. Where protected structures, such as heated glasshouses are used, then the energy loss is very dependent on wind speed and may be proportional to the wind speed squared or even to a higher power.

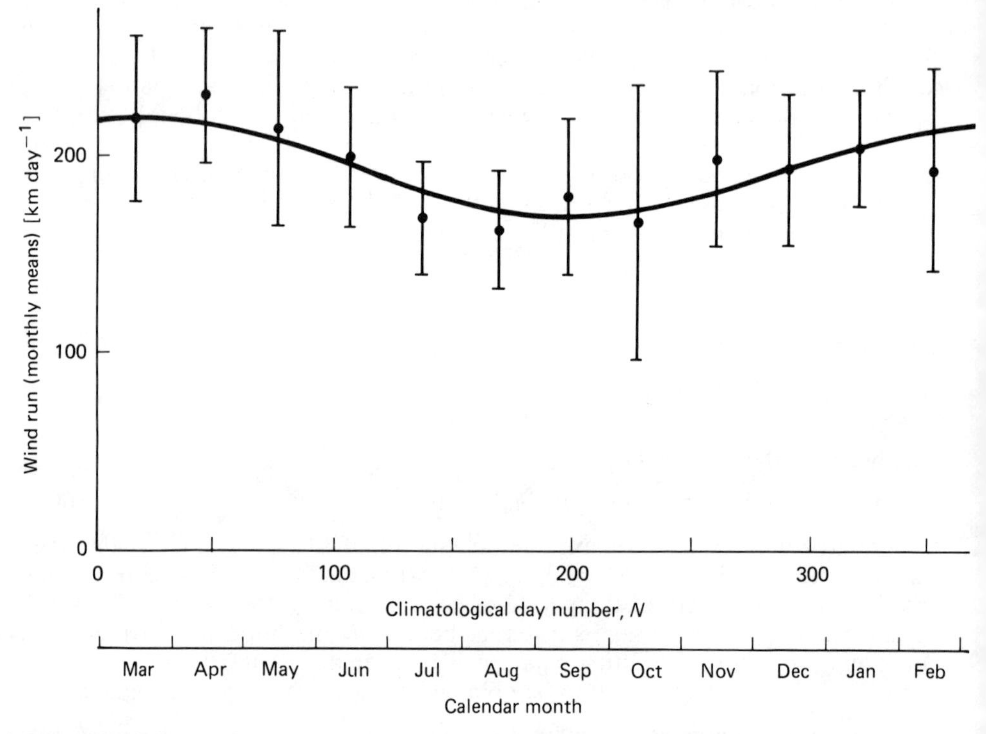


*Figure 6.8* Windrun at Littlehampton, Sussex. The points are 10-year averages (1967–76) of monthly means, expressed on a daily basis; the bars are two standard deviations. The solid line is a fitted sine wave, of the type of equation (6.34), with $a = 194\,\text{km day}^{-1}$, $b = -24\,\text{km day}^{-1}$, and $N_o = 108$ day

The monthly means of the daily windrun at Littlehampton, Sussex, averaged over 10 years, are shown in *Figure 6.8*. Although the winter and spring months are more windy than the summer and autumn months, the variation during the year is not very great, and there is considerable year-to-year variability.

## Exercises

### Exercise 6.1

Construct formulae converting the decimal part of the day, $t_d$ $(0 \leqslant t_d < 1)$, to hours, minutes and seconds. And also vice versa.

**Exercise 6.2**

Use equation (6.6), or otherwise, to find the climatological day number, $N$, of the first day of each calendar month, and tabulate these.

**Exercise 6.3**

For your location (i.e. latitude) write a computer program to calculate, at 10-day intervals throughout the year starting on 1 March: the day of the month $D$ and the calendar month $M$ using equations (6.7); the declination $\delta$ using equations (6.8) and (6.9); the zenith angle $z_n$ at solar noon (use equation (6.11) with $h = 0$ and $\phi$ equal to your latitude); the daylengths $g_N$ corresponding to the zenith-angle definitions of $z = 90°, 90.83°, 96°, 102°$ and $108°$ (equation (6.14)).

**Exercise 6.4**

Using equation (6.12) to eliminate $h_o$, derive an alternative equation (namely, $\cos a_o = -\sin \delta / \cos \phi$) to equation (6.15b) for the azimuth angle $a_o$ at sunrise or sunset.

**Exercise 6.5**

For your location and with a zenith angle of 96° (civil twilight), use your program from Exercise 6.3, or otherwise, to work out (using equations (6.8), (6.9) and (6.14)) the daylengths on 21 June ($N = 113$) and on 21 December ($N = 296$): $g_{113}$ and $g_{296}$. Construct a sine-wave approximation to the variation in daylength using

$$g_N = (g_{113} + g_{296}) + (g_{113} - g_{296}) \sin\left[\left(\frac{N - 21}{365}\right) 360\right]$$

Tabulate the results and compare these with the more accurate daylengths obtained in Exercise 6.3.

**Exercise 6.6**

For Kew, London (51.47°N, 0.32°W), a 10-year average of the monthly means of the daily receipt of photosynthetically active radiation (using the conversion factor 100 kilolux-hours = $400 \times 3600 = 1.44$ MJ m$^{-2}$ PAR, and for the period 1966–76 omitting 1973 for which the record is missing) is

| | Jan | Feb | Mar | Apr | May | Jun | Jul | Aug | Sep | Oct | Nov | Dec |
|---|---|---|---|---|---|---|---|---|---|---|---|---|
| 10-year average | 1.05 | 1.98 | 3.78 | 5.65 | 7.71 | 9.09 | 8.44 | 7.02 | 5.11 | 2.93 | 1.48 | 0.87 |
| *SD* | 0.13 | 0.36 | 0.64 | 0.85 | 0.89 | 1.31 | 1.10 | 0.89 | 0.61 | 0.45 | 0.27 | 0.14 |
| $\frac{SD}{\text{average}}$ | 0.12 | 0.18 | 0.17 | 0.15 | 0.12 | 0.14 | 0.13 | 0.13 | 0.12 | 0.15 | 0.18 | 0.16 |

The units for the mean and standard deviation ($SD$) are MJ $m^{-2}$ of PAR. Note that the coefficient of variation ($SD$/average) is fairly constant throughout the year at 0.15, with a slight suggestion that it might be lower in the summer months than in the winter. This supports the assumption in equations (6.17) and (6.18) that ln(daily light receipt) is normally distributed.

Assuming a relationship of type equation (6.16), namely

$$\bar{J}_N = a + b \sin\left[\left(\frac{N - 21}{365}\right) 360\right]$$

where $N$ is the climatological day number ($N$ = 1 on 1 March), estimate approximately the constants $a$ and $b$ from the average monthly means above, and plot a graph showing the equation and the average monthly means. This method only makes use of $\bar{J}_{113}$ and $\bar{J}_{296}$. Application of the method outlined in the Glossary for Fourier analysis makes use of all 12 data points, resulting in a better approximation.

**Exercise 6.7**

At Kew, London (51.47°N, 0.32°W), for the 21 days beginning on 11 June 1976 ($N$ = 103), the daily illumination (in kilolux-hours) took the following values: 711, 513, 947, 942, 807, 184, 389, 837, 303, 513, 496, 938, 957, 992, 944, 842, 794, 716, 955, 1038, 1004. Assume that at this time of year $\bar{J}_N$ is constant, that is, equation (6.16) does not vary much about $N$ = 113 (21 June). Assuming equations (6.17) to (6.19), estimate the two parameters $\sigma_1$ and $\rho_1$ which are used to characterize the daily radiation receipt.

**Exercise 6.8**

At Littlehampton, Sussex (50.82°N, 0.52°W), 10-year averages of the monthly means of the air temperature [½(maximum + minimum)], $T_a$, its standard deviation, $SD$, the soil temperature at a depth of 100 cm, $T_s$, and its standard deviation, $SD$, are as follows (*Figures 6.5* and *6.6*)

| | Jan | Feb | Mar | Apr | May | Jun | Jul | Aug | Sep | Oct | Nov | Dec |
|---|---|---|---|---|---|---|---|---|---|---|---|---|
| $T_a$ | 5.6 | 5.0 | 6.3 | 8.2 | 11.6 | 14.7 | 16.6 | 17.0 | 14.5 | 11.9 | 7.7 | 5.6 |
| $SD$ | 1.1 | 1.3 | 1.1 | 0.8 | 0.9 | 1.5 | 1.1 | 1.0 | 0.9 | 1.7 | 0.8 | 1.7 |
| $T_s$ | 7.4 | 7.0 | 7.1 | 8.8 | 11.4 | 14.1 | 16.3 | 17.0 | 16.1 | 14.3 | 11.3 | 8.6 |
| $SD$ | 0.7 | 0.5 | 0.6 | 0.7 | 0.4 | 0.7 | 1.1 | 1.0 | 0.7 | 0.7 | 0.7 | 0.7 |

For both $T_a$ and $T_s$, calculate the parameters $a_y$, $b_y$, and $N_o$ of the first-harmonic Fourier approximation of equation (6.31) (*see* Glossary).

**Exercise 6.9**

The total October rainfall for the 10 years (1968–77) at Littlehampton and the number of days on which the rainfall was above 0.2 mm are

| 95 | 1 | 35 | 59 | 18 | 55 | 86 | 35 | 145 | 76 | mm |
|---|---|---|---|---|---|---|---|---|---|---|
| 15 | 2 | 12 | 5 | 6 | 8 | 13 | 6 | 19 | 14 | days |

Calculate the arithmetic means and standard deviations of both. Using the binomial rainfall model of pp. 108–109, find the parameters $q$ and $n_p$, where $n_p$ is the length (number of days) of the periods of good or bad weather, and $q$ is the probability of rain falling during a given period of $n_p$ days. What is the probability of the month of October passing without rain?

Chapter 7

# Plant and crop processes

## Introduction

In this chapter, a critical survey is given of what in fact is a vast area. No attempt is made to be exhaustive in coverage, but on each topic, the most promising of current approaches are outlined. Mathematically complex treatments are eschewed, and the methods presented are simple enough for possible incorporation in a whole-plant or crop model. Some general references are cited in the Bibliography.

The division of material between this chapter and the next chapter on crop responses and models presented a problem. The intention is to discuss in the present chapter those processes where the plant or the crop has an important role, and where the time interval of interest is about a day or less. Thus, fertilizer response, where concern is usually with final yield, and development, where there are three or four events during a growing season, are considered in the next chapter. So is water uptake and use since, although this occurs on a daily or within-day time scale, the plant or crop fills a largely passive role, as an intermediary between soil and atmosphere.

A notable omission from the plant processes discussed below is translocation. There are many monographs on the topic (Canny, 1973; Moorby, 1981); our treatment for within-plant carbon fluxes is summarized in equations (7.57) and (7.62) in the section on partitioning—they represent a phenomenological approach that is finding increasing use.

## Light interception

The interaction of the crop or plant canopy with the light environment is the first of many sequential and parallel processes occurring in the growth and development of these systems. Although most crops begin life as an array of isolated plants, it is much easier to treat crops than plants, and this is the most convenient starting point for our discussion.

### Crops

The Monsi–Saeki equation (1953) is the standard treatment for this problem; Monteith's (1965) analysis gives an equivalent result (Exercise 7.2). Let $I_o$ be the

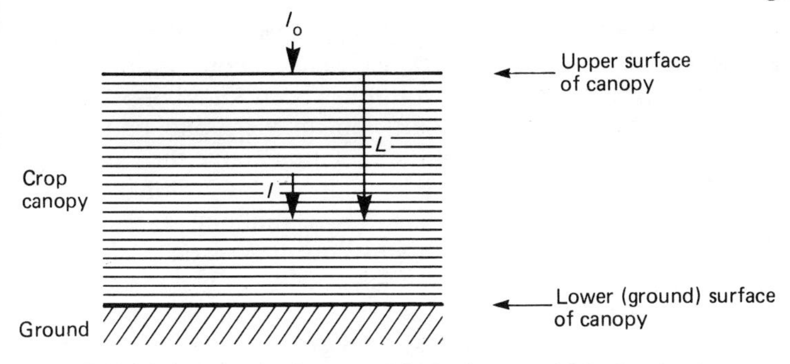


*Figure 7.1* Light interception in crops. $I_o$ is the downward light flux density on a horizontal plane above the canopy; $I$ is the downward light flux density on a horizontal plane at 'depth' $L$ within the canopy, where $L$ is the leaf area index measured from the upper canopy surface

downward light flux density [W m$^{-2}$] on a horizontal plane above the crop canopy (*Figure 7.1*), $I$ the downward light flux density on a horizontal plane within the canopy at a 'depth' $L$ from the upper canopy surface, where $L$ is the leaf area index (defined as the leaf area per unit ground area) measured downwards from the upper canopy surface, then the Monsi–Saeki equation is written

$$I = I_o e^{-kL} \tag{7.1}$$

where $k$ is a constant known as the extinction coefficient. The behaviour of equation (7.1) is shown in *Figure 7.2*.

For calculations of photosynthetic rates, the light flux density incident on a leaf surface, $I_\ell$, is required, and this is not necessarily the same as the light flux density incident on a horizontal plane, $I$. Saeki (1960) proposed, and it can be shown (Thornley, 1976, pp. 88–89) that $I_\ell$ and $I$ are related by

$$I_\ell = \left(\frac{k}{1-m}\right) I = \left(\frac{kI_o}{1-m}\right) e^{-kL} \tag{7.2}$$

where $m$ is the transmission coefficient of the leaf.

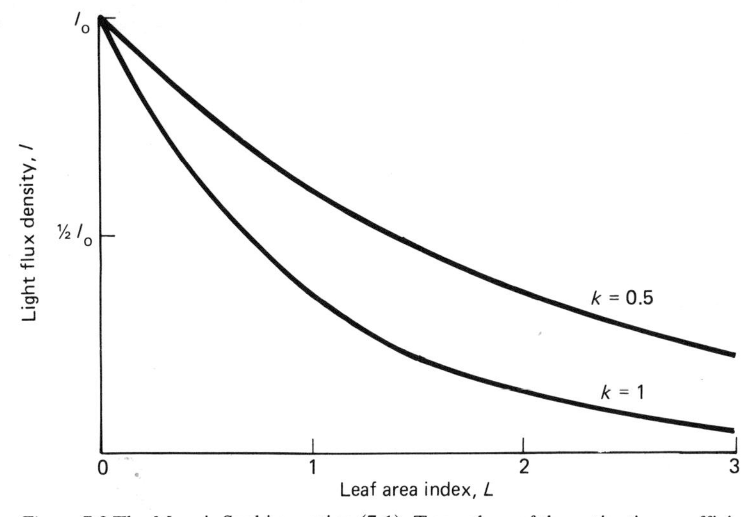


*Figure 7.2* The Monsi–Saeki equation (7.1). Two values of the extinction coefficient $k$ are shown

The Monsi–Saeki equation only applies rigorously to the rather unreal situation of a crop consisting of randomly distributed black leaves interacting with radiation only from above (i.e. vertical radiation). However, it has been found in practice to apply to a variety of crops. The extinction coefficient may depend upon many factors, for instance, the angular distribution of the radiation above the canopy (p. 103), the angular distribution of the leaves, the spatial distribution of the leaves (these may be clumped (tending to lie under each other), random, or distributed (tending to avoid each other) (Acock, Thornley and Warren Wilson, 1970)), and their optical properties. Thus, the extinction coefficient is usually regarded as an empirical constant, which is assigned a value from experimental data, but whose theoretical basis may be obscure. There are some detailed treatments of this problem (de Wit, 1965; Anderson, 1966; Chartier, 1966; Cowan, 1968; Idso and de Wit, 1970; Nilson, 1971; Norman and Jarvis, 1975; Bikhele, Moldau and Ross, 1980), and although these contribute to our understanding of the role of crop architecture in light interception, they are rather specialized for most crop physiologists and agronomists, whose objectives may lie elsewhere.

**Single plants or rows of plants**

No method has yet been developed for plants analogous to the simplicity of the Monsi–Saeki equation (7.1). An extension of the Monsi–Saeki approach which can be used for plants or rows with a defined canopy structure has been described by Charles-Edwards and Thornley (1973), and an outline is given of this technique.

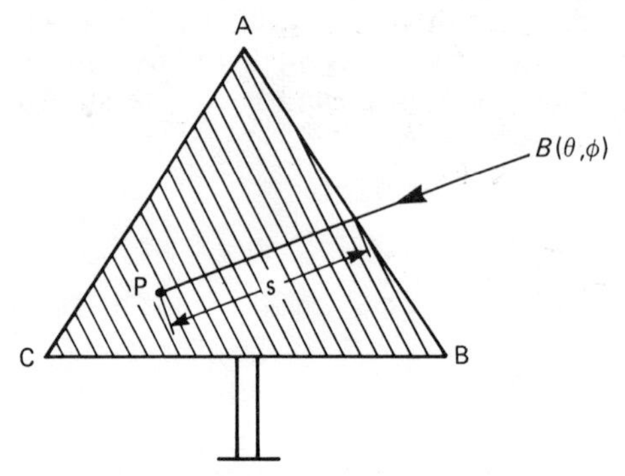


*Figure 7.3* Light interception in a single plant. ABC defines the canopy of the plant. Radiation from an element of sky, defined by the two angles, $\theta$ and $\phi$, and with luminosity $B(\theta,\phi)$, traverses a path length $s$ in order to reach the point P. The radiation is attenuated according to the Monsi–Saeki equation (7.1) as it travels along $s$ within the canopy. Contributions from the whole sky (all $\theta$ and $\phi$ above the horizon) are summed to give the total radiation reaching P

In *Figure 7.3* an example is shown for a plant whose canopy is an isosceles triangle of revolution. If $dI_p$ is the contribution at P to the light flux density on a horizontal plane, from an element of sky at an angular position defined by angles $\theta$ and $\phi$ with luminosity $B(\theta, \phi)$ [$W\,m^{-2}\,steradian^{-1}$], then assuming that the Monsi–Saeki equation (7.1) can be used to give the attenuation along the pathlength $s$ which the radiation traverses in order to reach P, and the leaf area density is constant and equal to $F$ ($F$ is the leaf area per unit volume of space), therefore

$$dI_p = B(\theta, \phi)e^{-kFs} \cos\theta \sin\theta \, d\theta \, d\phi \quad (7.3)$$

where $k$ is the extinction coefficient, $Fs$ is the effective leaf area index traversed over pathlength $s$, $\cos\theta$ resolves the radiation onto the horizontal plane, and $\sin\theta \mathrm{d}\theta \mathrm{d}\phi$ is the element of solid angle. The pathlength $s$ varies with the angles $\theta$ and $\phi$, and the total radiation at P, $I_p$, is obtained by integrating equation (7.3) over all angles $\theta$ and $\phi$ above the horizon. This can only be achieved numerically, even for simple canopy shapes such as parts of a sphere or an ellipsoid, but computer programs can be written so that application of the method is fairly routine (Charles-Edwards and Thorpe, 1976; Palmer, 1977; Whitfield, 1980; Whitfield and Connor, 1980).

In the same way that a factor $k/(1-m)$ was applied in the Monsi–Saeki theory in going from the light flux density falling on a plane to that falling on a leaf surface (equations (7.1) and (7.2)), the light flux density on a horizontal plane at point P, $I_p$, must be corrected to give the light flux density falling on a leaf surface at point P, $I_\ell$, by means of

$$I_\ell = \left(\frac{k}{1-m}\right) I_p \tag{7.4}$$

## Discontinuous canopies

While the method described in the last section can be used for any canopy whose shape can be geometrically defined (an uneven crop, or hedgerow, or any array of plants or trees), it has not been found to be well suited for use in crop models, possibly because of the large amount of computing involved in integrating equation (7.3), and also the difficulty in defining biologically plausible geometrical forms that are mathematically and computationally tractable. For these reasons, an interesting and promising alternative approach has been proposed by Jackson and Palmer (1979), which is now described.

They suggest that it is advantageous to view the light transmission, $T$, through a discontinuous canopy as consisting of two discrete and additive components (*Figure 7.4*)

$$T = T_b + T_c \tag{7.5}$$

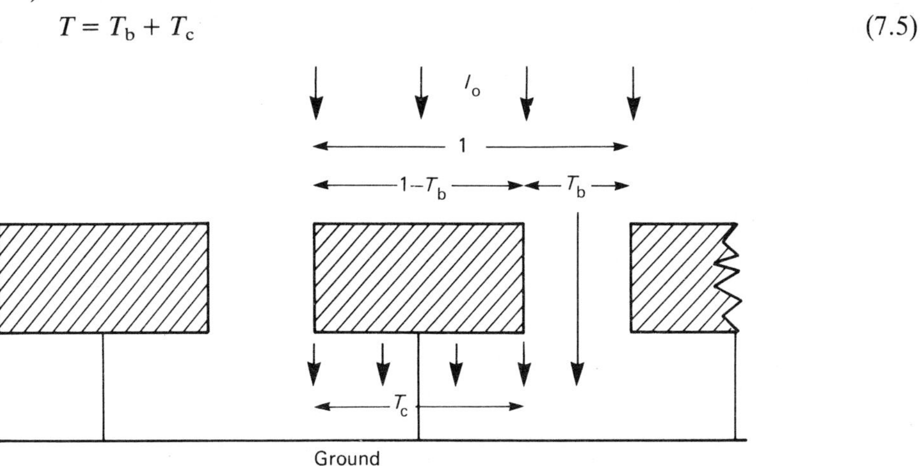


*Figure 7.4* Discontinuous canopy model (Jackson and Palmer, 1979), illustrated for vertical radiation $I_o$. Per unit area, a fraction $T_b$ (b denotes black canopy) reaches the ground without traversing the canopy. The remaining fraction $1 - T_b$ passes through the canopy, where it is attenuated, and of this, $T_c$ ($T_c \leq 1 - T_b$) reaches the ground. The total transmission $T = T_b + T_c$

where $T_b$ is the fraction of light that reaches the ground when the canopy is black or completely non-transmitting, and $T_c$ is the fraction of the light that has passed through the canopy. As shown in *Figure 7.4*, the canopy fraction $T_c$ has a maximum value of $1 - T_b$ if the canopy is transparent, and in general $T_c$ is given by

$$T_c = (1 - T_b) \times \text{attenuation by the canopy} \tag{7.6}$$

It is then assumed that the attenuation factor in equation (7.6) can be calculated by means of the Monsi–Saeki equation (7.1). If $L$ is the leaf area index (with respect to the total ground area), then the effective leaf area index $L'$ is given by (*Figure 7.4*)

$$L' = \frac{L}{1 - T_b} \tag{7.7}$$

and the attenuation by the canopy is given by

$$e^{-kL'} \tag{7.8}$$

where $k$ is the extinction coefficient. Combining (7.5) to (7.8) gives

$$T = T_b + (1 - T_b)\exp[-kL/(1 - T_b)] \tag{7.9}$$

There are various ways of estimating $T_b$ (Jackson and Palmer, 1979), some involving measurement, or alternatively by calculation. For some crops, a simple overhead estimate of the visible ground area might suffice. Jackson and Palmer (1979) tested equation (7.9), comparing its predictions with observations on hedgerow apple orchards, giving non-systematic discrepancies of order 2–4%, suggesting that equation (7.9) is an accurate and useful model for appropriate cases.

Both equations (7.2) and (7.4) give the light incident on the leaf surface, and in order to obtain the photosynthetic rate of the plant or crop, one must sum the contributions of the leaves, which are differently illuminated. On the other hand, equation (7.9) gives the light intercepted by the canopy as a whole, $I_c$, as

$$I_c = I_o(1 - T) \tag{7.10}$$

The calculation of canopy photosynthesis from canopy light interception is discussed in the next section.

## Photosynthesis

Photosynthesis provides the driving force for most of the processes that concern the agricultural scientist, and is one of the more intensely researched areas of biology today. While the basic principles still elude us, it is a phenomenon that can be observed at levels from the subcellular to the whole-crop, and many of the responses that interest the crop physiologist have been determined. In this section, a simple outline is given of how photosynthesis may be analysed at the leaf, plant and crop levels of organization. Respiration and photorespiration are ignored. More comprehensive treatments can be found in Thornley (1976, Chapter 4); Acock *et al.* (1976); and Hesketh and Jones (1980).

The problem is as follows. If $P_c$ denotes the photosynthetic rate of a plant or a crop canopy, then the crop physiologist wishes to be able to calculate $P_c$ given the leaf area index $L$ (or the leaf area $A$ of the plant), the light flux density above the crop $I_o$, the ambient $CO_2$ level $C_a$, and the temperature $T$. Formally

$$P_c = P_c(I_o, C_a, T, L) \tag{7.11}$$

In practice, direct calculations along the lines of (7.11) have not proved fruitful, possibly because of the wide variety of canopy architectures that are encountered. The more usual approach is to define a function that enables the leaf photosynthetic rate $P_\ell$ to be calculated:

$$P_\ell = P_\ell(I_\ell, C_a, T) \tag{7.12}$$

where $I_\ell$ is now the light flux density incident on the leaf. Equations (7.11) and (7.12) are now connected by summing over the leaf area elements in the crop, to give

$$P_c = \int_0^L P_\ell \, dL \tag{7.13}$$

**Leaf photosynthesis**

Many different equations have been proposed for leaf photosynthetic response to light and $CO_2$ (for a survey and original references see Thornley, 1976, p. 94; Hesketh and Jones, 1980, p. 125), but perhaps the most frequently used for the leaf photosynthetic rate $P_\ell$ is

$$P_\ell = \frac{\alpha I_\ell P_{max}}{\alpha I_\ell + P_{max}} \tag{7.14}$$

where $\alpha$ is a constant and $P_{max}$ is the value of $P$ at saturating light levels ($I_\ell \rightarrow \infty$). Equation (7.14) is drawn in *Figure 7.5*. Mathematically, the curve is known as a rectangular hyperbola, as it has two asymptotes that intersect at right-angles, at $\alpha I_\ell$

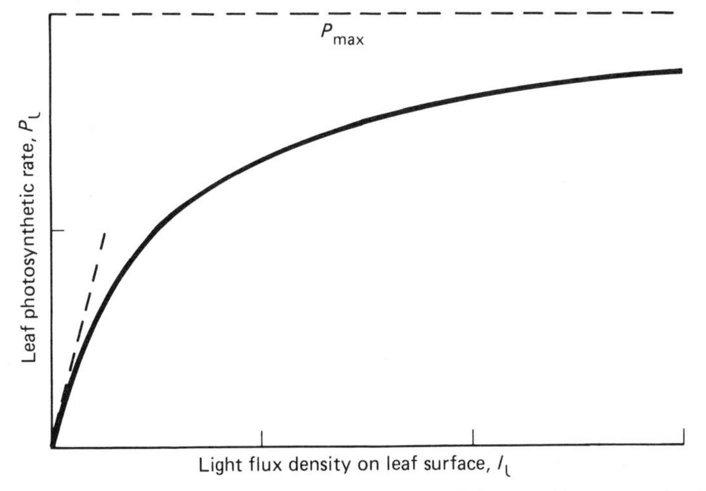


*Figure 7.5* Leaf photosynthetic response curve to light according to equation (7.14). The dashed line at $P_{max}$ denotes the light-saturated value of photosynthetic rate $P$; $\alpha$ is the initial slope of the response curve (equation (7.15))

$= -P_{max}$ and $P_\ell = P_{max}$. The initial slope of the response curve is equal to the constant $\alpha$:

$$\frac{dP_\ell}{dI_\ell}(I_\ell = 0) = \alpha \tag{7.15}$$

and $\alpha$ is known as the photosynthetic efficiency. An explicit $CO_2$ dependence is sometimes put into equation (7.14) by assuming that $P_{max}$ is a function of the ambient $CO_2$ concentration $C_a$, for instance according to

$$P_{max} = \tau C_a \tag{7.16}$$

where $\tau$ is a constant, often referred to as the $CO_2$ conductance. Typical values for the parameters are (cf. Acock *et al.*, 1976, who analysed photosynthesis in green peppers) $\tau = 0.002\ m\,s^{-1}$ and $C_a = 0.0006\ kg\,CO_2\,m^{-3}$ (c. 330 vol. $\times\ 10^{-6}$), giving with equation (7.16)

$$P_{max} = 1.2 \times 10^{-6}\ kg\,CO_2\,m^{-2}\,s^{-1}$$

The photosynthetic efficiency $\alpha$ is, for mature leaves of $C_3$ plants, often approximately

$$\alpha = 13 \times 10^{-9}\ kg\,CO_2\,(J\,PAR)^{-1}$$

where PAR denotes photosynthetically active radiation.

## Crop photosynthesis

Evaluation of the photosynthetic rate, $P_c$, of a crop (or a plant) requires summation over the elements of leaf area using equation (7.13); this is not trivial because the leaf photosynthetic rate $P_\ell$ depends on the light flux density incident on the leaf $I_\ell$ (cf. equation (7.14)), which varies within the canopy (equation (7.2)). This combination of a canopy structure and light interception model with a leaf photosynthesis model means that many different crop photosynthesis models are required, e.g. for an apple orchard, a lettuce crop, a spruce forest, or a stand of corn. However, the approach can be nicely illustrated with a case that yields analytic solutions by combining the Monsi–Saeki theory with the rectangular hyperbola for leaf photosynthetic response.

Substitution of equation (7.14) into equation (7.13) gives

$$P_c = \int_0^L \frac{\alpha I_\ell P_{max}}{\alpha I_\ell + P_{max}}\,dL \tag{7.17}$$

Substitution of equation (7.2) of the Monsi–Saeki theory into equation (7.17) leads to

$$P_c = \int_0^L \frac{kI_o\,\alpha P_{max}\,e^{-kL}}{kI_o\,\alpha e^{-kL} + P_{max}(1-m)}\,dL \tag{7.18}$$

This can be integrated to give

$$P_c = -\frac{P_{max}}{k}\left\{\ln[kI_o\,e^{-kL} + P_{max}(1-m)]\right\}_0^L$$

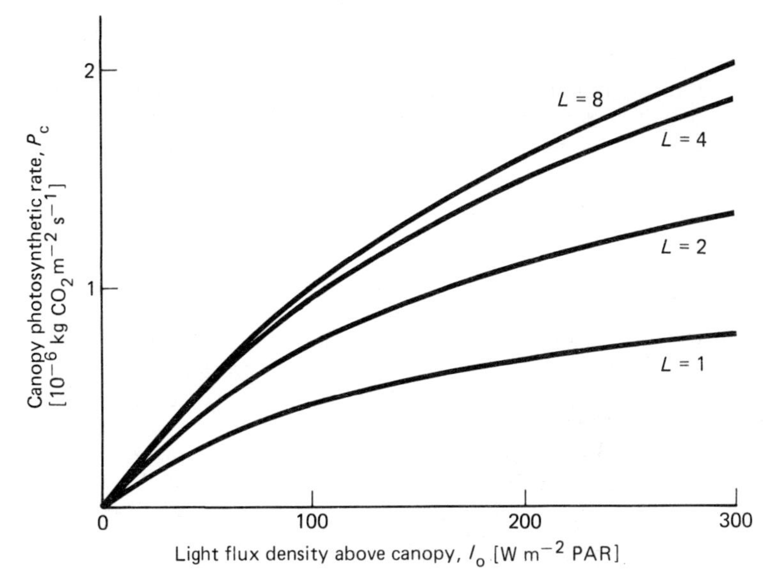


*Figure 7.6* Crop photosynthetic response curve to light according to equation (7.19), for the leaf area indices $L$ shown. Parameter values are $P_{max} = 1.2 \times 10^{-6}$ kg $CO_2$ $m^{-2}$ $s^{-1}$, $\alpha = 13 \times 10^{-9}$ kg $CO_2$ (J PAR)$^{-1}$, $k = 0.8$ and $m = 0.1$

and therefore

$$P_c = \frac{P_{max}}{k}\left\{\ln\left[\frac{k\alpha I_o + P_{max}(1-m)}{kI_o\alpha e^{-kL} + P_{max}(1-m)}\right]\right\} \tag{7.19}$$

The initial slope of equation (7.19) is

$$\frac{dP_c}{dI_o}(I_o = 0) = \frac{\alpha(1 - e^{-kL})}{1 - m} \tag{7.20}$$

and the light-saturated photosynthetic rate is

$$P_c(I_o \rightarrow \infty) = P_{max} L \tag{7.21}$$

The crop response equation (7.19) is drawn in *Figure 7.6*. It is similar in shape to the leaf response in *Figure 7.5*, the principal difference being that the approach to the asymptotic value is rather slower for crops than for single leaves.

## Respiration

The energetics of crop growth and yield are closely bound up with respiratory activity and the production of $CO_2$. It is generally considered that respiration in plants is concerned with two metabolic functions: conversion and maintenance. Conversion includes processes such as the synthesis of plant material from the sugars produced by photosynthesis; the conversion of stored carbohydrates from the stem of the wheat plant into grain material; and the conversion of sugars and nitrogen into plant proteins. Respiration accompanying conversion is sometimes referred to as 'growth' respiration, or 'constructive' respiration. The maintenance function is concerned with maintaining the *status quo* of already existing plant material. At the biochemical level, many of the biochemical processes involved in carrying these two metabolic functions are likely to be the same—it is the way in which the biochemistry is integrated within the organ or plant that is different.

Crop modelling and attempts to predict crop yields accurately have focused attention on the importance of treating respiration correctly, and especially maintenance respiration. There is still no consensus on how this is to be achieved. Below, four approaches to the problem are outlined.

## Phenomenological approach

McCree (1970) reported the results of experiments performed on white clover plants, in which the respiration $R$ [kg $CO_2$ day$^{-1}$ plant$^{-1}$] and gross photosynthesis during the light period $P$ [kg $CO_2$ day$^{-1}$ plant$^{-1}$] were measured or estimated for plants of dry weight $W$. Plant dry weight $W$ is conveniently expressed in $CO_2$ equivalents [kg $CO_2$ plant$^{-1}$]. Both plant weight $W$ and photosynthesis $P$ were varied over a range, as a result of which respiration $R$ also varied. McCree found that his data could be summarized by the single equation

$$R = kP + cW \tag{7.22}$$

where $k$ and $c$ are constants. For white clover

$$k = 0.25 \quad \text{and} \quad c = 0.0125 \text{ day}^{-1} \tag{7.23}$$

implying that the plants lose by respiration 25% of photosynthesis, and per day 1.25% of their dry weight. This equation divides respiration into two components, corresponding to the two metabolic functions of conversion and maintenance mentioned earlier. The two ideas, that there is a loss in converting photosynthate into average plant material, and that some basal metabolism is required to maintain the current status of the plant, both appear physiologically plausible. However, it has often been found that crop models that assume a fixed maintenance requirement as in equation (7.22), fail to perform satisfactorily. Other approaches to respiration may be viewed as attempting to put some understanding into the coefficients $k$ and $c$ of equation (7.22), which might enable them to be treated as variables, or to find some alternative approach to respiration which might be more satisfactory in a crop model.

## Biochemical approach

This method involves using biochemical knowledge about pathways of reactions, their stoichiometry and energetics, to predict the efficiencies of the growth and maintenance processes. The technique has largely been developed by Penning de Vries (1975a, b). For the synthesis of plant material, *Table 7.1*, derived from Penning de Vries (1975b, Table 20.1), illustrates typical results. The main differences between the yields in *Table 7.1* are due to the different specific energy contents (heats of combustion) of the components, but modified by the efficiencies of the biochemical pathways involved.

The conversion yields of *Table 7.1* can be used to construct the substrate requirements for the biosynthesis of different types of plant material; results of this type are shown in *Table 7.2*, taken from Penning de Vries (1975b, Table 20.2), where the organic substrates are assumed to be sucrose, which is the principal translocated sugar, and amino acids.

**TABLE 7.1. Conversion of glucose into the principal components of plant dry matter in the dark. Each component is a natural mixture of different molecular species. The nitrogenous compounds comprise amino acids, proteins and nucleic acids. (From Penning de Vries, 1975b, Table 20.1)**

| *Chemical component* | *Yield* [kg product (kg glucose)$^{-1}$] | *$O_2$ consumed* [kg $O_2$ (kg glucose)$^{-1}$] | *$CO_2$ produced* [kg $CO_2$ (kg glucose)$^{-1}$] | *Comment* |
|---|---|---|---|---|
| Nitrogenous | 0.616 | 0.137 | 0.256 | + $NH_3$ |
| component | 0.404 | 0.174 | 0.673 | + $NO_3$ |
| Carbohydrates | 0.826 | 0.082 | 0.102 | |
| Lipids | 0.330 | 0.116 | 0.530 | |
| Lignin | 0.465 | 0.116 | 0.292 | |
| Organic acids | 1.104 | 0.298 | −0.050 | |

**TABLE 7.2. Substrate requirements of the biosynthesis of 1 kg dry matter of tissues with various chemical compositions; the compositions refer to per cent of carbohydrates, nitrogenous compounds (amino acids, proteins, nucleic acids), lignin, lipids and minerals, respectively. (From Penning de Vries, 1975b, Table 20.2)**

| *Type of biomass* | *Composition* [%] | | | | | *Sucrose* [kg] | *Amino acids* [kg] | *$CO_2$ produced* [kg] | *$O_2$ consumed* [kg] |
|---|---|---|---|---|---|---|---|---|---|
| Leaves | 66.5 | 25 | 4 | 2.5 | 2 | 1.06 | 0.305 | 0.333 | 0.150 |
| Non-woody stem | 74 | 12.5 | 8 | 2.5 | 2 | 1.15 | 0.153 | 0.278 | 0.135 |
| Woody stem | 45 | 5 | 40 | 5 | 5 | 1.52 | 0.061 | 0.426 | 0.176 |
| Bean seeds | 55 | 35 | 2 | 5 | 3 | 1.01 | 0.427 | 0.420 | 0.170 |
| Rice seeds | 90 | 5 | 1 | 2 | 2 | 1.14 | 0.061 | 0.186 | 0.110 |
| Peanut seeds | 21 | 20 | 6 | 50 | 3 | 1.92 | 0.245 | 1.017 | 0.266 |
| Bacteria | 25 | 60 | 2 | 5 | 8 | 0.80 | 0.732 | 0.573 | 0.208 |

It should be noted that the numbers in *Tables 7.1* and *7.2* are dependent to some extent on assumptions about pathways, which have not always been experimentally substantiated. In addition, the transport costs of moving materials around the plant are not known reliably. In photosynthetic organs, it is possible for light energy to be used to drive other reactions, and not only the reduction of $CO_2$ to glucose; for example, in some plants a substantial amount of nitrate reduction occurs in the leaves, being driven directly by photophosphorylation. Thus, the results of biochemical analysis are a valuable guide to energy costs of plant growth, but do not yet provide the last word.

Penning de Vries (1975a) attempted to estimate the maintenance costs of plant tissue. This is a more difficult exercise than the synthetic costs of plant growth, because there is less knowledge about the processes involved. Maintenance of molecular structures requires energy where breakdown and resynthesis occur, for instance with proteins, and messenger-RNA, lipids, starch and some cell-wall components. Given the turnover rates and the energy released usefully in breakdown, the maintenance requirement can be calculated from the energy requirements for synthesis. However, turnover rates are notoriously variable, depending strongly upon type of tissue and age. There are also costs from maintaining ionic and other concentrations against leakage through membranes. Penning de Vries (1975a, Table 3) lists many measured maintenance respiration rates, and in *Table 7.3*, some of the entries from his table are given. The wide range of values is immediately apparent. Crop models using equation (7.22) to describe respiration costs, with $c$ proportional to the maintenance coefficient, indicate that crop yields are very sensitive to the maintenance rate, highlighting the importance of a better understanding of this process.

**TABLE 7.3. Maintenance respiration rates of plant tissue. (From Penning de Vries, 1975a, Table 3)**

| *Tissue and conditions* | *Maintenance respiration rate* [$10^{-3}$ kg glucose (kg dry matter)$^{-1}$ day$^{-1}$] |
|---|---|
| *Pisum sativum*, seeds | 0.0039 |
| Conifer stem wood, bark, 15 °C | 1.3 |
| *Trifolium repens*, plants, 20 °C, low light | 15.0 |
| *Gossypium* sp., bolls | 6 ± 10 |
| *Helianthus annuus*, plants, 25 °C | 41.0 |
| *Helianthus annuus*, leaves, 25 °C | 60.0 |
| *Phaseolus vulgaris*, 25 °C, high light | 80.0 |
| *Phaseolus vulgaris*, 25 °C, low light | 12.0 |

## Substrate-balance analysis of respiration

Pirt (1965) showed experimentally that the respiration of some bacterial populations can be divided into two components, identified with growth respiration and maintenance respiration, and he analysed bacterial growth in these terms. This approach was extended to plants, principally by Thornley (1970, 1971, 1976), and an account of this is now given.

It is assumed that a plant (or a part of a plant) is supplied with a quantity of substrate $\Delta s$ during a time interval $\Delta t$ and during $\Delta t$, $\Delta s$ is completely utilized. If it is assumed that the substrate can only be used for two purposes, growth or maintenance, then one can write

$$\Delta s = \Delta s_g + \Delta s_m \tag{7.24}$$

where $\Delta s_g$ is used for growth and $\Delta s_m$ is used for maintenance. All the substrate in the term $\Delta s_m$ is respired, whereas only part of $\Delta s_g$ is respired, the rest being turned into an increment of plant material $\Delta w$. The substrate components and plant material $w$ are all measured in the same units, for instance kg $\{CH_2O\}$. If $\Delta s_r$ is the part of growth substrate $\Delta s_g$ that is respired, then

$$\Delta s_g = \Delta w + \Delta s_r \tag{7.25}$$

The yield of the conversion of substrate into plant material can be defined by

$$Y_g = \frac{\Delta w}{\Delta s_g} = \frac{\Delta w}{\Delta w + \Delta s_r} \tag{7.26}$$

where $Y_g$ is the conversion yield of the growth process itself. Note that $Y_g$ will depend upon the type of plant material under consideration, as in *Table 7.2*. The overall conversion yield, denoted by $Y$, is less than the growth conversion yield $Y_g$, owing to the maintenance losses. $Y$ is defined by

$$Y = \frac{\Delta w}{\Delta s} = \frac{\Delta w}{\Delta w + \Delta s_r + \Delta s_m} \tag{7.27}$$

Combining equations (7.26) and (7.27) gives

$$\frac{1}{Y} = \frac{1}{Y_g} + \frac{\Delta s_m}{\Delta w}$$

It is usual to define a maintenance coefficient $m$ by

$$\frac{\Delta s_m}{\Delta t} = mw \tag{7.28}$$

and therefore

$$\frac{1}{Y} = \frac{1}{Y_g} + mw\frac{\Delta t}{\Delta w} \tag{7.29}$$

Finally, the specific growth rate, $\mu$, of the plant or organ is defined by

$$\mu = \frac{1}{w}\frac{\Delta w}{\Delta t} \tag{7.30}$$

and thus equation (7.29) becomes

$$\frac{1}{Y} = \frac{1}{Y_g} + \frac{m}{\mu} \tag{7.31}$$

Where it is possible to manipulate the specific growth rate $\mu$ experimentally, and determine the overall conversion yield $Y$, then one can test the hypothesis that the growth conversion yield $Y_g$ and the maintenance coefficient $m$ are constant, by plotting the data as in *Figure 7.7*. If the data lie on a straight line, then $Y_g$ and $m$ are easily obtained. For fast rates of growth, maintenance becomes unimportant, and

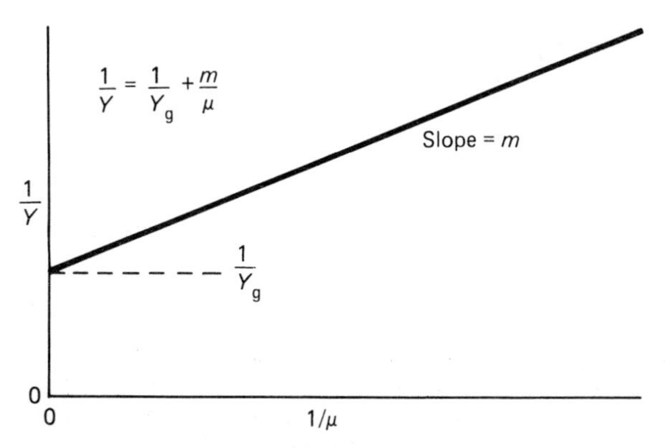


*Figure 7.7* Relationship between the overall conversion yield $Y$ and the specific growth rate $\mu$, as in equation (7.31). $Y_g$ is the growth conversion yield and $m$ the maintenance coefficient; the line drawn illustrates the situation where these are constant

the conversion yield $Y$ approaches its maximum value of $Y_g$; for slower rates of growth, the conversion yield falls progressively as maintenance losses become relatively more important.

The respiration rate $R$ is given by

$$R = \frac{\Delta s_m}{\Delta t} + \frac{\Delta s_r}{\Delta t} \tag{7.32}$$

With equations (7.24) and (7.25) this becomes

$$\frac{\Delta s}{\Delta t} = \frac{\Delta w}{\Delta t} + R \tag{7.33}$$

From equation (7.26)

$$\frac{\Delta w}{\Delta t} = \frac{Y_g}{1 - Y_g}\frac{\Delta s_r}{\Delta t}$$

which with equations (7.32) and (7.28) gives

$$\frac{\Delta w}{\Delta t} = \frac{Y_g}{1 - Y_g}(R - mw)$$

which becomes

$$R = \left(\frac{1 - Y_g}{Y_g}\right)\frac{\Delta w}{\Delta t} + mw \tag{7.34}$$

Substitution for $\Delta w/\Delta t$ in equation (7.33) then gives after rearrangement

$$R = (1 - Y_g)\frac{\Delta s}{\Delta t} - Y_g mw \tag{7.35}$$

Both equations (7.34) and (7.35) are expressions for the respiration rate $R$, the first in terms of net dry matter increment $\Delta w$ and the second in terms of gross supply of substrate $\Delta s$. The latter is equivalent to equation (7.22) used by McCree (1970).

The specific respiration rate $r$ is defined by

$$r = \frac{R}{w} \tag{7.36}$$

and equation (7.34) becomes

$$r = \left(\frac{1 - Y_g}{Y_g}\right)\frac{1}{w}\frac{\Delta w}{\Delta t} + m \tag{7.37}$$

If by experiment the time-course of dry matter $w$ and specific respiration rate $r$ are measured, giving

$$w = w(t) \quad \text{and} \quad r = r(t) \tag{7.38}$$

then equation (7.37) can be used to test the applicability of the above analysis and assumptions to the data, and if equation (7.37) does describe the data adequately, then estimates of the parameters $Y_g$ and $m$ can be obtained. Thornley and Hesketh (1972) analysed cotton boll growth and respiration in this way.

### Recycling model of growth and maintenance respiration

The usual whole-plant approach to maintenance respiration is to treat it as a fixed requirement that must be met obligatorily. This can give rise to difficulties, and an approach has been proposed (Thornley, 1977) in which both components of respiration (growth and maintenance) are produced by the same synthetic process, but a component of the structural dry matter of the plant is degraded and can be resynthesized, producing a maintenance-like respiration component.

This model is illustrated in *Figure 7.8*, where all the quantities appearing in the model are defined. The three differential equations corresponding to the three state variables are immediately written down:

$$\frac{dW_s}{dt} = P - k_g W_s + k_d W_d \tag{7.39a}$$

$$\frac{dW_d}{dt} = Y_d Y_g k_g W_s - k_d W_d \tag{7.39b}$$

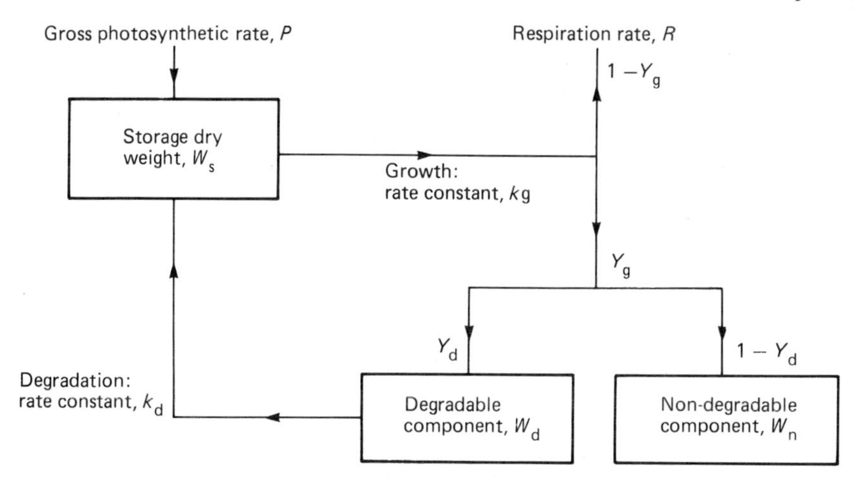


*Figure 7.8* A recycling model of growth and maintenance respiration. Plant dry weight $W$ is divided into the three components shown. $Y_g$ and $Y_d$ are conversion yields, $0 \leqslant Y_g, Y_d \leqslant 1$

$$\frac{dW_n}{dt} = (1 - Y_d)Y_g k_g W_s \tag{7.39c}$$

First-order kinetics are assumed with rate constants $k_g$ and $k_d$ for growth and degradation. $Y_g$ is the conversion yield of the growth process, and $Y_d$ gives the fraction of the newly synthesized material that becomes part of the degradable pool. The respiration rate $R$ is given by

$$R = k_g(1 - Y_g)W_s$$

which can be shown to be equal to (Barnes and Hole, 1978)

$$R = \frac{1 - Y_g}{Y_g}\left[\frac{d(W - W_s)}{dt} + k_d W_d\right] \tag{7.40}$$

For a plant of constant composition, the ratios of the storage to total dry weight, $f_s$, and of the degradable component to total dry weight, $f_d$, defined by

$$f_s = \frac{W_s}{W} \quad \text{and} \quad f_d = \frac{W_d}{W}$$

are constant, so that

$$\frac{dW_s}{dt} = f_s \frac{dW}{dt}$$

Substitution into equation (7.40) gives

$$R = \frac{1}{Y_g}(1 - Y_g)\left[(1 - f_s)\frac{dW}{dt} + f_d k_d W\right] \tag{7.41}$$

Comparison of this equation with equation (7.34) gives a substantially altered interpretation of respiration coefficients: the plant composition now enters into the apparent coefficients, which may vary as plant composition varies, although the underlying parameters (the growth conversion yield $Y_g$ and the degradation rate $k_d$) may be constant (Thornley, 1982).

## Partitioning of dry matter and its components

It is self-evident that the partitioning of dry matter between different plant organs and between different chemical components is relevant to crop productivity and to the nutritional value of the crop. Few topics give the crop physiologist or agronomist so much difficulty. There is still no agreed best or satisfactory approach, and it is therefore our intention to introduce the reader briefly to the methods that have been applied to this problem (Thornley, 1977, reviews this area in more detail).

It is the dynamics of partitioning that is crucial, since this determines, after integration over time, the observed partitioning pattern. The general problem is easily stated: let $X$ denote a chemical substance (this may be an element, compound, or group of compounds), then $n$ substances are denoted by

$$X = \{X_1, X_2, \ldots, X_n\} \tag{7.42}$$

These substances $X$ are located in compartments (tissues, organs, or groups of organs) $L$; there are $m$ compartments, so that

$$L = \{L_1, L_2, \ldots, L_m\} \tag{7.43}$$

Let $W_{i,j}$ be the weight of substance $X_i$ in location $L_j$; these are state variables. Then, as described in Chapter 2 for dynamic deterministic models, the assumptions that lead to the equations ($n \times m$ in number)

$$\frac{\mathrm{d}W_{i,j}}{\mathrm{d}t} = \text{function of state variables and other quantities} \tag{7.44}$$

define the system completely and also the ensuing partitioning pattern. This general approach allows for changing (adaptive) partitioning patterns between organs and also between chemical substances. For biological partitioning problems, carbon and nitrogen are the two elements of most interest.

For a plant (or crop), let $\Delta C_{\text{gross}}$ be the gross amount of carbon supplied to the plant by photosynthesis (for example, this might be calculated by equation (7.19)) during time interval $\Delta t$, then one may write

$$\Delta C_{\text{gross}} = \text{losses (respiration, senescence, etc.)} + \Delta C_{\text{net}} \tag{7.45}$$

The net carbon supply may be partitioned over organs by

$$\Delta C_{\text{net}} = \sum_{\text{organ}} \Delta W_{\text{C,organ}} \tag{7.46}$$

The main drive behind plant partitioning studies is to describe and understand the determinants of equation (7.46), where $\Delta W_{\text{C,organ}}$ is the incremental weight of carbon assigned to each organ.

### Empirical approaches

At the simplest level, this method uses directly partitioning measurements to set the coefficients in equation (7.46). For instance, with shoot : root : inflorescence partitioning, (7.46) takes the form

$$\Delta C_{\text{net}} = \Delta W_{\text{C,shoot}} + \Delta W_{\text{C,root}} + \Delta W_{\text{C,inflorescence}} \tag{7.47}$$

Partitioning fractions $\phi_{shoot}$, $\phi_{root}$ and $\phi_{inflorescence}$ may be defined by

$$\phi_{shoot} = \Delta W_{C,shoot}/\Delta C_{net}, \text{ etc.} \tag{7.48}$$

with

$$\phi_{shoot} + \phi_{root} + \phi_{inflorescence} = 1 \tag{7.49}$$

Finally, the $\phi$'s are assigned numerical values, usually appealing directly to experiment; for instance, in this example these may be 0.4, 0.1 and 0.5. The circularity of this approach, in which experimental data are used in a model which is used to describe possibly the same experimental data, may have dangers for the unwary. Essentially, this method has been used by Patefield and Austin (1971), Scaife and Smith (1973), Holt *et al.* (1975), Curry, Baker and Streeter (1975), and others.

**Priorities; dynamic partitioning**

The values assigned to the $\phi$'s of equations (7.48) and (7.49) reflect priorities between the organs. In the simplest approach, the $\phi$'s do not change with time, although sometimes the $\phi$'s are assumed to be (empirical) functions of time, and in other cases, the $\phi$'s depend on the substrate supply ($\Delta C_{net}$) and its relation to the potential maximum rate of use by the organs.

Consider the case where the shoot and inflorescence are given an absolute priority over the root until their requirements are satisfied. A maximum possible rate of growth is assigned to each organ, denoted by

$$\Delta W_{C,shoot,max} \quad \text{and} \quad \Delta W_{C,inflorescence,max} \tag{7.50}$$

which may depend on the sizes of the organs ($W_{shoot}$ and $W_{inflorescence}$) and possibly their stage of development; $\Delta C_{net}$ may be assigned with (say) $\phi_{shoot} = 0.3$ and $\phi_{inflorescence} = 0.7$ to the shoot and the inflorescence; if either shoot or inflorescence reaches its maximum rate of growth (7.50), then that organ cannot absorb more carbon, and the extra goes to the other organ until both shoot and inflorescence are growing at their maximum rate. For still higher values of $\Delta C_{net}$, the carbon which the shoot and inflorescence cannot absorb is assigned to the root.

There are many ways of varying this approach, which is, however, still essentially empirical, in that the data used in fixing partitioning coefficients and priorities are still at the same hierarchical level as the predictions of the partitioning model.

Another method of setting priorities which is quite frequently encountered is to use Michaelis–Menten equations for the growth response to some substrate pool of concentration $S$, but to assign different parameter values to different parts of the plant. For the two organs A and B, the growth rates $G_A$ and $G_B$ may be written

$$G_A = \frac{k_A S}{K_A + S} \quad \text{and} \quad G_B = \frac{k_B S}{K_B + S} \tag{7.51}$$

where the $k$'s and $K$'s are constants. These equations can be used to model both a fixed system of priorities and changing priorities, as illustrated in *Figure 7.9*. The initial slopes and asymptotes are given by $k_A/K_A$, $k_B/K_B$ and $k_A$, $k_B$ respectively. If these quantities are in the same ratio, then the priorities may be the same for different substrate levels; alternatively, and as illustrated in *Figure 7.9*, these equations can be used to give priority to organ A at low substrate level, but changing over to priority for organ B at higher values of substrate level $S$.

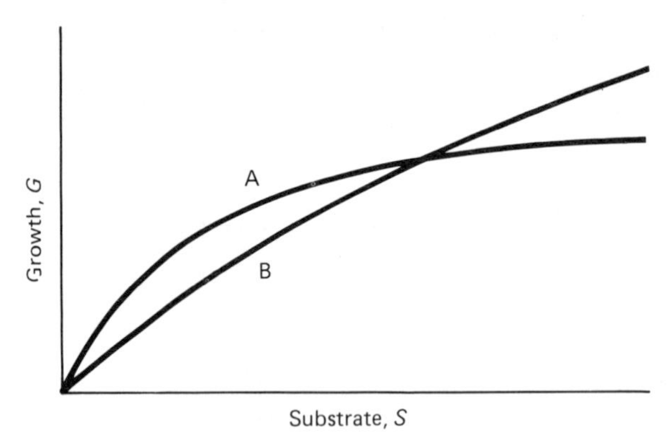


*Figure 7.9* Growth responses, $G_A$ and $G_B$, to a substrate pool of concentration $S$ for two organs, A and B, according to equations (7.51)

**Functional approaches**

White (1937), Brouwer (1962) and subsequently Davidson (1969) have been associated with the hypothesis that a functional equilibrium exists between the size and activity of the shoot, supplying carbohydrate, and the size and activity of the root, supplying water and essential nutrients. This may be formally expressed as

$$(\text{shoot mass} \times \text{shoot specific activity}) \propto (\text{root mass} \times \text{root specific activity}) \tag{7.52}$$

If $W_{shoot}$ is the shoot mass, $P_g$ is the gross photosynthetic rate of the shoot, and $\sigma_{shoot}$ is the shoot specific activity, then $\sigma_{shoot}$ is defined by

$$\sigma_{shoot} = P_g / W_{shoot} \tag{7.53a}$$

Considering root activity in terms of nitrogen uptake rate $U_N$, then the root specific activity $\sigma_{root}$ is given by

$$\sigma_{root} = U_N / W_{root} \tag{7.53b}$$

where $W_{root}$ is the root mass.

Equation (7.52) can only apply to a plant that is adapted to its environment. For instance, if shoot activity or root activity is suddenly changed by altering the light flux density incident on the shoot or the root temperature, then equation (7.52) will no longer be satisfied. In the functional approach to partitioning, it is presumed that the partitioning of newly synthesized dry matter occurs in such a way that equation (7.52) is re-established. One must then discuss how the dynamic partitioning coefficients, the $\phi$'s of equations (7.48) and (7.49), respond to a disequilibrium in the plant, which is extant as a deviation from equation (7.52).

*Figure 7.10* illustrates how dynamic shoot : root partitioning may be achieved for the case where water is the limiting nutrient supplied by the root, and it is assumed that the dynamic partitioning coefficients $\phi_{shoot}$ and $\phi_{root}$ respond to the relative water content of the plant.

It is clear that the functional equilibrium concept of equation (7.52) can be applied to any organ which provides an activity or function that is essential to the well-being of the plant, including processes like the synthesis of hormones.

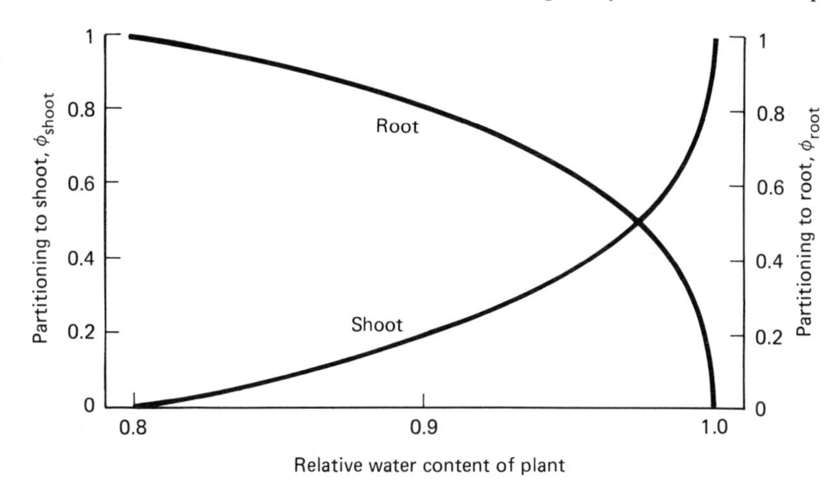


*Figure 7.10* Dynamic partitioning of dry matter to shoot and root in response to the relative water content of the plant. $\phi_{shoot}$ and $\phi_{root}$ are the dynamic partitioning coefficients of equations (7.48) and (7.49). (After de Wit *et al.* 1978, p. 57)

For carbon and nitrogen in relation to shoot and root partitioning, Reynolds and Thornley (1982) proposed a partitioning function of the following sort:

$$\frac{\Delta W_{shoot}/W_{shoot}}{\Delta W_{root}/W_{root}} = \text{constant} \times \left(\frac{N}{C}\right)^q \qquad (7.54)$$

where $N$ and $C$ are the nitrogen and carbon concentrations in the storage pools of those substances, and $q$ is a number. The use of equation (7.54) in a plant growth model causes the shoot : root ratio to stabilize at a certain value that depends on the $N:C$ ratio, and hence the specific activity with which shoot and root take up these quantities from the environment. The parameter $q$ determines the sensitivity with which this response occurs.

The functional approach to shoot : root partitioning is also discussed by Charles-Edwards (1976) and by Thornley (1977). It may perhaps be described as a phenomenological approach, as it is a step beyond simple empiricism, but does not yet speak explicitly of mechanism.

## A mechanistic approach to partitioning

In considering the effects of carbohydrate and nitrogen supply on shoot : root partitioning in the pondweed *Lemna*, White (1937) outlined concepts that were formalized mathematically by Thornley (1972a, b). The method is based on the assumption that the only processes that contribute towards plant growth are chemical conversion and transport. Thus, the growth pattern of a plant depends upon the processes by which substances are transported around the plant, and the processes by which the substances are used for growth of the plant parts.

For a two-substrate (carbon, nitrogen) two-compartment (shoot, root) system, the method is illustrated in *Figure 7.11*, which is taken from Cooper and Thornley (1976), who applied a mechanistic partitioning model with some success to experimental data on the tomato plant. To illustrate the method, the initial

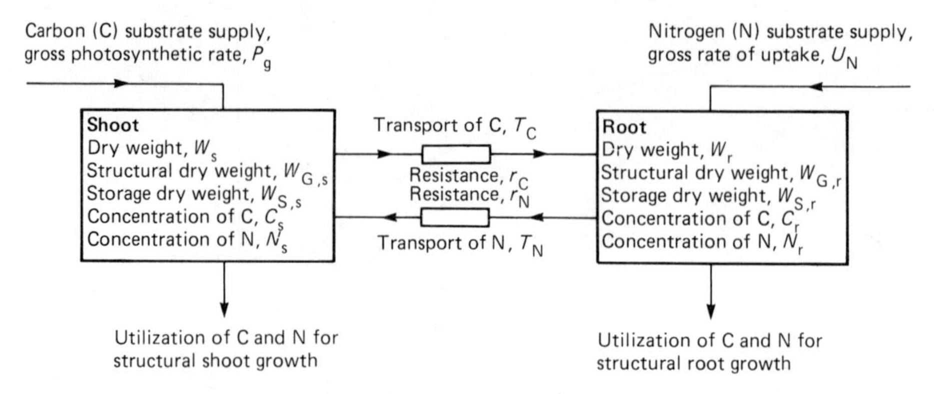


*Figure 7.11* A shoot:root partitioning model dependent on the supply of carbon by the shoot and nitrogen by the root. (From Cooper and Thornley, 1976)

mathematical statement of the model is outlined, although reference should be made to Thornley (1972a, b), Thornley (1976), and Cooper and Thornley (1976) for a detailed account.

It is assumed that

$$\text{rate of use of carbon substrate} = kCN \tag{7.55}$$

in units of kg mol carbon $m^{-3}\,s^{-1}$, where $k$ is a constant, and $C$ and $N$ are the concentrations of carbohydrate and nitrogen. Of this utilized material, a fraction $Y_G$ appears in plant material, $Y_G$ being a conversion efficiency. If $\lambda$ is the ratio of nitrogen atoms to carbon atoms in the structural material of the plant, then

$$\text{rate of use of nitrogen substrate} = \lambda Y_G CN \tag{7.56}$$

The fluxes of carbon, $T_C$, and nitrogen, $T_N$, between shoot and root are taken to be

$$T_C = \frac{\beta(C_s - C_r)}{r_C} \tag{7.57}$$

and

$$T_N = \frac{\beta(N_r - N_s)}{r_N} \tag{7.58}$$

in units of kg mol carbon or nitrogen $s^{-1}$, where $\beta$ is a scaling factor, $r_C$ and $r_N$ are resistances to the movement of carbon and nitrogen, and the subscripts r and s refer to the concentrations in the root and the shoot.

It is assumed that the gross rates of uptake of carbon and nitrogen, $P_g$ and $U_N$, are proportional to the shoot and root structural dry weights $W_{G,s}$ and $W_{G,r}$ respectively. The subscripts G and S are used to denote the structural and storage components of dry weight. This gives rise to a double-subscript notation which, although clumsy, has the advantage of being explicit. Thus $W_{G,s}$ refers to the structural dry weight in the shoot, and $W_{G,r}$ to the storage dry weight in the root. The constants of proportionality, $k_C$ and $k_N$, are defined by

$$P_g = k_C W_{G,s} \quad \text{and} \quad U_N = k_N W_{G,r} \tag{7.59}$$

in units of kg mol carbon or nitrogen $s^{-1}$.

The model has six unknowns ($W_{G,s}$, $W_{G,r}$, $C_s$, $C_r$, $N_s$ and $N_r$) and the six differential equations of the model are

$$\begin{aligned}
\frac{d}{dt}\left(\frac{W_{G,s}\,C_s}{\rho}\right) &= k_C W_{G,s} - T_C - \frac{W_{G,s}\,kC_s N_s}{\rho}\\
\frac{d}{dt}\left(\frac{W_{G,s}\,N_s}{\rho}\right) &= T_N - \frac{W_{G,s}\,\lambda Y_G\,kC_s N_s}{\rho}\\
\frac{d}{dt}\left(\frac{W_{G,r}\,C_r}{\rho}\right) &= T_C - \frac{W_{G,r}\,kC_r N_r}{\rho}\\
\frac{d}{dt}\left(\frac{W_{G,r}\,N_r}{\rho}\right) &= k_N W_{G,r} - T_N - \frac{W_{G,r}\,\lambda Y_G\,kC_r N_r}{\rho}\\
\frac{dW_{G,s}}{dt} &= \frac{\gamma Y_G\,W_{G,s}\,kC_s N_s}{\rho}\\
\frac{dW_{G,r}}{dt} &= \frac{\gamma Y_G\,W_{G,r}\,kC_r N_r}{\rho}
\end{aligned} \tag{7.60}$$

In these equations $\rho$ is the density of structural plant tissue with units of kg (m$^3$ fresh tissue)$^{-1}$, $\gamma$ is a conversion factor relating kg mol carbon to kg of structural material [units: kg structural dry weight (kg mol of carbon in structural material)$^{-1}$], and $t$ is the time variable [s]. These equations can be solved by integrating numerically from a given set of initial conditions (the values of the six unknowns at $t = 0$).

Cooper and Thornley (1976) went on to solve these equations for balanced exponential growth, and compared the predictions of the model directly with experiment.

It should be noted that the utilization function of equation (7.55) represents perhaps the simplest of possible assumptions. A more general two-substrate function for the rate of carbohydrate utilization $U_C$ is

$$U_C = \frac{k'CN}{1 + C/K_C + N/K_N + CN/K_{CN}} \tag{7.61}$$

where $k'$ and the $K$'s are constants. If these parameters are assigned different values in the shoot and the root, then this may be viewed as giving a priority to the shoot or root, over and above the partitioning effects caused by the transport resistances in equations (7.57) and (7.58); this point is discussed by Thornley (1977, pp. 372–373). Hormones, growth factors and additional substrates can be handled either by varying the parameters of equation (7.61), or by extending it for more variables.

The transport equations (7.57) and (7.58) describe transport down concentration gradients, with the transport fluxes vanishing when, for instance, $C_s = C_r$. Thornley (1977, pp. 376–379) demonstrated that a formalism which can be used for active transport fluxes is to use equations of the type

$$T_C = k_1 C_s - k_2 C_r \tag{7.62}$$

where $k_1$ and $k_2$ are constants. The method may then be applied to systems such as sugar beet, where sugars appear to move up concentration gradients.

## Nutrient uptake

The role of the plant and crop canopies in intercepting light and carrying out photosynthesis is analysed on pp. 114–121 in terms of several simple models, which often lead to analytical solutions. There is no analogous analytical treatment for nutrient uptake by the root, which takes account of root size, structure and activity in relation to the soil environment. However, much work has been done in this area. An early significant contribution is by Passioura (1963); two recent monographs are by Nye and Tinker (1977) and Russell (1977); a recent and succinct review is by Clarkson (1981). Our account of the problem is necessarily incomplete, and is written from the point of view of the physiologist more interested in representing whole-plant or crop response, than in a rigorous delineation of mechanisms.

### A simple model

In *Figure 7.12*, the essentials of the system are defined, in a scheme that treats the root as a homogeneous compartment and ignores spatial aspects of root structure and soil environment. Amongst the state variables in the root compartment, the root surface area, $A$, over which uptake can take place is clearly important; the carbohydrate concentration, $C$, is included because this is the energy source for processes such as active uptake.

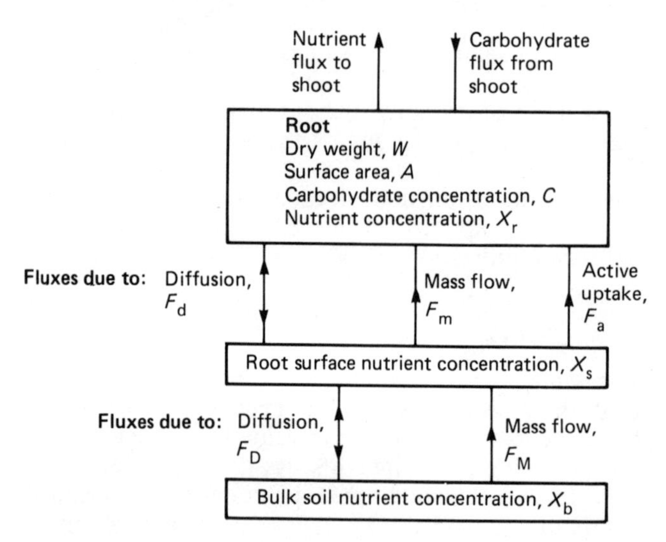


*Figure 7.12* Nutrient uptake by the root: a homogeneous model

The nutrient concentration at the root surface $X_s$ is defined by equating the fluxes into and out of this compartment (which is of negligible physical size), giving

$$F_D + F_M = F_d + F_m + F_a \tag{7.63}$$

The diffusion flux $F_D$ may be written as

$$F_D = g_{bs}(X_b - X_s) \tag{7.64a}$$

where $g_{bs}$ [$m^3 day^{-1}$] is a conductance, and is given by an expression of the type

$$g_{bs} = \frac{D_b A}{\ell_{bs}} \times \text{(dimensionless geometrical factors)} \tag{7.64b}$$

where $D_b$ is a diffusion constant [$m^2 day^{-1}$], and $\ell_{bs}$ is a characteristic length. If $v$ [$m^3 day^{-1}$] is the volume flow of soil solution, then the mass flow flux is

$$F_M = vX_b \tag{7.64c}$$

Similarly, two of the fluxes from the root surface into the root are given by

$$F_d = g_{sr}(X_s - X_r) \quad \text{and} \quad F_m = vX_s \tag{7.65}$$

where $g_{sr}$ is a conductance. The active-uptake flux $F_a$ may be described by a Michaelis–Menten equation:

$$F_a = \frac{kX_s}{K + X_s} \tag{7.66}$$

where $k$ and $K$ are constants. Note that the constant $k$ of this equation may be a function of the carbohydrate concentration in the root, $C$, since this process is driven by energy released from the respiration of sugars. Substituting equations (7.64)–(7.66) into equation (7.63) gives

$$g_{bs}(X_b - X_s) + vX_b = g_{sr}(X_s - X_r) + vX_s + \frac{kX_s}{K + X_s} \tag{7.67}$$

which leads to a quadratic in the root surface nutrient concentration $X_s$:

$$\begin{aligned} 0 = {} & X_s^2(g_{sr} + g_{bs} + v) + X_s[K(g_{sr} + g_{bs} + v) + k - g_{sr}X_r - g_{bs}X_b] \\ & - K(g_{sr}X_r + g_{bs}X_b) \end{aligned} \tag{7.68}$$

A simpler approximate expression may be obtained by linearizing the active uptake term from equation (7.66), replacing it with

$$F_a = \ell X_s \tag{7.69}$$

where $\ell$ is a constant. The solution to the modified equation (7.67) is

$$X_s = \frac{g_{bs}X_b + vX_b + g_{sr}X_r}{g_{bs} + g_{sr} + v + \ell} \tag{7.70}$$

When two processes occur in series, as in *Figure 7.12*, it is often found that the overall result is a small modification of the second process, and the effective parameter values are changed. Thus, a semi-empirical approximation to the total flux $F$ may be obtained by taking the right side of equation (7.67), replacing $X_s$ by $X_b$, and modifying the parameters:

$$F = g'_{sr}(X_b - X_r) + vX_b + \frac{k'X_b}{K' + X_b} \tag{7.71}$$

This last equation is in a convenient form for using in a plant or crop growth model.

Quantities such as root volume $V$, total root length $L$, and mean root radius $r$ are obtained by

$$V = W/\rho \tag{7.72a}$$

where $\rho$ is a dry matter density,

$$\pi r^2 L = V \tag{7.72b}$$

and

$$2\pi r L = A \tag{7.72c}$$

It is necessary to make an assumption about the $L/r$ ratio (for instance, that this is a constant in a growing root system) in order to evaluate the root surface area $A$ in terms of dry weight $W$ using equations (7.72).

Finally, it is pointed out that one cannot always separate the diffusive and mass flow fluxes, as in equations (7.63) and (7.71), since these can, in some circumstances, be strongly interdependent.

## Outline of a one-dimensional crop model

A typical scheme is sketched in *Figure 7.13*, where some of the symbols needed are defined. Considering diffusion (*see* Glossary) into a small element at depth $z$, continuity requirements give

$$D\frac{\partial^2 X}{\partial z^2} = \frac{\partial X}{\partial t} + \text{sinks} - \text{sources} \tag{7.73}$$

The term on the left is diffusion into the element with diffusion coefficient $D$. The sinks removing $X$ from the soil into the root system are as illustrated in *Figure 7.12* and discussed above, and in this case may be written

$$\text{sinks} = g\sigma(z)[X(z) - X_r(z)] + v(z)\sigma(z)X(z) + \sigma(z)\frac{kX(z)}{K + X(z)} \tag{7.74}$$

where $g$ is a conductance, and $k$ and $K$ are uptake activity parameters. $g$, $k$ and $K$ may vary with depth $z$, and $k$ and $K$ may depend on the carbohydrate status of the root at depth $z$. The sources term in equation (7.73) may arise from exudation from the root, diffusion or leakiness of the root, or root senescence; it may also include the release of nutrient $X$ from the solid phase of the soil into the soil solution.

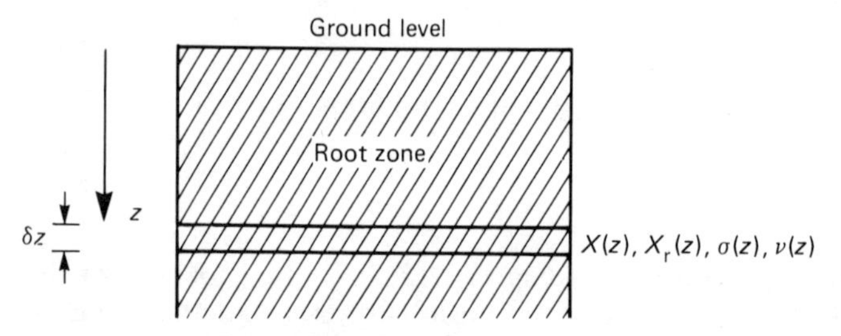


*Figure 7.13* A one-dimensional nutrient uptake model for a crop root system. $z$ is the distance measured downwards from ground level; $X(z)$ is the mean soil concentration of nutrient $X$ in the horizontal plane at depth $z$; $X_r(z)$ is the mean concentration of nutrient $X$ within the root on the horizontal plane at depth $z$; $\sigma(z)$ is the root surface area per unit volume at depth $z$; $v(z)$ is the volume flux density [$m^3(m^2$ root surface$)^{-1}$ day$^{-1}$] at depth $z$

Even with highly simplifying assumptions, equations (7.73) and (7.74) can usually only be solved numerically, and the lower boundary conditions can present a considerable problem.

## Exercises

### Exercise 7.1

Assuming a canopy extinction coefficient $k = 0.6$ (typical of grasses) in equation (7.1), what is the leaf area index $L$ of a canopy which only allows 10% of the incident light to reach the ground?

If the intercepted light is assumed to be uniformly distributed over the leaves in the canopy, what is the mean light level incident on the leaves as a fraction of that above the canopy?

### Exercise 7.2

Monteith (1965) proposed an equation for light interception:

$$I(L) = [s + (1 - s)m]^L I_o$$

where $I(L)$ is the downward light flux density under leaf area $L$, $I_o$ is that above the canopy, $s$ is the fraction of radiation that passes through a layer of unit leaf area index without interception by foliage, and $m$ is the transmission coefficient of foliage. This binomial expression (*see* Glossary) assumes that the unit layers of foliage act independently.

Derive, by using $(1 - x)^n \rightarrow \exp(-nx)$ as $x \rightarrow 0$, $n \rightarrow \infty$ and $nx$ = constant, the Monsi–Saeki result of equation (7.1).

### Exercise 7.3

Assuming that equation (7.14) describes leaf photosynthetic response to light $I_\ell$ with $P_{max} = 1.2 \times 10^{-6}$ kg $CO_2$ m$^{-2}$ s$^{-1}$ and $\alpha = 13 \times 10^{-9}$ kg $CO_2$ (J PAR)$^{-1}$, calculate the light level $I_\ell$ for 90% saturation of leaf response, and the light level where the linear approximation $P_\ell = \alpha I_\ell$ is 10% in error.

### Exercise 7.4

Write a computer program to calculate canopy photosynthetic rate by equation (7.19) for $P_{max} = 1.2 \times 10^{-6}$ kg $CO_2$ m$^{-2}$ s$^{-1}$, $\alpha = 13 \times 10^{-9}$ kg $CO_2$ (J PAR)$^{-1}$, $m$ = 0.1, examining the values when $I_o$ = 10, 100, 1000 J m$^{-2}$ s$^{-1}$, $k$ = 0.4, 1.2, and $L$ = 1, 6. In particular, consider which values of the canopy extinction coefficient $k$ give high canopy photosynthetic rates for various light : leaf area index combinations.

### Exercise 7.5

By comparison of equation (7.22) and equation (7.35), from the white clover parameters measured by McCree and given in equation (7.23), obtain numerical values for the conversion efficiency $Y_g$ and the maintenance coefficient $m$.

Chapter 8

# Crop responses and models

## Introduction

The emphasis in the last chapter was on the physiological, chemical and physical processes that contribute to plant and crop growth, and their definition in terms of mechanistic equations. In the present chapter, the main thrust is towards a higher level in the organizational hierarchy, towards more direct relationships between crop growth and productivity, environmental variables, and agronomic inputs such as fertilizer and water. To make such short-cuts, approximations are necessary, with concomitant losses in precision and realism. Whether or not the latter are acceptable is a matter for continual reassessment.

## Water use and uptake

It must first be said that the phenomena involved in water use and uptake by plants and crops are so complex that they have not yet been completely unravelled. Our account of the problem is therefore incomplete, and should be regarded as an introductory guide to this important area.

The major part of the water taken up by a crop flows straight through the plants and into the atmosphere. Less than 1% is generally used in tissue growth, and still less in photosynthesis in the reaction

$$H_2O + CO_2 \rightarrow \{CH_2O\} + O_2$$

However, the presence of water is essential for the physiological and biochemical functions of the plant, and for transport processes in the soil.

Transpiration, which is the flow of water through the crop, is often limited by either

1. the supply of water to be evaporated, or
2. the supply of energy to evaporate the water.

If there is no water to be evaporated, then to prevent dehydration of plant tissue, the stomata close, and photosynthesis all but ceases. On the other hand, if water is in plentiful supply and evaporation is energy limited, often the same energy supply is limiting photosynthesis. Thus, simple relations between transpiration and dry matter accumulation are often observed.

**Water use efficiency**

The water use efficiency $w$ is defined as

$$w = \frac{\text{kg dry matter of crop}}{\text{kg of water transpired}} \tag{8.1}$$

For $C_3$ species, $w$ typically lies in a range about a value of 0.0016 kg dry matter (kg water)$^{-1}$, whereas for $C_4$ species it is about twice as large at 0.0032 kg dry matter (kg water)$^{-1}$ (Larcher, 1980, p. 231). Water use efficiency has proved to be a useful concept because it is often relatively constant, and it is easy to understand why this is so in simple terms.

Let $R$ [J m$^{-2}$] be the radiation receipt by a crop, and let $R_p$ be the photosynthetically active part of this radiation, given by

$$R_p = fR \tag{8.2}$$

where $f$ is a (dimensionless) fraction. If $\lambda$ [J kg$^{-1}$] is the latent heat of evaporation of water, and all the incident energy is used to evaporate a quantity of water $q$ [kg water m$^{-2}$], then

$$q = \frac{R}{\lambda} \tag{8.3}$$

Assuming a net photosynthetic efficiency of $\alpha$ [kg $CO_2$ (J PAR)$^{-1}$], then the dry matter produced, $W$ [kg], is

$$W = \frac{30}{44}\,\alpha f R \tag{8.4}$$

where it is assumed that crop material is carbohydrate with a molecular weight of 30 and the molecular weight of $CO_2$ is 44. Thus, the water use efficiency $w$ is given by

$$w = \frac{W}{q} = \frac{30}{44}\alpha f \lambda \tag{8.5}$$

Substituting reasonable numbers into this expression, $f = 0.5$, $\lambda = 2.5 \times 10^6$ J kg$^{-1}$, and for net photosynthesis including respiration losses (p.120, p.122) $\alpha = 7 \times 10^{-9}$ kg $CO_2$ (J PAR)$^{-1}$, gives

$$w = 0.006 \text{ kg dry matter (kg water)}^{-1} \tag{8.6}$$

This estimate is about two to four times as large as the experimental values given above; this is acceptable in view of the many simplifying assumptions that have been made; possibly the assumption with the greatest error is that the photosynthetic efficiency of the canopy is a constant, given by $\alpha$. At high irradiance, the efficiency is likely to fall well below $\alpha$.

**Transpiration models**

Penman (1948) applied energy conservation and water transport equations to the problem of water loss from a crop canopy, and was able to eliminate the requirement for leaf temperature. Monteith (1965) added to this theoretical framework, incorporating an additional resistance to water loss associated with the leaf stomata and the pathways by which water molecules escape from the leaf.

Jarvis, Edwards and Talbot (1981) give a useful survey of this area. A simple derivation of the Penman–Monteith equation (equation (8.7)) is given separately at the end of the section (equations (8.17) to (8.27)); this is based on an account by Jarvis (1981).

Let $E$ be the transpiration rate of a crop, in units of kg water (m$^2$ ground area)$^{-1}$ s$^{-1}$, then the Penman–Monteith formula for $E$ may be written

$$E = \frac{sA + c_p \rho [p_s(T_a) - p_a] g_a}{\lambda[s + \gamma(1 + g_a/g_c)]} \qquad (8.7)$$

$s$ = rate of change of saturated vapour pressure with temperature at air temperature $T_a$ [Pa K$^{-1}$]
$A$ = available energy = net radiation − soil heat flux [J m$^{-2}$ s$^{-1}$]
$c_p$ = specific heat capacity of air at constant pressure [J kg$^{-1}$ K$^{-1}$]
$\rho$ = density of air [kg m$^{-3}$]
$p_s(T_a) - p_a$ = vapour pressure deficit of air [Pa], $p_s(T_a)$ being the saturated water vapour pressure at air temperature $T_a$, and $p_a$ being the actual water vapour pressure in the air
$g_a$ = boundary layer conductance [m s$^{-1}$]
$\lambda$ = latent heat of evaporation of water [J kg$^{-1}$]
$\gamma$ = psychrometric constant [Pa K$^{-1}$]
$g_c$ = canopy conductance [m s$^{-1}$]

The psychrometric constant $\gamma$ is given by

$$\gamma = \frac{c_p P}{\lambda \epsilon} \qquad (8.8)$$

where $P$ is the total atmospheric pressure [Pa], and $\epsilon$ is the ratio of the molecular weights of water vapour and air (0.622).

The boundary layer conductance of the canopy $g_a$ is the sum of the conductances of the water vapour pathways from the surfaces of the canopy elements (leaves, stems, etc.) lying above 1 m$^2$ of ground area to a reference level above the canopy. The canopy conductance $g_c$ is the conductance from an assumed free water surface within the canopy tissues, through the stomata and along associated transfer pathways, to the surface of the canopy elements; again this is with reference to 1 m$^2$ of ground area.

$s$ is quite strongly temperature-dependent, with values of 44.3, 82.5, 145, and 244 Pa K$^{-1}$ at temperatures of 0, 10, 20 and 30 °C, respectively. A measurement of air temperature is therefore needed. The parameters $c_p$, $\rho$, $\lambda$ and $\gamma$ are comparatively insensitive to temperature. The available energy $A$ can often be acceptably approximated by the net radiation $R_n$, since the soil heat flux is usually small.

There are two components of water loss—these may be termed 'energy-driven' and 'diffusion-driven'—and they are clearly reflected in the structure of equation (8.7). Writing these two terms as $E_e$ and $E_d$, from equation (8.7), therefore

$$E_e = \frac{sA}{\lambda[s + \gamma(1 + g_a/g_c)]} \qquad (8.9a)$$

$$E_d = \frac{c_p \rho [p_s(T_a) - p_a] g_a}{\lambda[s + \gamma(1 + g_a/g_c)]} \qquad (8.9b)$$

Many authors make approximations from equations (8.7) or (8.9) to obtain simpler expressions that might be applicable to particular conditions of canopy and environment.

### 1. STOMATAL CONDUCTANCE HIGH: $g_c >> g_a$

In this case, equation (8.7) becomes

$$E_1 = \frac{sA + c_p\rho[p_s(T_a) - p_a]g_a}{\lambda(s + \gamma)} \tag{8.10}$$

Clearly $E_1 > E$ and this gives an upper limit to the estimated transpiration. Equation (8.10) may be valid when the stomata are wide open and the boundary layer conductance is low.

### 2. DIFFUSION TERM NEGLIGIBLE: $[p_s(T_a) - p_a]g_a \approx 0$

The energy-driven term of equations (8.9) alone remains, so that

$$E_2 = E_e \tag{8.11}$$

Humid still environments with low crops are best likely to satisfy these requirements.

### 3. DIFFUSION TERM DOMINANT: $E_d >> E_e$

Transpiration rate is now given by

$$E_3 = E_d \tag{8.12}$$

This approach may be justified for tall rough crops (such as trees) in windy conditions (giving a large $g_a$) where the incident radiation may be low.

Another approach in which there has been some interest is to assume that the diffusion contribution is some empirical fraction of the radiation term, so that transpiration rate $E_4$ is described by

$$E_4 = \alpha \frac{sA}{\lambda(s + \gamma)} \tag{8.13}$$

where $\alpha$ is an empirical constant. This method works best in conditions where the diffusion term is relatively small, but when applied over a wider range of environments, it has considerable shortcomings.

### APPLICATION OF THE PENMAN–MONTEITH TRANSPIRATION EQUATION

Although equation (8.7) is itself based on certain assumptions (for instance, the canopy is treated as consisting of a single layer), it is generally thought that these do not lead to serious error. Moreover, the equation has a sound basis in physics, and the inputs required to use the model do not present insuperable difficulties. The inputs required are

1. available energy, $A$;
2. vapour pressure deficit, $p_s(T_a) - p_a$;
3. air temperature, $T_a$;

4. wind speed, $u$, because this affects the boundary layer conductance, $g_a$, which can then be calculated given some knowledge of the aerodynamic properties of the canopy;
5. canopy conductance, $g_c$, whose calculation may involve a quite complex submodel.

The last two items in the list are far from straightforward in practical application, and there is therefore still much interest in less-demanding more-empirical approaches of sufficient accuracy.

### Soil water balance

For the estimation of irrigation needs, the calculation of water balances is required. The soil moisture deficit $S_d$ is defined as the amount of water [$m^3$] that must be added to unit area of ground to reach the point where run-off occurs; $S_d$ therefore has units of $m^3$ water $(m^2 \text{ ground})^{-1}$, or m, the depth of water that has to be added.

Defining $R$, $E_v$ and $L$ as the rates of rainfall, evaporation and run-off (loss), in m $day^{-1}$, then since $S_d$ is always greater or equal to zero,

if $S_d > 0$ or $R < E_v$

$$S_d = S_d(t = 0) - \int (E_v - R)\, dt \quad \text{and} \quad L = 0 \tag{8.14a}$$

and if $S_d = 0$ and $R \geqslant E_v$

$$L = R - E_v \tag{8.14b}$$

$E_v$ is the actual transpiration rate, and this may or may not be the same as the potential transpiration rate $E_{vpot}$. $E_{vpot}$ may be estimated by the Penman–Monteith equation (8.7), or more commonly by some simpler derivative of it, such as equation (8.13). In many applications, the actual transpiration $E_v$ is reduced below the potential transpiration $E_{vpot}$ by some empirical factor depending upon the soil moisture deficit. For example, a simple linear decline with $S_d$ gives

$$E_v = E_{vpot}(1 - \beta S_d) \tag{8.15}$$

where $\beta$ is a constant. Sometimes, a threshold value of $S_d$ is used, possibly the wilting point, followed by a linear decline, giving

$$S_d \leqslant S_d\,(\text{threshold}) \quad E_v = E_{vpot}$$

$$S_d > S_d\,(\text{threshold}) \quad E_v = E_{vpot}\{1 - \beta[S_d - S_d\,(\text{threshold})]\} \tag{8.16}$$

### A simple derivation of the Penman–Monteith equation

This derivation follows Jarvis (1981).

Using the notation defined after equation (8.7), then equating the available energy $A$ with the energy loss in latent heat $\lambda E$ and the energy loss in sensible heat $C$ [J $m^{-2}$]

$$A = \lambda E + C \tag{8.17}$$

This energy balance equation for the biomass covering unit ground area neglects changes in metabolic energy and heat content of the biomass, which are generally very small.

Using Thom (1975), the equations for heat transfer from the canopy element are

$$\lambda E = \frac{c_p\rho}{\gamma}[p_s(T_\ell) - p_a]g_w \tag{8.18a}$$

and

$$C = c_p\rho(T_\ell - T_a)g_h \tag{8.18b}$$

where $T_\ell$ is the leaf temperature [°C], and $g_w$, $g_h$ are the transfer conductances from the leaves [m s$^{-1}$] (w refers to water vapour and h to sensible heat). $s$, the rate of change with temperature of the saturated vapour pressure, is defined by

$$s = \frac{dp_s}{dT} \tag{8.19}$$

and if the difference between leaf and air temperature is small, therefore

$$p_s(T_\ell) - p_s(T_a) \approx s(T_\ell - T_a) \tag{8.20}$$

Thus

$$\begin{aligned} p_s(T_\ell) - p_a &= p_s(T_\ell) - p_s(T_a) + p_s(T_a) - p_a \\ &\approx s(T_\ell - T_a) + p_s(T_a) - p_a \end{aligned} \tag{8.21}$$

We now use equation (8.21) to eliminate $p_s(T_\ell) - p_a$ from equation (8.18a), then equation (8.18b) to substitute for $T_\ell - T_a$ in the resulting equation, and finally equation (8.17) to eliminate $C$. The equation for $\lambda E$ is

$$\lambda E = \frac{sA + c_p\rho[p_s(T_a) - p_a]g_h}{s + \gamma g_h/g_w} \tag{8.22}$$

If the foliage is regarded as wet, and the pathways of sensible heat and water vapour transfer are the same, so that $g_h = g_w$, therefore

$$\lambda E = \frac{sA + c_p\rho[p_s(T_a) - p_a]g_h}{s + \gamma} \tag{8.23}$$

For dry foliage, the water vapour transfer pathway includes movement out of the substomatal cavities to the leaf surface (with conductance $g_c$), and then transfer from the leaf surface across the boundary layer (conductance $g_a$), so that

$$\frac{1}{g_w} = \frac{1}{g_c} + \frac{1}{g_a} \tag{8.24}$$

Substituting this expression into equation (8.22) gives

$$\lambda E = \frac{sA + c_p\rho[p_s(T_a) - p_a]g_h}{s + \gamma g_h\left(\dfrac{1}{g_c} + \dfrac{1}{g_a}\right)} \tag{8.25}$$

For the quantity being transported (heat or water vapour), the conductance across a boundary layer is proportional to (molecular diffusivity in air)$^{2/3}$ (Thom, 1968). For heat and water vapour in air at 15 °C, these are 20.8 and 23.4 mm$^2$ s$^{-1}$ (Monteith, 1973). Assuming an identical pathway across the boundary layer for heat and water vapour,

$$\frac{g_h}{g_a} = \left(\frac{20.8}{23.4}\right)^{2/3} = 0.924 \tag{8.26}$$

and equation (8.25) becomes

$$\lambda E = \frac{sA + c_p\rho[p_s(T_a) - p_a]0.924g_a}{s + 0.924\gamma(1 + g_a/g_c)} \tag{8.27}$$

For most purposes equation (8.7) is a sufficiently accurate approximation of equation (8.27).

## Fertilizer responses

Considerable benefit may accrue from the optimum application of fertilizers to crops, and as a result, there is much interest in the problem of estimating optimum applications in relation to crop, soil type, management strategy and weather. While we are still some way from a general solution to this problem, there are available simple practical methods that may easily be used by farmers and growers. Our account of the topic leans heavily on work described by Greenwood, Cleaver and Turner (1974), and Thornley (1978).

It must first be stated that there are two approaches to the problem, which may be termed static and dynamic. The static approach is to use response equations to describe crop yield $Y$ as a function of the level of fertilizer $X$; for instance, two typical response equations are

$$Y = a_0 + a_1X + a_2X^2 \tag{8.28a}$$

and

$$Y = \frac{aX}{X + b} \tag{8.28b}$$

where $a_0$, $a_1$, $a_2$, $a$ and $b$ are constants. Equation (8.28a), which is a quadratic, is statistically convenient, although biologically it is not the most appropriate function; a methodology based on such equations has been elaborated by Colwell (1974, 1977). Equation (8.28b) belongs to the class of functions known as inverse polynomials because they can be inverted to give a linear form, namely

$$\frac{1}{Y} = \frac{1}{a} + \frac{b}{aX} \tag{8.29}$$

We believe that equation (8.28b) and its derivatives provide a more useful framework than equation (8.28a) for a static analysis of the fertilizer problem, and this is the approach we develop later.

In our view, the dynamic modelling approach promises, in the long term, to give the definitive solution to the fertilizer problem. Here we do no more than indicate the method. As described in Chapter 2 (p. 18), a dynamic crop growth model may be written schematically as

$$\frac{dW}{dt} = f(W, P, E) \tag{8.30}$$

where $W$ is crop dry matter, $f$ denotes some function of the arguments, $P$ indicates a set of parameters and constants, and $E$ environmental variables and parameters; $E$

includes weather variables, soil nutrient status, and also possible management inputs. Yield $Y$ is some part of the dry matter $W$ at the end of or during the season. Equation (8.30) is a convenient way of indicating a large mechanistic crop model with submodels for photosynthesis, water status, and nutrient uptake. With inputs of mean weather variables, the model may be used to define fertilizer inputs for optimum results in an average year. However, few years are average, and one of the principal problems in using static response functions as in equations (8.28), is the strong dependence of fertilizer response on actual weather. In the dynamic approach, the model of equation (8.30) is run with actual weather up to the present moment of time, and then the best assumption is made about the weather for the rest of the season, which is probably that the weather will be average weather. Thus, on the basis of actual weather up to the present, and assuming average weather in the future, the model may be used to predict the management inputs (including fertilizer application) to give optimum results over the rest of the season.

## The general static response function approach

Denote the level of fertilizer $i$ by $X_i$ [kg m$^{-2}$] and the yield per unit area of the crop by $Y$ [kg m$^{-2}$], then it is assumed that at a particular site and in an average season a fertilizer response function $f$ can be defined by

$$Y = f(X_1, X_2, \ldots, X_n) \tag{8.31}$$

where $n$ fertilizers are considered.

### MAXIMUM RESPONSE

The fertilizer levels for maximum yield $Y$ are determined by equating the first partial derivatives of equation (8.31) to zero, giving

$$\frac{\partial Y}{\partial X_1} = 0, \frac{\partial Y}{\partial X_2} = 0, \ldots, \frac{\partial Y}{\partial X_n} = 0 \tag{8.32}$$

Solution of these $n$ equations for the $n$ quantities $X_1, \ldots, X_n$ gives the fertilizer treatment for maximum yield, namely, $X_{1,\max}, \ldots, X_{n,\max}$. Substitution of these values into equation (8.31) gives the maximum yield $Y_{\max}$. For many nutrients such as phosphate and potassium, the response is one of the diminishing-returns type, the maximum is at $X_i = \infty$, and the fertilizer levels for maximum yield are of very limited interest.

### ECONOMIC RESPONSE

Let $p$ [£ kg$^{-1}$] be the price obtained for the product, and $c_i$ [£ kg$^{-1}$] the unit cost of the fertilizer $X_i$, including the cost of application. The farmer's profit is maximized when

$$\text{marginal return} = \text{marginal cost} \tag{8.33}$$

giving

$$p\delta Y = c_i\, \delta X_i \tag{8.34}$$

where $\delta Y$ and $\delta X_i$ represent small increments in $Y$ and $X_i$. From equation (8.34), taking the limit, therefore

$$\frac{\partial Y}{\partial X_i} = \frac{c_i}{p} \quad (i = 1, \ldots, n) \tag{8.35}$$

Equations (8.35) are solved to give the most economical treatment $X_{i,\text{ec}}$, and the yield with this treatment, $Y_{\text{ec}}$ is given by

$$Y_{\text{ec}} = f(X_{1,\text{ec}}, X_{2,\text{ec}}, \ldots, X_{n,\text{ec}}) \tag{8.36}$$

## COMPOUND FERTILIZERS

It is assumed in the above treatment that the fertilizers $X_i$ are applied independently, and although the agricultural scientist usually prefers to examine crop response to single nutrients varied independently, it is more convenient for a farmer to apply a compound fertilizer or fertilizers to his crops. Consider a compound fertilizer $Z$ defined in terms of the fertilizers $X_i$ by

$$X_i = \beta_i Z \quad (i = 1, \ldots, n) \tag{8.37}$$

where the $\beta_i$ are coefficients. Substituting equations (8.37) into equation (8.31) gives a different response function $F$, with

$$Y = F(Z) \tag{8.38}$$

The function $F$ contains, in addition to the parameters of equation (8.31), the coefficients $\beta_i$. The maximum response is obtained by solving the equation $\mathrm{d}F/\mathrm{d}Z = 0$. More usefully, the most economical response $Y_{\text{ec}} = F(Z_{\text{ec}})$, is determined by solving

$$\frac{\mathrm{d}F}{\mathrm{d}Z} = \frac{c_Z}{p} \tag{8.39}$$

where $c_Z$ is the cost of the compound fertilizer $Z$. Since equation (8.37) is linear, the relationship

$$\frac{\mathrm{d}F}{\mathrm{d}Z} = \sum_i \frac{\partial f}{\partial X_i} \frac{\mathrm{d}X_i}{\mathrm{d}Z} = \sum_i \beta_i \frac{\partial f}{\partial X_i} \tag{8.40}$$

can be derived, which can be used in equation (8.39).

## NONLINEAR FERTILIZER APPLICATION COSTS

The analysis of the most economic fertilizer application in equations (8.34)–(8.36) assumes that the unit cost of buying and applying fertilizer is constant, and independent of the level of fertilizer applied. It is more realistic to assume a cost per unit area of fertilizer purchase and application of

$$d_i + e_i X_i + h_i X_i^2 \tag{8.41}$$

for fertilizer $X_i$, rather than $c_i X_i$ which leads to equation (8.34); $d_i$ represents fixed costs (of, for example, machinery and labour), $e_i$ a linear cost component (price of fertilizer, machinery wear and tear, and possibly labour), and $h_i$, which is usually negative, represents the savings that may be associated with bulk-buying or large-scale application. Equation (8.35) is now replaced by

$$\frac{\partial Y}{\partial X_i} = \frac{e_i + 2h_i X_i}{p} \quad (i = 1, \ldots, n) \tag{8.42}$$

These $n$ equations are solved for the fertilizer levels $X_{i,\mathrm{ec}}$. The profit is given by

$$\text{profit} = pY - \Sigma(d_i + e_i X_i + h_i X_i^2) \tag{8.43}$$

**Inverse polynomial response to $N$, $P$ and $K$**

Many authors have employed an inverse polynomial function to describe fertilizer responses, usually with a modification to allow for the adverse effects of high nitrogen levels on yield, typically

$$\frac{1}{Y} = \frac{1}{(1 - N/\alpha)}\left(\frac{1}{A} + \frac{1}{B_\mathrm{N} N} + \frac{1}{B_\mathrm{P} P} + \frac{1}{B_\mathrm{K} K}\right) \tag{8.44}$$

where $Y$ is the yield; $\alpha$ is a constant that accounts for the depression of yield at high nitrogen levels; $A$, $B_\mathrm{N}$, $B_\mathrm{P}$ and $B_\mathrm{K}$ are constants; and $N$, $P$ and $K$ are the levels of nitrogen, phosphate and potassium. The responses to $N$ and $P$ are shown in *Figure 8.1*; the $K$ response is similar to that for $P$. The constant $A$ represents the yield that

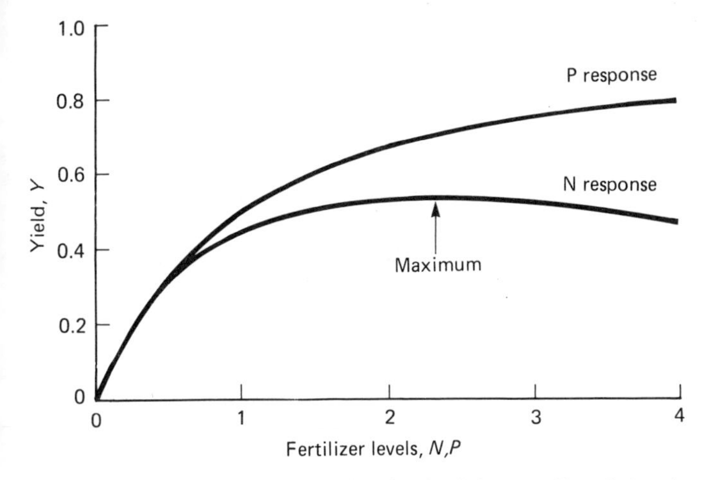


*Figure 8.1* Response of crop yield, $Y$, to levels of nitrogen, $N$, and phosphate, $P$, as given by the inverse polynomial response function, equation (8.44). For the $N$ response $A = 1$, $P = \infty$, $K = \infty$, $\alpha = 10$, $B_\mathrm{N} = 1$; for the $P$ response $A = 1$, $N = \infty$, $K = \infty$, $N/\alpha = 0$, $B_\mathrm{P} = 1$. The maximum in the $N$ response is given by equation (8.51). Units are arbitrary

would be obtained at high values of $N$, $P$ and $K$ in the absence of a nitrogen yield depression ($N/\alpha = 0$). The constants $B_\mathrm{N}$, $B_\mathrm{P}$ and $B_\mathrm{K}$ are the initial slopes of the $Y{:}N$, $Y{:}P$ and $Y{:}K$ curves for large $\alpha$; they might be viewed as conversion efficiencies (kg yield per kg of nitrogen, phosphate or potassium) and are possibly characteristic of the plant in question.

There is no definite physiological or biochemical meaning attached to the parameters of equation (8.44). However, it can be shown that, assuming the plant is of constant composition (with respect to nitrogen, phosphate and potassium), and the uptake rates of nitrogen, phosphate and potassium per unit of root mass are proportional to their soil levels, then the specific growth rate of the crop, $\mu = (1/W)(dW/dt)$, where $W$ is the dry weight, is given by an equation of this type (Thornley, 1977, equations (41) and (44)).

Equation (8.44) may be inverted to give

$$Y = \frac{AB_N NB_P PB_K K(1 - N/\alpha)}{B_N NB_P PB_K K + AB_N NB_P P + AB_N NB_K K + AB_P PB_K K} \tag{8.45}$$

Equation (8.45) is more easily considered if it is written in the form

$$Y = \frac{a_N N(1 - N/\alpha)}{N + a_N/b_N} = \frac{a_P P}{P + a_P/b_P} = \frac{a_K K}{K + a_K/b_K} \tag{8.46}$$

where

$$\begin{aligned} a_N &= AB_P PB_K K/(B_P PB_K K + AB_P P + AB_K K) \\ a_P &= AB_N N(1 - N/\alpha)B_K K/(B_N NB_K K + AB_N N + AB_K K) \\ a_K &= AB_N N(1 - N/\alpha)B_P P/(B_N NB_P P + AB_N N + AB_P P) \\ b_N &= B_N \qquad b_P = B_P(1 - N/\alpha) \qquad b_K = B_K(1 - N/\alpha) \end{aligned} \tag{8.47}$$

From equations (8.46) the initial slopes of the fertilizer response curves are given by

$$\frac{dY}{dX}(X \approx 0) = b_X \qquad X = N, P \text{ or } K \tag{8.48}$$

Also note that

$$\frac{a_N}{b_N} = \frac{A}{B_N}\left(1 + \frac{A}{B_P P} + \frac{A}{B_K K}\right)^{-1} \tag{8.49}$$

with similar expressions for $a_K/b_K$ and $a_P/b_P$. $a_X/b_X$ has a maximum value of $A/B_X$ for X = N, P or K.

MAXIMUM RESPONSE

Differentiating equations (8.46) and equating to zero as in equations (8.32), gives

$$\begin{aligned} \frac{\partial Y}{\partial N} &= \frac{a_N}{\alpha}\left(\alpha\frac{a_N}{b_N} - 2\frac{a_N}{b_N}N - N^2\right) \Big/ \left(N + \frac{a_N}{b_N}\right)^2 = 0 \\ \frac{\partial Y}{\partial P} &= a_P \frac{a_P/b_P}{(P + a_P/b_P)^2} = 0 \\ \frac{\partial Y}{\partial K} &= a_K \frac{a_K/b_K}{(K + a_K/b_K)^2} = 0 \end{aligned} \tag{8.50}$$

Substitution with equations (8.49) and (8.47), and solving for $N$, $P$ and $K$, therefore

$$N_{max} = \frac{A}{B_N}\left[\left(1 + \alpha\frac{B_N}{A}\right)^{1/2} - 1\right] \qquad P_{max} \rightarrow \infty \qquad K_{max} \rightarrow \infty \tag{8.51}$$

Substituting these values into equation (8.44), the maximum yield $Y_{max}$ is

$$Y_{max} = A\left\{1 - \frac{2A}{B_N \alpha}\left[\left(1 + \alpha\frac{B_N}{A}\right)^{1/2} - 1\right]\right\} \tag{8.52}$$

Note that $Y_{max}/A$ is always less than unity.

## ECONOMIC RESPONSE

To obtain the most economical fertilizer treatment for the inverse polynomial response, equations (8.35) and (8.50) are combined to give

$$\begin{aligned}
\frac{c_N}{p} &= \frac{a_N}{\alpha}\left(\alpha\frac{a_N}{b_N} - 2\frac{a_N}{b_N}N - N^2\right)\Bigg/\left(N + \frac{a_N}{b_N}\right)^2 \\
\frac{c_P}{p} &= a_P\,\frac{a_P/b_P}{(P + a_P/b_P)^2} \\
\frac{c_K}{p} &= a_K\,\frac{a_K/b_K}{(K + a_K/b_K)^2}
\end{aligned} \tag{8.53}$$

Solution of these three equations for the economic fertilizer treatment $N_{ec}$, $P_{ec}$ and $K_{ec}$, and substitution of these values into equation (8.44) give the economic yield $Y_{ec}$.

In general, these equations can only be solved numerically. However, for large $\alpha$ the $N$ equation takes the same form as the $P$ and $K$ equations, and an analytic solution is possible. Defining $\nu$, $\pi$, $\kappa$ and $\lambda$ by

$$\nu = \left(\frac{pB_N}{c_N}\right)^{1/2} \qquad \pi = \left(\frac{pB_P}{c_P}\right)^{1/2} \qquad \kappa = \left(\frac{pB_K}{c_K}\right)^{1/2} \tag{8.54}$$

$$\lambda = \nu\pi\kappa - \nu\pi - \pi\kappa - \nu\kappa$$

it can be shown that

$$N_{ec} = \frac{A}{B_N}\frac{\lambda}{\pi\kappa} \qquad P_{ec} = \frac{A}{B_P}\frac{\lambda}{\nu\kappa} \qquad K_{ec} = \frac{A}{B_K}\frac{\lambda}{\nu\pi} \tag{8.55}$$

Substituting these expressions into equation (8.45) gives

$$Y_{ec} = A\left(1 - \frac{1}{\nu} - \frac{1}{\pi} - \frac{1}{\kappa}\right) = \frac{A\lambda}{\nu\pi\kappa} \tag{8.56}$$

Note that if $\lambda$ is negative, then there is no gain in applying any fertilizer treatment. If, say, the nitrogen and phosphate fertilizers are very cheap, with $c_N$ and $c_P$ small so that $\nu$ and $\pi$ are large, then approximately

$$\lambda \approx \nu\pi(\kappa - 1)$$

so that (to give positive $\lambda$), we must have

$$pB_K \geqslant c_K \tag{8.57}$$

Thus, the initial slope of the fertilizer response curve multiplied by the price for the product must be larger than the unit cost of fertilizer.

APPLICATION TO SUMMER CABBAGE

The parameters defining the relative response of cabbage to fertilizers N, P and K are (Greenwood, Cleaver and Turner, 1974, Table 1)

$$\frac{A}{B_N} = 210\,\text{kg N ha}^{-1} \qquad \frac{A}{B_P} = 5\,\text{kg P ha}^{-1} \qquad \frac{A}{B_K} = 18\,\text{kg K ha}^{-1}$$

$$\alpha = 2300\,\text{kg N ha}^{-1} \tag{8.58}$$

Using equations (8.51), the maximum response is

$$N_{max} = 516\,\text{kg N ha}^{-1} \qquad P_{max} \rightarrow \infty \qquad K_{max} \rightarrow \infty \tag{8.59}$$

and with equation (8.52), the relative yield is

$$\frac{Y_{max}}{A} = 0.55 \tag{8.60}$$

It is assumed that the asymptotic yield parameter $A$, the product price $p$, and the costs of fertilizer (1976 prices) are

$$A = 50\,000\,\text{kg fresh weight ha}^{-1} \qquad p = 0.10\,£\,(\text{kg fresh weight})^{-1}$$

$$c_N = 0.20\,£\,(\text{kg N})^{-1} \qquad c_P = 0.63\,£\,(\text{kg P})^{-1} \qquad c_K = 0.29\,£\,(\text{kg K})^{-1} \tag{8.61}$$

First, assuming that $\alpha$ is large, equations (8.54), (8.55) and (8.56) give $\nu = 10.9$, $\pi = 39.8$, $\kappa = 31.0$ and $\lambda = 11\,442$, so that

$$N_{ec} = 1950\,\text{kg N ha}^{-1} \qquad P_{ec} = 169\,\text{kg P ha}^{-1} \qquad K_{ec} = 475\,\text{kg K ha}^{-1}$$

$$Y_{ec}/A = 0.85 \tag{8.62}$$

Equations (8.62) are used to give initial estimates for $N$, $P$ and $K$ for a numerical solution to equations (8.53) with $\alpha = 2300$, giving

$$N_{ec} = 466\,\text{kg N ha}^{-1} \qquad P_{ec} = 115\,\text{kg P ha}^{-1} \qquad K_{ec} = 321\,\text{kg K ha}^{-1}$$

$$Y_{ec}/A = 0.51 \tag{8.63}$$

Comparing equations (8.63) and (8.60), it is seen that the economic treatment gives a yield that is as high as 93% of the maximum. Comparing equations (8.63) and (8.62) shows that the effects of the nitrogen depression of yield is to reduce all the computed treatment levels, but with a very marked effect on the nitrogen level, with smaller effects on the phosphate and potassium levels.

DIFFERENT CROPS AND DIFFERENT SOILS

For a given fertilizer or nutrient X, the soil itself will contain a certain level of X, $X_s$, irrespective of the application of X by fertilizer treatment, $X_a$. It is assumed that the indigenous $X_s$ and the applied $X_a$ act additively, so that we may write

$$X = X_s + X_a \tag{8.64}$$

Note that the level of soil nutrient $X_s$ is likely to be a property of both the plant as well as of the soil, since the extent to which the plant root system explores the soil and is able to take up nutrients is important.

For a given crop growing in a given soil, and using equation (8.44) as the response function but neglecting yield depression effects, the parameters needed to define the problem are

$$A/B_{\mathrm{X}} \quad \text{and} \quad X_{\mathrm{s}} \tag{8.65}$$

There are many different crops and different soil types; it would be very expensive to determine experimentally the parameters of (8.65) for all possible combinations. We describe briefly a method of estimating these parameters for cases where they have not been directly determined, suggested by Greenwood, Cleaver and Turner (1974).

Consider a parameter $C$ (either $A/B_{\mathrm{X}}$ or $X_{\mathrm{s}}$), and let $C_{ij}$ be the parameter appropriate to crop $i$ growing in soil $j$. For $k$ crops growing in $l$ soils, a matrix can be constructed

$$\begin{array}{lc|ccc} & & & \textit{Soils} & \\ & & 1\ldots & j\ldots & l \\ \hline & 1 & C_{11} & C_{1j} & C_{1l} \\ & \vdots & & & \\ \textit{Crops} & i & C_{i1} & C_{ij} & C_{il} \\ & \vdots & & & \\ & k & C_{k1} & C_{kj} & C_{kl} \end{array} \tag{8.66}$$

There are $k \times l$ values of the parameter $C$. Greenwood, Cleaver and Turner (1974) suggested that it is reasonable to assume that

$$\frac{C_{ij}}{C_{1j}} = \frac{C_{i1}}{C_{11}} \tag{8.67}$$

If this method is valid, then a considerable saving in experimentation is possible, since now $k + l - 1$ crop–soil combinations define the $k \times l$ crop–soil combinations in the matrix of (8.66).

## Development

The way in which a plant or crop develops can greatly affect yield, and this is therefore a topic with which physiologists and agronomists are much concerned (Landsberg, 1977). Although efforts to quantify growth have had considerable success over the years, attempts to similarly quantify development have not made the same progress, and we still largely fall back on a qualitative characterization.

The life cycle of a plant is punctuated by events which take place effectively at points in time (in reality, over periods of time short compared with the length of the

**TABLE 8.1. Scheme for plant development**

| *Markers* (*events occurring at points in time*) | *Processes* (*these occur over a period of time*) |
|---|---|
| Sowing | |
| | Processes of germination, G |
| Germination, emergence | |
| | Vegetative growth, V |
| Floral initiation | |
| | Reproductive growth, R |
| Maturity | |
| | Senescence, S |
| Death | |

life cycle); these events are used as markers to separate phases of the plant's development which occur over longer periods of time. A typical scheme is shown in *Table 8.1*. The markers should correspond with readily observable changes in the plant's morphology; for different plant species different sets of markers might be appropriate.

The four phases of plant development in *Table 8.1* are denoted by G, V, R and S. A variable giving a measure of development, $D_i$, with $i$ = G, V, R or S, may notionally be associated with each phase. The $D_i$ have arbitrary units, and it is convenient to normalize the $D_i$ to lie between 0 and 1, so they are 0 at the beginning of a phase, and 1 when the phase is complete, and the plant is about to move into the next phase. We use $D$ to denote any of the $D_i$.

The rate of development is the rate at which $D$ changes, and by definition this is equal to the developmental rate constant $k$, so that

$$\frac{\mathrm{d}D}{\mathrm{d}t} = k \tag{8.68}$$

If the phase begins when time $t = t_i$ and $D = 0$, then

$$D = \int_{t_i}^{t} k \, \mathrm{d}t \tag{8.69}$$

gives the value of $D$ at time $t$. At time $t_f$, $D = 1$, and this signifies the end of the current phase and the beginning of the next phase. Note that it has been assumed that development can be measured by a single-valued quantity. The topic is not yet well enough understood to know if this is a reasonable assumption.

The crux of many agronomic studies of crop and plant development is how the developmental rate constant $k$ depends on environment. The environmental quantities that affect development are temperature, daylength, radiation, nutritional status and water status; however, interest is usually centred on the first two of these, with temperature having a dominant role.

## Heat sums or the day–degree rule

If it is assumed that the developmental rate constant $k$ is proportional to temperature $T$ above a threshold temperature $T_c$, then

$$k = aH(T - T_c)(T - T_c) \tag{8.70}$$

where $a$ is a constant, and $H$ is the unit step (or Heaviside) function:

$$H(T - T_c) = 0 \quad \text{for} \quad T < T_c \qquad H(T - T_c) = 1 \quad \text{for} \quad T \geqslant T_c \tag{8.71}$$

It is customary to form a heat sum, $h_{sum}$ [day °C], ignoring the normalization constant $a$, and using mean temperatures, to give

$$h_{sum} = \sum_i H(\bar{T}_i - T_c)(\bar{T}_i - T_c) \tag{8.72}$$

where $\bar{T}_i$ is the mean temperature on the $i$th day. To illustrate how this works, consider a sequence of 20 days, with a threshold temperature of $T_c = 10$ °C, to give

| $i$ | 1 | 2 | 3 | 4 | 5 | 6 | 7 | 8 | 9 | 10 | 11 | 12 | 13 | 14 | 15 | 16 | 17 | 18 | 19 | 20 |
|---|---|---|---|---|---|---|---|---|---|---|---|---|---|---|---|---|---|---|---|---|
| $\bar{T}_i$ | 8 | 8 | 9 | 11 | 10 | 12 | 9 | 10 | 11 | 12 | 9 | 10 | 14 | 13 | 12 | 14 | 13 | 15 | 14 | 14 |
| $h_{sum}$ | 0 | 0 | 0 | 1 | 1 | 3 | 3 | 3 | 4 | 6 | 6 | 6 | 10 | 13 | 15 | 19 | 22 | 27 | 31 | 35 |

(8.73)

For a particular crop, and a particular phase of development, it is found that the attainment of a certain value for the heat sum (50 day °C, say) corresponds quite well with the end of that phase of development. Thus, using the heat-sum method predictively, one can examine whether the phenology of a crop is suited by the average temperatures experienced at a location.

## Daylength and other environmental factors

Daylength is an environmental variable that is often next in importance after temperature in its effects on development. Some crop plants are 'day-neutral' or unaffected by daylength, but for many plants daylength is critical, especially for the reproductive markers. Because of the dominant effect of temperature, it may be convenient to write the influence of daylength in terms of day-degree equivalents, for instance by replacing equation (8.72) by

$$h_{sum} = \sum_i [H(\bar{T}_i - T_c)(\bar{T}_i - T_c) + bH(g_i - g_c)(g_i - g_c)] \tag{8.74}$$

where $b$ is a constant, $g_i$ the daylength on the $i$th day, and $g_c$ is the critical value of the daylength for development. Equation (8.74) is written for a 'long-day' plant, where days of length greater than $g_c$ promote development. For a 'short-day' plant, the subscripts $i$ and c in the daylength term are reversed.

Other environmental factors may be added to equation (8.74) in a similar manner. Equation (8.74) illustrates linear addition of the environmental factors. Some authors have found it useful to include higher-order terms, and combine these differently. For example, Robertson (1968) considered the developmental rate of a cereal crop, and its dependence on day and night temperature and photoperiod; he obtained good results using an equation equivalent to

$$\begin{aligned} k = \sum_i \{ & H(g_i - g_c)[a_1(g_i - g_c) + a_2(g_i - g_c)^2] \\ & + H(T_{i\ell} - T_c)[b_1(T_{i\ell} - T_c) + b_2(T_{i\ell} - T_c)^2] \\ & + H(T_{id} - T_c)[d_1(T_{id} - T_c) + d_2(T_{id} - T_c)^2]\} \end{aligned} \tag{8.75}$$

where $a_1$, $a_2$, $b_1$, $b_2$, $d_1$ and $d_2$ are constants, $g_i$, $g_c$ and $T_c$ are as before, and $T_{i\ell}$ and $T_{id}$ are the daily maximum and minimum temperatures on the $i$th day. It will be clear that many alternatives are possible when formulating empirical relations between developmental rate and environmental factors.

## Vernalization

Vernalization is the application of a duration of cold to make possible the development of a flower. Such treatment may be required for young plants, seeds, bulbs, etc. The phenomenon was first studied in cereals, where there are some varieties ('winter' varieties) which if sown in the spring will not flower at all in that year.

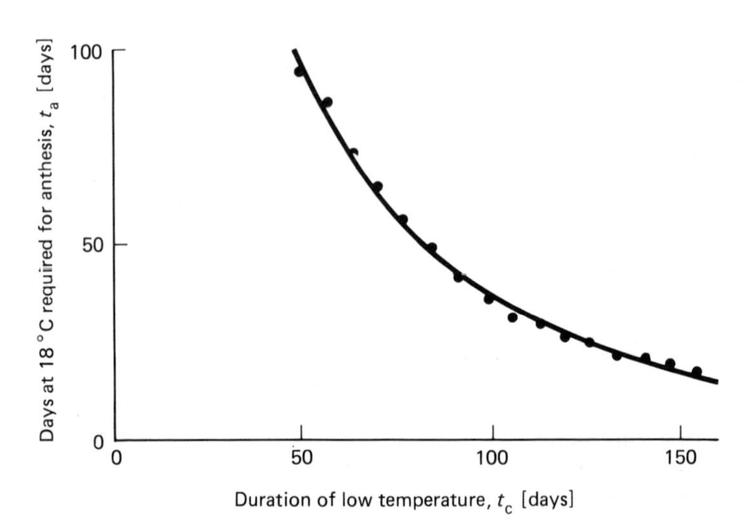


*Figure 8.2* Vernalization response, showing the days required at 18 °C to reach anthesis ($t_a$), as influenced by the duration of the cold temperature (9 °C) treatment ($t_c$). The continuous line shows the best fit to the data achieved with equation (8.76), with parameter values of $p_a = -24$ (standard error = 5), $p_c = -1$ (se = 5), and $p = 79$ (se = 5). The data are taken from Rees, Turquand and Briggs (1972)

A typical vernalization response is illustrated in *Figure 8.2*; in tulips, the duration needed for the development of a flower can be greatly reduced by the prior application of a period of low temperature. An attempt to model mechanistically the cold requirement of the tulip was made by Charles-Edwards and Rees (1974), who considered the forward and backward conversions between starch and labile sugars, and the use of the latter for growth and development of the young bulbs. The model functioned by assigning to these three rate constants a particular temperature dependence.

However, vernalization is usually treated in a very practical manner, and it does not fit readily into the heat-sum approach of equation (8.72). The data shown in *Figure 8.2* can be reasonably approximated by the empirical equation

$$(t_a - p_a)(t_c - p_c) = p^2 \qquad (8.76)$$

where $t_a$ is the time [days] required at the normal growing temperature for anthesis, $t_c$ is the time [days] of cold temperature vernalization treatment, and $p_a$, $p_c$ and $p$ are parameters. $p_a$ and $p_c$ give the asymptotes of this hyperbolic relationship (if $t_c \to \infty$, then $t_a \to p_a$ and vice versa); the parameter $p$ determines how closely the $t_a : t_c$ graph is to the two asymptotes—for example, equation (8.76) is satisfied at $t_a = p + p_a$, $t_c = p + p_c$.

The actual objective is often to minimize the total developmental time, denoted by $t_{dev}$, which is equal to

$$t_{dev} = t_c + t_a \tag{8.77}$$

It can be shown (Exercise 8.4) that $t_{dev}$ has its minimum value at

$$t_{dev}\,(\text{minimum}) = p_a + p_c + 2p \tag{8.78}$$

**Temperature dependence of development and the Arrhenius equation**

An equation attributed to Arrhenius is often used to describe the temperature dependence of the rate constant $k$ of a chemical reaction, namely

$$k = A\exp[-E_a/(RT)] \tag{8.79a}$$

where $A$, $E_a$ and $R$ are constants and $T$ is the absolute temperature. The constant $E_a$ is known as the activation energy of the reaction, and $R$ is the familiar gas constant. Equation (8.79a) is drawn in *Figure 8.3a*. Frequently, when examining

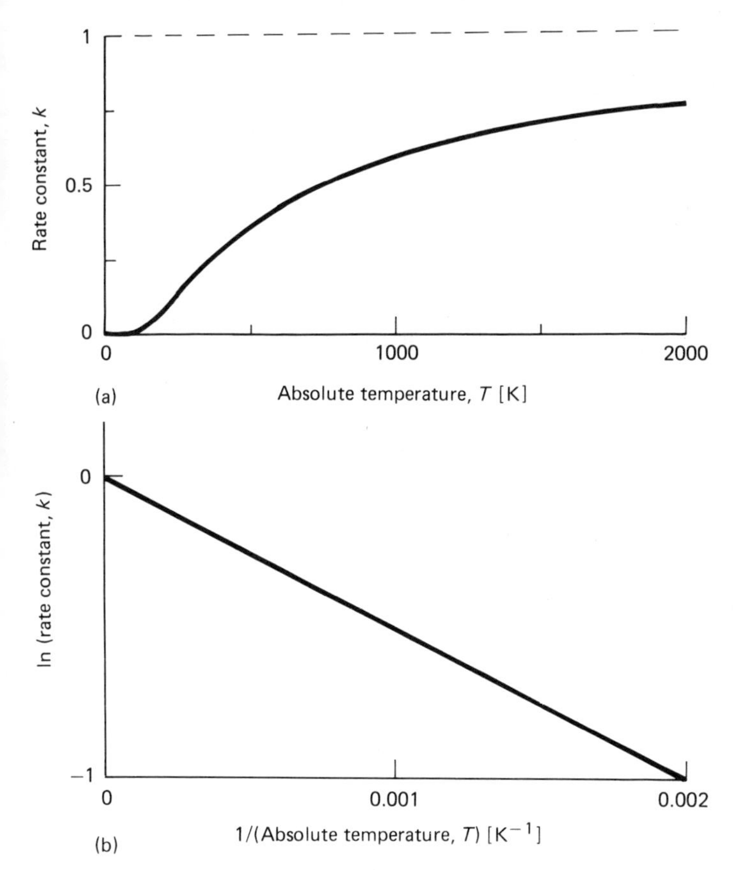


*Figure 8.3* Arrhenius equation for the rate constant, $k$, of a simple chemical reaction, according to equation (8.79a); $T$ is the absolute temperature. In (a), the temperature dependence of $k$ is shown directly; in (b), the linearized form of equation (8.79b) is shown. The dashed line in (a) is the asymptotic value of $k$. Parameters are $A = 1$ and $E_a/R = 500$ K throughout

reaction rate : temperature data, a linearized form of equation (8.79a) is used; this is

$$\ln k = \ln A - (E_a/R)/T \tag{8.79b}$$

This is shown in *Figure 8.3b*.

By differentiating equation (8.79a) twice, and equating the result to zero, it may be shown that the inflexion point is given by

$$T(\text{inflexion}) = E_a/(2R) \tag{8.80}$$

and in the region of this temperature, the response of rate constant $k$ to temperature $T$ is fairly linear. The equation of the straight line through the inflexion point with the same slope as equation (8.79a) is

$$k = \frac{4AR}{E_a e^2}[T - E_a/(4R)] \tag{8.81}$$

Comparison with equation (8.70) gives a threshold temperature $T_c = E_a/(4R)$.

Thus, the Arrhenius equation is compatible with the heat-sum concept.

## Crop yield : planting density responses

The relationship between crop yield $Y$ [kg m$^{-2}$] and planting density $\rho$ [number of plants m$^{-2}$] is an important topic that has received a good deal of attention. The problem was thoroughly reviewed by Willey and Heath (1969), and more recent contributions include those of Pant (1979) and Thornley (1983).

Essentially, there are two types of yield : density responses observed (*Figure 8.4*). In the hyperbolic responses, the yield curve approaches an asymptote, whereas in the parabolic responses, an optimum is reached followed by a subsequent decline; the latter are termed parabolic because a parabola is often used to fit this type of data.

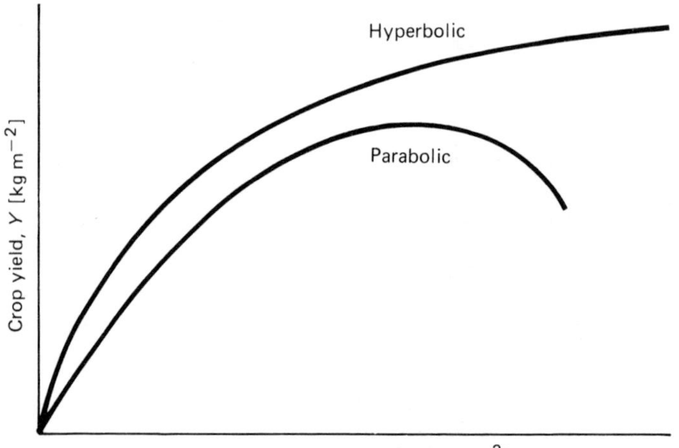


*Figure 8.4* Crop yield, $Y$: planting density, $\rho$ responses. The two main types of response are shown: the hyperbolic response approaches on asymptote; the parabolic response reaches an optimum followed by a decline, and is often fitted by a parabola

## Survey of response functions

The functions that have been used usually fall into one of the two categories: hyperbolic or parabolic; but there are intermediate types which can have either behaviour depending upon parameter values, and also (but rarely used) functions that possess neither an asymptote nor an optimum.

Hudson (1941) used a quadratic

$$Y = a + b\rho + c\rho^2$$

where, of the constants $a$, $b$ and $c$, $a$ and $b > 0$, and $c < 0$. The quadratic is symmetric about the optimum. This symmetry constraint was overcome by Sharpe and Dent (1968) who modified the above quadratic to

$$Y = a + b\rho + c\rho^{1/2}$$

where $a$ and $c > 0$, and $b < 0$.

Duncan (1958) proposed the gamma curve equation

$$Y = a\rho e^{b\rho}$$

with the constants $a > 0$ and $b < 0$.

These three empirical equations all exhibit an optimum. The so-called power (allometric) equation (Warne, 1951; Kira, Ogawa and Sakazaki, 1953), with

$$Y = a\rho^b$$

where the constants $a > 0$ and $0 < b < 1$, possesses neither an optimum nor an asymptote.

Asymptotic functions are frequently more appropriate to biological situations than the four functions so far discussed. The rectangular hyperbola

$$Y = \frac{a\rho}{\rho + b} \tag{8.82}$$

has been widely used (Shinozaki and Kira, 1956). It has been suggested (Holliday, 1960a) that this equation does not allow for a constant yield per plant $w$ [kg], where

$$w = \frac{Y}{\rho} = \frac{a}{\rho + b} \tag{8.83}$$

at low densities where there is little or no competition. However, this is mostly an algebraic illusion: if $A$ [m$^2$] is the area per plant, then

$$A = \frac{1}{\rho} \tag{8.84}$$

and substituting for $\rho$ in equation (8.83), therefore

$$w = \frac{aA}{1 + bA} \tag{8.85}$$

For large values of $A$ (little competition), weight per plant $w$ is constant.

The asymptotic equation (8.82) can be modified to give an optimum (Holliday, 1960b)

$$Y = \frac{a\rho}{1 + b\rho + c\rho^2} \tag{8.86}$$

Different values of the parameters $a$, $b$ and $c$ allow this equation to describe responses ranging from parabolic to asymptotic.

Bleasdale and Nelder (1960) proposed an equation

$$Y = \frac{\rho}{(a + b\rho^c)^d} \tag{8.87}$$

with constants $a$, $b$, $c$ and $d$. This equation is based on the Richards growth function (p. 85). It is usually satisfactory to take $c = 1$. By taking $d = 1$ or $d > 1$, the asymptotic and parabolic responses can be obtained.

Farazdaghi and Harris (1968) discuss and use a function similar to equation (8.87) but with $d = 1$; this can describe an asymptotic or parabolic yield depending upon the value of $c$.

With some row crops, one is concerned with within-row spacing ($s_1$) and between-row spacing ($s_2$), rather than a simple scalar planting density $\rho$. Berry (1967) suggested taking account of plant rectangularity by using for plant weight $w$ the expression

$$\frac{1}{w^d} = a + b\left(\frac{1}{s_1} + \frac{1}{s_2}\right) + \frac{c}{s_1 s_2} \tag{8.88}$$

and $Y = \rho w$ for the yield, with $1/\rho = s_1 s_2$.

## Mechanistic basis for yield : density responses

An equation can usually be found amongst those discussed above, which will describe successfully the data from a particular experimental situation. However, there are benefits to be derived from a less empirical approach, and efforts have been made to provide a biological or mechanistic basis for some of these relationships (for example, see Willey and Heath, 1969; Pant, 1979), including for instance factors such as within-row and between-row spacing (Berry, 1967). In this section, we give a derivation (Thornley, 1983) using the established responses of plants to photosynthetic activity and to nutrients from the soil, of a crop yield : planting density response function with parameters that are of some biological significance.

The area $A$ [$m^2$] associated with each plant is defined as

$$A = \frac{1}{\rho} \tag{8.89}$$

and it is assumed that the soil volume $V$ [$m^3$] associated with each plant is

$$V = (1/\rho)^{3/2} \tag{8.90}$$

The assumption that a volume can be assigned to each plant is more questionable than for area, due to factors such as root density, activity and depth, which may be variable, responding to environmental factors. However, equations (8.89) and (8.90) are to be regarded as a first approximation to a complex situation.

Let $C$ [kg carbon per plant] be the quantity of carbon substrate available to each plant; to be specific, it is assumed that the soil nutrient limiting growth is nitrogen, and $N$ [kg nitrogen per plant] is defined as the quantity of nitrogen substrate available to each plant. The dependence of plant dry matter $w$ [kg] upon the carbon and nitrogen substrates $C$ and $N$ is, by assumption,

$$w = w_m \left( \frac{1}{1 + k_c/C + k_n/N + k_{cn}/CN} \right) \tag{8.91}$$

where $k_c$, $k_n$ and $k_{cn}$ are constants, and $w_m$ is the maximum value of $w$ obtained when both carbon and nitrogen are present in saturating quantities. There are several possible justifications for equation (8.91), none of them rigorous. Fertilizer responses of this type have been widely observed (Greenwood, Cleaver and Turner, 1974); two-substrate biochemical reactions obey this equation (Dixon and Webb, 1964, p. 71); for a tomato plant the response of relative growth rate to endogenous carbon and nitrogen levels is consistent with equation (8.91) (Cooper and Thornley, 1976).

It is assumed that the carbon and nitrogen available to each plant, $C$ and $N$, are related to the area $A$ and volume $V$ associated with each plant of equations (8.89) and (8.90) by

$$C = b_c A = \frac{b_c}{\rho} \quad \text{and} \quad N = b_n V = \frac{b_n}{\rho^{3/2}} \tag{8.92}$$

$b_c$ is a constant, possibly proportional to photosynthetic efficiency and light receipt; the constant $b_n$ may depend on the level of available soil nitrogen and a root uptake parameter. Equations (8.92) will break down at low planting densities since the plants will be unable to fully exploit the available space; however, for present purposes this is not important, as equation (8.91) will give a plant dry matter $w$ at its asymptotic value $w_m$ for the large $C$ and $N$ values obtaining at low densities. Substituting equations (8.92) into equation (8.91),

$$w = w_m \left( \frac{1}{1 + \dfrac{k_c \rho}{b_c} + \dfrac{k_n \rho^{3/2}}{b_n} + \dfrac{k_{cn} \rho^{5/2}}{b_c b_n}} \right)$$

and rewriting this with newly defined parameters $g_1$, $g_2$ and $g_3$, therefore

$$w = w_m \left( \frac{1}{1 + g_1 \rho + g_2 \rho^{3/2} + g_3 \rho^{5/2}} \right) \tag{8.93}$$

Crop yield $Y$ is given by

$$Y = \rho w \tag{8.94}$$

The range of responses that can be described by equations (8.93) and (8.94) for individual plant weight $w$ and for crop yield $Y$ versus planting density $\rho$ is shown in *Figure 8.5*. For crop yield $Y$ this includes the asymptotic response and the 'parabolic' response. For the same crop, different responses may be obtained depending upon season and management. A rationale for such behaviour is contained in the above analysis. Similarly, the plant dry matter responses may cover a range that includes sigmoidal and non-sigmoidal behaviour.

To outline explicitly one way in which environmental effects may be incorporated into equation (8.93), it could be assumed that $b_c$ and $b_n$ of equations (8.92) are given by

$$b_c = \alpha J \quad \text{and} \quad b_n = \beta N_s \tag{8.95}$$

where $J$ is energy receipt or light receipt, and $N_s$ is the level of soil nitrogen (endogenous and added); $\alpha$ and $\beta$ are constants. In equation (8.93) this then gives

$$g_1 = \frac{k_c}{\alpha J} \qquad g_2 = \frac{k_n}{\beta N_s} \qquad g_3 = \frac{k_c k_n}{\alpha J \beta N_s} \tag{8.96}$$

Equations (8.93) and (8.94) with parameters given by equations (8.96) could be tested in density : yield experiments in which treatments of fertilizer nitrogen and shading are applied.

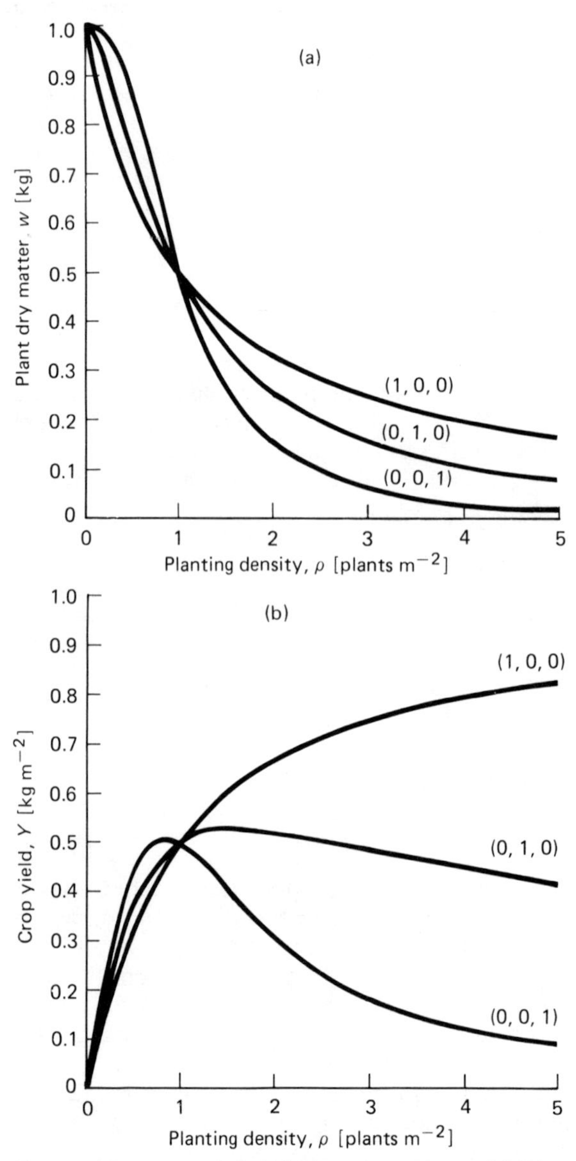


*Figure 8.5* Responses of plant dry matter $w$ and crop yield $Y$ to planting density $\rho$, according to equations (8.93) and (8.94). Parameter values are $w_m = 1$, and $(g_1, g_2, g_3)$ as shown

Frequently, only some part of the plant, with dry matter $w_h$, is harvested. Equation (8.94) is then replaced by

$$Y = \rho w_h \tag{8.97}$$

The harvestable part of the plant $w_h$ is some fraction $f_h$ of the total plant dry matter $w$, giving

$$w_h = f_h w \tag{8.98}$$

Some authors (see Willey and Heath, 1969, for references) have used the allometric equation to relate $w_h$ to $w$, taking $w_h \propto w^\gamma$, or $f_h \propto w^{\gamma-1}$, where $\gamma$ is the allometric constant. However, the environment may have considerable effects on plant partitioning (and the fraction $f_h$), even at a constant plant dry matter $w$. This may be denoted by

$$f_h \equiv f_h(E, w) \tag{8.99}$$

where $E$ indicates a set of environmental parameters. The use of equations (8.97), (8.98) and (8.91), with the allometric or some other expression for $f_h$, may modify the responses of *Figure 8.5*.

## Models of crop growth

This section is concerned with crop models which can be operated over a whole growing season to predict growth and especially yield. Some of the models already described in this chapter fulfill this function, for example equation (8.4) on water use, equation (8.44) on response to fertilizer, and equation (8.93) on response to planting density. These simple models, which often deal with a single factor and are expressed in a single equation, are of considerable value, but there is a need for more accurate and comprehensive models which include the principal environmental factors.

The number of published crop models is such that a complete review of the area is not attempted here; recent surveys are by Baker (1980) and Legg (1981). We discuss the general principles and give some examples. The categorization of crop models into empirical or mechanistic types is, of course, an approximation, and many models fall somewhere in between.

### Empirical models

Many attempts have been made to relate crop growth and yield directly to various aspects of climate, weather and environment; the objectives are to account for observed yield variation, to discover which factors affect yield most greatly, and finally, armed with this information, to consider whether management can manipulate some of these factors so as to increase yields or decrease costs.

The most important factors are rainfall, temperature and solar radiation, and usually of lesser importance are daylength, fertilizer or nutrient status and $CO_2$ concentration. Rainfall affects plant water status, which is crucial for many crops in certain parts of the world, although water status may also depend on humidity, wind speed, irrigation and soil characteristics. Temperature may well include soil temperature, and maximum and minimum air temperatures.

**TABLE 8.2. Some published empirical crop models. Environmental or management factors included: *I* = irradiance; *W* = water status (rainfall, humidity, wind); *T* = temperature; *F* = fertilizer, nutrient status; *D* = daylength; *C* = $CO_2$ concentration**

| *Crop* | *Factors* | *Method* | *Time interval* | *Reference* |
|---|---|---|---|---|
| Grass | *I,W,T,D* | Growth and development rates are modified | 7 days | Angus *et al.* (1980) |
| Rice | *I,T* | Regression of climatic factors on yield | Phases of development | Murata (1975) |
| Wheat | *W,T* | Stepwise multiple regression on yield | Day | Bridge (1976) |
| Wheat | *W* | Linear regression of yield on seasonal rainfall | Season | Seif and Pederson (1978) |
| Wheat | *T,W* | Regression of yield on a derived index | Month | Sakamoto (1978) |
| Wheat | *T,W* | Multiple curvilinear regression | Month | Pitter (1977) |
| Wheat | *T,W* | Regression equations for daily growth rate | Day | Haun (1974) |

*Table 8.2* summarizes some published accounts of empirical crop models, revealing a wide variety of approaches. Let $W$, the crop dry matter, be the crop variable of interest, and denote the environment by $E$. The simplest approach is to assume that $W$ is related to the environment averaged or accumulated over the whole growing season, $E_s$. Formally, this could be denoted

$$W = f_s(E_s) \tag{8.100a}$$

where $f_s$ is a function of the seasonal environmental data $E_s$. However, this approach is not very successful because weather fluctuations are considerable, and often have large effects on crop growth and development; also, the weather during certain phases of the growing season can be much more important than during others.

An obvious refinement of this method is to use environment data for weekly or monthly periods, or sometimes over phases of the crops developmental life cycle. Denoting such weekly, monthly or phasic environmental data by $E_w$, $E_m$ or $E_p$, then equation (8.100a) is replaced by

$$W = f_w(E_w) \qquad W = f_m(E_m) \quad \text{or} \quad W = f_p(E_p) \tag{8.100b}$$

where $f_w$, $f_m$ and $f_p$ denote different functions of the environmental data.

Rather than regressing yield or crop dry weight directly on environment, it is becoming more popular to regress growth rate or development rate on the environmental factors. This approach appears to be more effective, perhaps because it moves the empirical models some way towards considerations of mechanism. Let the growing season consist of $h$ equal time intervals $\Delta t$, denoted by the index $i = 1, 2, \ldots, h$. It is common to choose a day or a week for $\Delta t$. Let $E_i$ represent the environmental factors, accumulated or meaned, over the $i$th time interval. Considering just the single attribute, crop dry weight $W$, it is assumed that the growth rate $\Delta W/\Delta t$ over the $i$th interval can be written

$$\frac{\Delta W}{\Delta t} = g_i(W_i, E_i) \tag{8.101a}$$

so that $W_{i+1}$, the crop dry weight at the beginning of the $(i + 1)$th time interval is given by

$$W_{i+1} = W_i + g_i(W_i, E_i)\Delta t \tag{8.101b}$$

$g_i$ represents some function of arguments ($W_i$ and $E_i$) appropriate to the $i$th period.

In predicting wheat yields, Haun (1979) used essentially a different function $g_i$ for each week of the growing season, using equations that were regressions on environmental and management factors such as soil moisture, minimum and maximum temperatures, rainfall, fertilizer inputs, planting date, etc. Idso *et al.* (1979) were able to predict wheat yields from crop temperature and reflectance data alone, both of which are amenable to remote-sensing instrumentation carried out in satellites. For forecasting yields in grain sorghum, Arkin, Richardson and Maas (1978) combine a deterministic crop model with stochastic weather data, although they give few details. Fitzpatrick and Nix (1970) proposed an approach in which a quantity called a growth index, GI, is expressed as a simple multiplicative function of a light index, LI, a thermal index, TI, and a moisture index, MI, giving

$$\mathrm{GI} = \mathrm{LI} \times \mathrm{TI} \times \mathrm{MI} \tag{8.102}$$

The indices may all be expressed as hyperbolic relationships lying between 0 and 1, giving the total effect on crop growth through the growth index. Although this method appears attractive, it has not so far proved possible to refine and develop it into a practical technique.

A possible semi-empirical approach which has not yet been fully explored is to take one of the biologically appropriate growth equations (Chapter 5), and to allow the environmental factors to modify the parameters of the growth equation. For instance, the Gompertz growth equation for finite increments is (equation (5.29), p. 83)

$$\frac{1}{W}\frac{\Delta W}{\Delta t} = \mu \mathrm{e}^{-Dt} \tag{8.103}$$

where $W$ is crop dry weight, $\mu$ is the specific growth rate parameter and $D$ a developmental rate parameter. If $\mu$ and $D$ are constant, then this equation can be integrated to give the Gompertz growth curve (*Figure 5.5*, p. 84). It is assumed that the effects of environment are to modify the parameters $\mu$ and $D$. There are many ways in which this might be accomplished, but as an example, it is assumed that, with a time interval of 1 day, the radiation receipt $J_i$ and soil moisture deficit $S_i$, on the $i$th day, are the important environmental factors. The dependence of the parameter $\mu$ on these factors could be expressed as

$$\mu = \mu_m\left(\frac{J_i}{K_J + J_i}\right)\left(\frac{K_S}{K_S + S_i}\right) \tag{8.104}$$

where $\mu_{\mathrm{m}}$, $K_J$ and $K_S$ are constants. A separate equation is needed to keep account of the soil moisture deficit (p. 142). The incorporation of temperature, which greatly affects development, into equation (8.103) is not so straightforward as with radiation and soil moisture deficit; it may be accomplished by treating the product $Dt$ as proportional to a developmental index, which increases at a temperature-dependent rate. Equation (8.103) may then be integrated day-by-day given a matrix of environmental inputs, which may represent historic, current or simulated environmental data.

Clearly, there is no shortage of environmental variables to regress against growth or yield, and the method in its various forms can be made to account for crop yield variation with a high degree of success. Ideally, an empirical model should account for yield variation from site to site without change in the parameters of the model. However, these models are usually less successful when used outside the range of sites and weather for which the models have been developed and tested, and such use is not recommended. Despite limitations of lack of generality, and a failure to give understanding into how the environment is having the observed effects on yield, empirical models currently provide the most effective method of accounting for crop yield variation. These models continue to be valuable for such purposes as yield prediction, the consideration of land use possibilities, application of fertilizers, irrigation, and the management of grass. However, working downwards towards mechanism, as in equations (8.103) and (8.104), may produce models more suited to application than those obtained by the approach described in the next section, where one begins with complexity, and attempts to integrate it into a model that will make simple predictions from given environmental inputs.

## Mechanistic models

Mechanistic models of crop growth are constructed by assuming that the system has a certain structure, and assigning to the components of the system properties and processes which can be assembled within a mathematical model. In *Table 8.3* processes that are often considered in such models are listed, although most models only incorporate some of these processes—those relevant to the system being studied. Many crop models have been developed, taking different combinations of

**TABLE 8.3. Processes that may be important in mechanistic models of plant and crop growth**

| | |
|---|---|
| 1 Phot | Light interception and photosynthesis: canopy architecture; radiation characteristics; leaf characteristics |
| 2 Nutr | Root activity and nutrient uptake: root system architecture; soil nutrient status; root status and characteristics |
| 3 Part | Partitioning: substrate pools of carbon compounds and nutrients replenished by 1 and 2; transport between pools; utilization of pool substances for growth; priorities |
| 4 Transp | Transpiration: water balance of plant and soil; water status of plant |
| 5 Gr & R | Growth of structural dry matter and the recycling of structural components; respiration |
| 6 LA exp | Leaf area expansion |
| 7 Dev | Development and morphogenesis: initiation, growth and development of new organs (stems, leaves, flowers, fruits, storage organs etc.) |
| 8 Sen | Senescence |

the processes in *Table 8.3*, and treating these processes in various ways. A survey of some of the published models is given in *Table 8.4*, and also by Legg (1981, p. 142) and Baker (1980). Here we give outline accounts of a simple model of grass growth (Johnson, Ameziane and Thornley, 1983), and a much more comprehensive model of soyabean (Meyer *et al.*, 1979), in order to illustrate the principles and practice of mechanistic modelling; for these two examples, the published papers give clearly stated assumptions for the models, and compare experimental measurement with model predictions.

**TABLE 8.4. Some published mechanistic crop growth models. See *Table 8.3* for submodel key; emp = empirical, mech = mechanistic. See *Table 8.2* for key to environmental factors**

| *Crop* | *Submodels* | | | | | | | | *Environmental factors* | *Reference* |
|---|---|---|---|---|---|---|---|---|---|---|
| | *Phot* | *Nutr* | *Part* | *Transp* | *Gr & R* | *LA exp* | *Dev* | *Sen* | | |
| Alfalfa | emp | | mech | | emp | emp | | | *I,W,T,D* | Holt *et al.* (1975) |
| Barley | mech | | | | mech | | | | *I,W,T* | Legg *et al.* (1979) |
| Chrysanthemum | mech | | | | emp | emp | | | *I,C* | Charles-Edwards and Acock (1977) |
| Clover (sub) | emp | | | | emp | emp | | | *I,T* | Fukai and Silsbury (1978) |
| Cotton | emp | emp | | | emp | | | | *I,T,D* | McKinion *et al.* (1975) |
| Grass | mech | | emp | | | emp | | emp | *I,T* | Johnson *et al.* (1983) |
| Grass | emp | | emp | | emp | emp | | emp | *I,T* | Sheehy *et al.* (1980) |
| Lettuce | emp | | | | mech | | | | *I,C,T* | Sweeney *et al.* (1981) |
| Maize | mech | emp | emp | mech | mech | emp | | emp | *I,W,T* | de Wit *et al.* (1978) |
| Sorghum | mech | emp? | emp? | mech | mech | emp | emp | | *I,W,T* | Arkin *et al.* (1976) |
| Soyabean | mech | emp | mech | mech | mech | emp | emp | emp | *I,W,T,F,D* | Meyer *et al.* (1979) |
| Sugar beet | mech | | emp | mech | mech | emp | emp | | | Fick *et al.* (1975) |
| Tobacco | mech | | mech | | mech | | | emp | *I,T* | Wann *et al.* (1978) |

## A GRASS GROWTH MODEL

This model was developed for a vegetative, irrigated, fertilized, disease-free grass sward. The processes treated are light interception and photosynthesis, growth and respiration, leaf area expansion and senescence; the processes ignored include root activity and nutrient uptake, partitioning, transpiration and water relations, and morphogenesis; the latter are deemed to be of lesser importance for a grass crop grown under the circumstances specified.

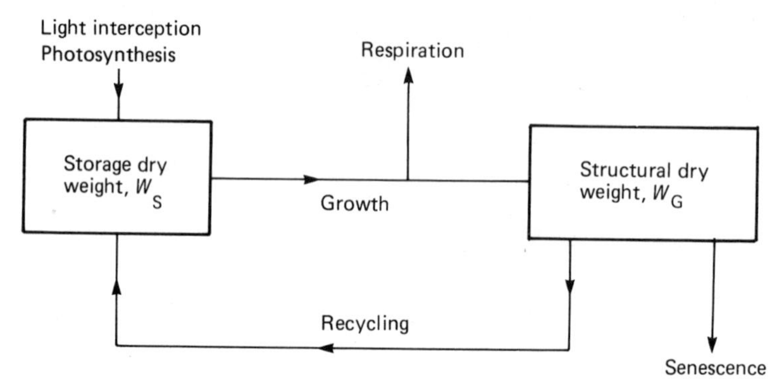


*Figure 8.6* A simple mechanistic grass growth model

The model is shown schematically in *Figure 8.6*. It is a carbon budget model, with crop dry weight $W$ expressed as carbon per unit ground area [kg carbon $m^{-2}$]. The dry weight $W$ comprises two components: storage dry weight $W_S$ and structural dry weight $W_G$, with

$$W = W_G + W_S \tag{8.105}$$

The experimentally measured dry weight $W_{ex}$ is related to $W$ by

$$W = f_C W_{ex} \tag{8.106}$$

where $f_C$ is the fractional carbon content of dry matter. It is assumed that the leaf area index $L$ is related to the structural dry weight $W_G$ by

$$L = aW_G \tag{8.107}$$

where $a$ is a constant [$m^2$ leaf area (kg carbon in structural dry weight)$^{-1}$].

*1. Light interception and photosynthesis*

For instantaneous canopy photosynthetic rate $P_c$ [kg $CO_2$ $m^{-2}$ $s^{-1}$], equation (7.19) is used, namely

$$P_c = \frac{P_m}{k}\left\{\ln\left[\frac{k\alpha I_o + (1 - m)P_m}{k\alpha I_o e^{-kL} + (1 - m)P_m}\right]\right\} \tag{8.108}$$

where $P_m$ and $\alpha$ are the asymptote and initial slope of the leaf light response curve (equation (7.14)), $m$ is the leaf transmission coefficient, $k$ is the canopy extinction coefficient and $I_o$ is the instantaneous light flux density above the canopy. To find the daily photosynthetic rate $P$, the instantaneous canopy rate $P_c$ must be integrated with respect to the within-day time variable $\tau$ over daylength $h$ [s]:

$$P = \int_0^h P_c \, d\tau \tag{8.109}$$

It is assumed that the irradiance $I_o$ varies throughout the day according to (p. 102)

$$I_o(\tau) = \frac{2J}{h} \sin^2\left(\frac{\pi\tau}{h}\right) \quad \text{for} \quad 0 \leqslant \tau \leqslant h \tag{8.110}$$

where $J$ is the daily light receipt. Substitution of equation (8.108) and then equation (8.110) into equation (8.109), followed by integration gives

$$P = \frac{P_m h}{k} \ln \left\{ \frac{\dfrac{\alpha kJ}{h} + (1-m)P_m + \left[\dfrac{2\alpha kJ}{h}(1-m)P_m + (1-m)^2 P_m^2\right]^{1/2}}{\dfrac{\alpha kJ}{h} e^{-kL} + (1-m)P_m + \left[\dfrac{2\alpha kJ}{h}(1-m)P_m e^{-kL} + (1-m)^2 P_m^2\right]^{1/2}} \right\} \tag{8.111}$$

This equation gives the daily photosynthetic integral in terms of single leaf, canopy and environmental parameters, based on assumptions about light attenuation through the canopy, single leaf photosynthesis and daily light distribution.

The photosynthetic input $P$ of equation (8.111) is in units of kg $CO_2$ $m^{-2}$ $day^{-1}$; to convert this to kg carbon $m^{-2}$ $day^{-1}$, $P$ is multiplied by $\theta$, with $\theta = 12/44 = 0.273$, so that the photosynthetic input to the model is

$$\theta P \tag{8.112}$$

*2. Growth and respiration*

The growth rate of the structural component $W_G$ occurs (*Figure 8.6*) at the expense of storage material, and it is assumed that it is given by

$$\frac{dW_G}{dt} = \mu_m \frac{W_G W_S}{W_G + W_S} \tag{8.113}$$

where $\mu_m$ is a parameter giving the maximum value of the specific growth rate $(1/W_G)(dW_G/dt)$, obtained when $W_S$ is large. Thornley and Hurd (1974) found that this type of hyperbolic relationship worked well in tomato.

If $Y$ is the conversion yield of growth (pp. 124–126), then the rate of utilization of storage is

$$\frac{1}{Y}\frac{dW_G}{dt} \tag{8.114}$$

and the respiration rate $R$ is given by

$$R = \frac{(1-Y)}{Y}\frac{dW_G}{dt} \tag{8.115}$$

The maintenance requirement is treated using the recycling model described on pp. 126–127 and in *Figure 7.8*. This assumes first-order kinetics for the degradation of structural dry matter, so $W_G$ is lost at a rate

$$\gamma W_G \tag{8.116}$$

where $\gamma$ is a constant [$\text{day}^{-1}$]; assuming no losses, storage material is added to at the same rate.

*3. Senescence*

The model is concerned with an established vegetative grass crop, and therefore as a simple approximation senescence is modelled as a linear loss of structural material,

$$\beta W_G \tag{8.117}$$

where $\beta$ is a rate constant.

*4. Partitioning, and root growth and maintenance*

A constant fraction $\phi$ of the photosynthetic input is allocated to shoot growth, the remainder being partitioned to the root for root growth, root maintenance and any other root activities. Thus from (8.112), the photosynthetic input to the shoot is

$$\phi\theta P \tag{8.118}$$

*5. Effects of temperature*

The parameters $\mu_m$, $\gamma$ and $\beta$ of (8.113), (8.116) and (8.117) may be regarded as quasi-chemical rate constants; they are therefore assumed to have the usual chemical $Q_{10}$-type behaviour with a $Q_{10}$ of $Q$. If $X(T)$ is the value of $X$ at temperature $T$, then

$$X(T) = X_o Q^{(T - T_o)/10} \tag{8.119}$$

where $X_o$ is the value of $X$ at the reference temperature $T_o$.

For the single-leaf photosynthetic parameters of equation (8.111), it is assumed that the photosynthetic efficiency $\alpha$ is independent of temperature, but the maximum photosynthetic rate $P_m$ is assigned a linear dependence:

$$P_m = P_0 + P_1 T \tag{8.120}$$

where $P_0$ and $P_1$ are constants.

*6. Mathematical summary of the model*

The two state variables $W_S$ and $W_G$ of *Figure 8.6* are described by two ordinary first-order differential equations:

$$\frac{dW_S}{dt} = \phi\theta P - \frac{\mu W_G}{Y} + \gamma W_G \tag{8.121a}$$

and

$$\frac{dW_G}{dt} = \mu_m \frac{W_G W_S}{W_G + W_S} - \gamma W_G - \beta W_G \tag{8.121b}$$

Equation (8.111) is used to calculate $P$; the temperature dependence is given by equations (8.119) and (8.120). The environmental inputs are daily light integral $J$, temperature $T$ and daylength $h$. The model has thirteen parameters: $a$, $k$, $m$, $\alpha$, $P_0$, $P_1$, $\theta$, $\mu_{mo}$, $Y$, $\gamma_o$, $\beta_o$, $\phi$ and $Q$. For comparison of model predictions with experiment the parameter $f_C$ (equation (8.106)) must be supplied. Initial conditions, $W_S$ ($t = 0$) and $W_G$ ($t = 0$), are also required so that equations (8.121) can be integrated.

*7. Solutions*

In general, equations (8.121) can only be solved numerically. However, with many models there are special or limiting cases which can be solved analytically to give a valuable check on the biology built into the model and also on the mathematical formulation of that biology. In this case, there are two such limiting situations, both for a constant environment.

Balanced exponential growth occurs at low values of leaf area index $L$ where equation (8.111) can be approximated

$$P = \zeta L \tag{8.122}$$

where $\zeta$ is a constant. A quadratic equation for the specific growth rate can be derived which will only give a positive root when the physiological parameters satisfy certain conditions.

At high leaf area indices, steady-state non-growing solutions can be obtained where the losses due to respiration and senescence are exactly balanced by the photosynthetic input.

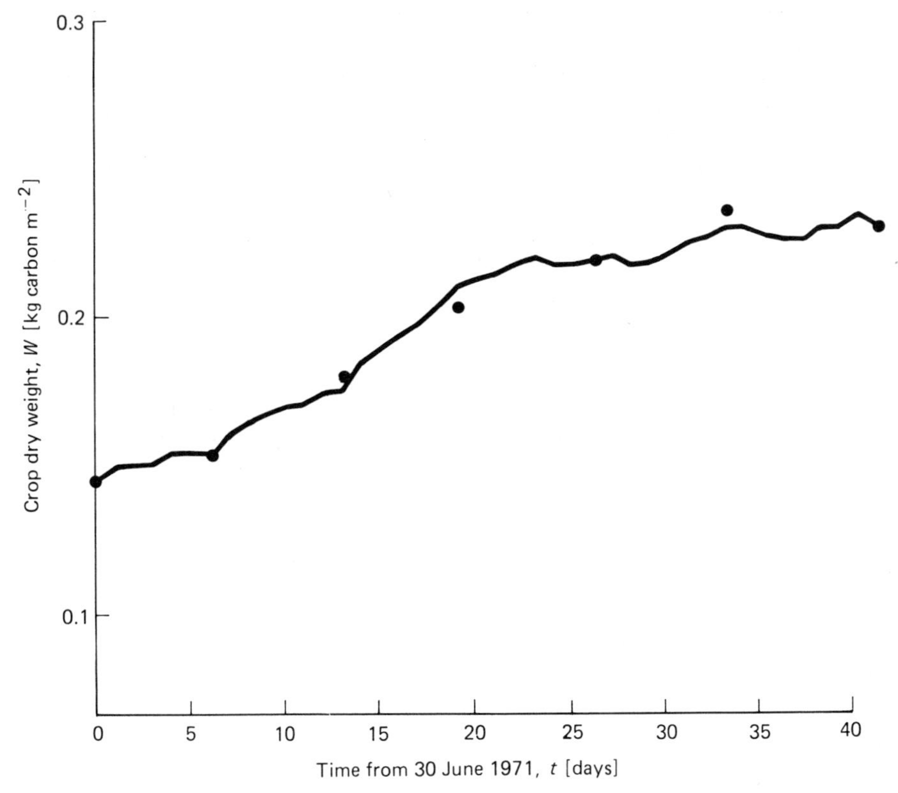


*Figure 8.7* Comparison of model (continuous line) with experiment (●). The data are on a perennial ryegrass crop (Leafe, Stiles and Dickinson, 1974). The plant parameters used were $a = 21\ m^2$ (kg carbon)$^{-1}$, $f_c = 0.4$ kg carbon (kg dry matter)$^{-1}$, $k = 0.5$, $m = 0.12$, $Q = 2$, $P_0 = 0.5 \times 10^{-6}$ kg $CO_2\ m^{-2}\ s^{-1}$ (°C)$^{-1}$, $P_1 = 0.05 \times 10^{-6}$ kg $CO_2\ m^{-2}\ s^{-1}$, $Y = 0.75$, $\alpha = 12 \times 10^{-9}$ kg $CO_2\ J^{-1}$, $T_o = 20$ °C, $\beta_o = 0.05$ day$^{-1}$, $\gamma_o = 0.1$ day$^{-1}$, $\theta = 12/44$ kg carbon (kg $CO_2$)$^{-1}$, $\phi = 0.9$, $\mu_{mo} = 0.4$ day$^{-1}$. The daily environmental variables were taken from local records. For integration, Euler's method (p. 21) with a time step of one day was used

Not only does consideration of limiting cases give an invaluable check on the model, but often useful physiological insights may also be obtained.

In *Figure 8.7* the model is applied to data on perennial ryegrass, using the parameter values given, and meteorological records from Hurley for mean daily temperatures, daily light integrals, and daylengths. A full account is given by Johnson, Ameziane and Thornley (1983).

## A SOYABEAN MODEL

This model, which bears the title 'SOYMOD/OARDC—A dynamic simulator of soybean growth, development, and seed yield', is comprehensively described in an excellent 36-page research bulletin (Meyer *et al.*, 1979). It is a large model, attempting to account as mechanistically as possible for all the processes in *Table 8.2*, and it is not possible to give a detailed description of it here. We attempt to give enough of an account to convey the flavour, the promise and the problems of this approach to crop growth simulation.

An overview of SOYMOD is given in *Table 8.5*, where the processes treated, the input data required, and the output data produced, are listed. The conceptual, mathematical and computational approach is analogous to that used in the grass growth model just described; there is simply more of everything: more submodels, more input data, more computing time and power, and more output. The way in which the computing problem was tackled by means of a FORTRAN program is

**TABLE 8.5. Structure of SOYMOD/OARDC—a dynamic simulator of soyabean growth, development and seed yield**

| *Processes* | *Input data* | *Output data* |
|---|---|---|
| Photosynthesis | Plant data | Physiological events |
| Light interception | Variety coefficients | Echo of some input data |
| Structural growth | Rooting density | Dry matter: total, leaves, petioles, |
| | characteristics for a given | root, fruit |
| | soil type | Plant height, stem diameter |
| Respiration | Planting data | Number of fruit |
| | Emergence date | Flowering and podfill events |
| Leaf reserves | Row width | Total leaf area |
| | Plant Spacing | Irrigation events |
| Phloem loading | Climatic data | Leaf abscission events |
| Phloem transport | Date, daylength | Seed yield |
| | Maximum air temperature | Summary by nodes of dry matter, |
| Phloem unloading | Minimum air temperature | protein, soluble sugar, starch, |
| Senescence | Dewpoint | and fibre contents; rates of |
| Nitrogen assimilation | Solar radiation | photosynthesis, respiration, |
| | Rainfall | phloem loading, |
| Root and soil moisture | Windrun | translocation, storage and |
| | Soil data | growth |
| | Soil type | Crop summary: total rainfall, |
| Discrete processes | Soil water retention curve | irrigation, solar radiation, |
| Development of | Bulk density | heat units, |
| vegetative shoot | Hydraulic conductivity | evapotranspiration; seed |
| Flowering and fruiting | Initial soil water content | yield, grams per 100 seed, |
| Leaf abscission | | total fruit dry matter per plant |

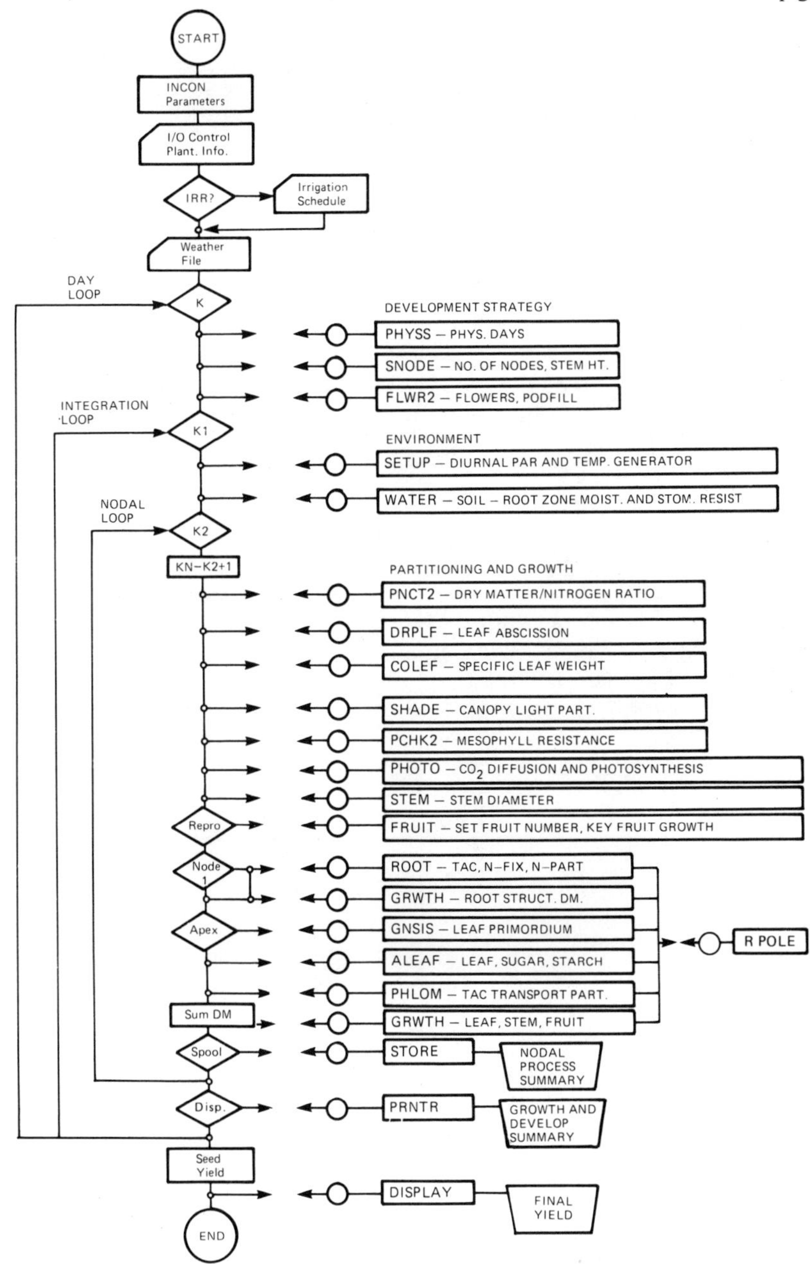


*Figure 8.8* Simplified flow diagram of SOYMOD/OARDC. (After Meyer *et al.*, 1979)

indicated in *Figure 8.8*, and in *Figure 8.9* some of the predictions of the model are shown alongside comparable experimental data.

Instead of the two categories, structure and storage, of the grass model (*Figure 8.6*), SOYMOD describes dry matter as the sum of four entities: structural carbohydrate, available carbohydrate (this comprises labile carbohydrates that can

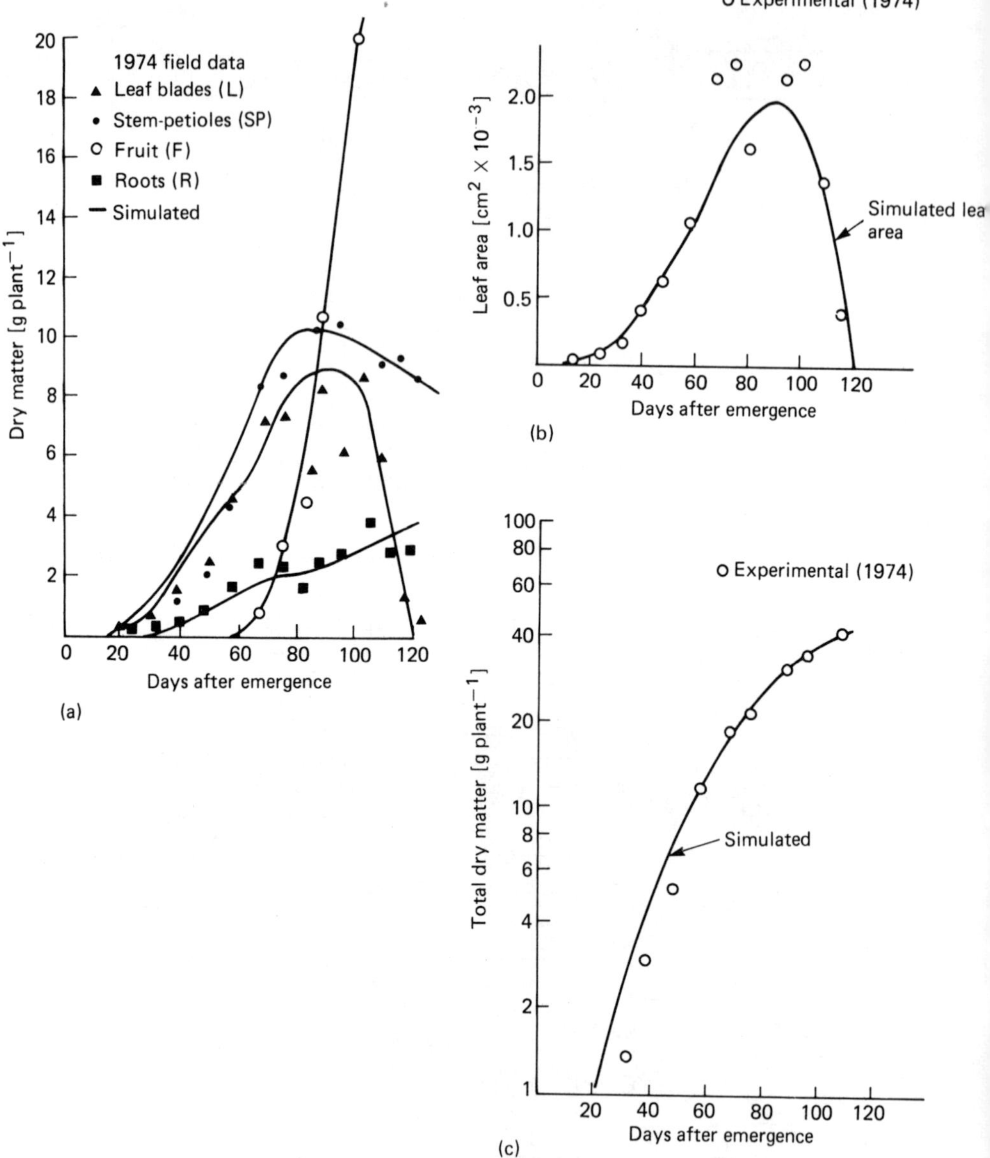


*Figure 8.9* Simulated and experimental data for SOYMOD (after Meyer *et al.*, 1979). (a) Dry matter for leaf blades, stem, petiole, fruit and roots. (b) Total leaf area. (c) Total dry matter

be transported around the plant), starch and protein. These substances are partitioned amongst many morphological parts: leaf blades, petioles, stems, fruit and roots, whereas in the grass model, effectively just the single morphological entity 'shoot' is considered. The model is of the carbon–nitrogen type, with materials being transported between the morphological entities, and utilized within those entities for growth or other purposes. Carbon/nitrogen ratios are assigned a control function for the growth of different plant parts.

In SOYMOD the aim has been to incorporate sufficient physiological detail so that the different levels of information for use by agronomists, plant breeders and farmers can be obtained. In the summary, the authors state that the model is not available as an extension (advisory) tool, and is most suitable for research and teaching about the physiological processes of soyabean.

### Empiricism and mechanism in crop modelling

There are three fairly distinct communities of people to whom a crop model or simulator may be of value: first, farmers, including advisory and extension services, who are concerned primarily with current production; second, applied scientists, including agronomists and plant breeders, whose objective is to improve the efficiency of production techniques, for the most part using current knowledge; and third, research scientists, whose aim is to extend the bounds of present-day knowledge. There is no way in which a single model can satisfy the differing objectives of these three groups, which are, at least in part, incompatible. There are, it seems, three types of model that correspond approximately with the requirements of these three communities: empirical models, sound mechanistic models based on current ideas, and more speculative research-oriented mechanistic models.

For the farming community, utility, ease of practical application, and a certain level of accuracy within a tested (though restricted) range are essential. These needs can be most usually met by an empirical model, although a carefully validated mechanistic model may be suitable if sufficiently powerful small computers are available, and good software is provided.

For the applied scientist, some representation of mechanism and physiological detail is usually required, and also a higher level of generality than can be provided by an empirical model. This may be obtained by a mechanistic model in which well-understood processes are represented mechanistically, and areas of uncertainty are described using the best empirical functions. Such a model is, compared to the empirical model, likely to be more general, less accurate, more complex, but possibly is still reasonably manageable.

The research scientist using models will, if he is worth his salt, construct speculative models extending into areas where there is uncertainty and lack of understanding. Such models are far too unreliable for practical application by the farmer or the applied scientist.

## Exercises

### Exercise 8.1

In summer in the UK a typical total daily radiation receipt is 18 MJ m$^{-2}$ (*see Figure 6.2*, p. 100). Calculate (a) the amount of water evaporated per unit area per day using equation (8.9a) and making the approximation $s >> \gamma$; (b) the quantity of dry matter produced per unit area per day assuming a water use efficiency of 0.0015 kg dry matter (kg water)$^{-1}$.

**Exercise 8.2**

Using equation (8.8) and $c_p$ (20 °C, atmospheric pressure) = 1006 J (kg air)$^{-1}$ °C$^{-1}$, $P = 1.013 \times 10^5$ Pa, $\lambda(20\,°C) = 2.453 \times 10^6$ J (kg water)$^{-1}$ and $\epsilon = 0.622$, calculate the psychrometric constant $\gamma$.

Using equation (8.9a) and $s(20\,°C) = 145$ Pa K$^{-1}$ (*see* p. 140), evaluate the correction factor to be applied to the estimate of Exercise 8.1 assuming $g_a/g_s = 0$.

**Exercise 8.3**

Assuming that the response of crop yield $Y$ to nitrogen level $N$ and phosphate level $P$ is given by the quadratic expression

$$Y = a_0 + a_1 N + a_2 N^2 + b_0 + b_1 P + b_2 P^2 + hNP$$

where $a_0$, $a_1$, $a_2$, $b_0$, $b_1$, $b_2$ and $h$ are constants, apply equations (8.32) and (8.35) to obtain equations for $N$ and $P$ in terms of the constants, giving the maximum crop yield and the economic crop yield.

**Exercise 8.4**

Derive equation (8.78) by differentiating equation (8.77) with respect to $t_c$ (the length of the cold temperature treatment), and assuming that equation (8.76) relating $t_c$ and $t_a$ (the time to anthesis at normal growing temperatures) is valid.

**Exercise 8.5**

Derive equations (8.80) and (8.81) from equation (8.79a), as outlined in the text.

Chapter 9

# Plant diseases and pests

## Introduction

An objective of the sciences concerned with crop protection is a good quantitative understanding of the significant problems in this area: that is, the host–pathogen–environment complex for a plant disease, and the host–pest–predator/parasite–environment complex for a pest. However, the main task of the farmer is the raising of crops and the production of food, and he will only be interested in applying control measures where they are effective and economically warranted.

Assessment of crop losses from pests and diseases is an essential requirement of management schemes aiming at economic control, although such assessment is very difficult, and usually preventative measures are carried out without any rigorous economic appraisal. Frequently, plant diseases and pests have been controlled without an understanding of the system and its mechanisms, and sometimes even without recognition of the causal agent. However, a thorough understanding is likely to lead to the most efficient methods of control.

In this chapter, some applications of mathematics to pest and disease problems are discussed, without attempting to be comprehensive. Van der Plank (1963) is one of the pioneers of the mathematical approach to plant disease epidemics and their control; more recent accounts and surveys are given by Kranz (1974), Krause and Massie (1975) and James (1974). Plant pest modelling has been largely the preserve of entomologists and applied ecologists. There is no wholly satisfactory general text, but Maynard Smith (1974) and Conway (1977) both contain material that will give a useful orientation.

## Plant diseases

### Estimation of crop losses

This problem is invariably attacked using empirical methods (James, 1974). The growth of healthy and diseased plant populations is monitored over the growing season. The different stages of plant development are identified using a key (e.g. Large, 1954). A disease assessment method is developed to measure the amount of disease at the different growth stages of the plant (James, 1971). Various regression methods (*see* Glossary) may be used to relate the dependent variable, $Y$ (the percentage yield loss), to the independent variables $X_i$, which give the severity of the disease at stage $i$ of the plant's development. For example, Burleigh, Roelfs and Eversmeyer (1972), estimating the losses in wheat caused by leaf rust, used

$$Y = 5.3788 + 5.5260X_2 - 0.3308X_5 + 0.5019X_7 \tag{9.1}$$

where $X_2$, $X_5$ and $X_7$ are the percentage leaf-rust severity at the boot, early berry and early dough stages respectively; this equation accounted for 79% of the variation. Similar work has been done on potato late blight and loss of tuber yield (James *et al.*, 1972) and many other crops.

Equation (9.1) is sometimes referred to as a 'multiple-point' model, since it incorporates disease at several points during the crop growth period. The so-called 'critical-point' models may be applicable for disease of short duration, and which affect the physiology of the growing plant at a particular and crucial phase. Katsube and Koshimizu (1970) used

$$Y = 0.57X \tag{9.2}$$

for estimating percentage loss of yield in rice due to rice blast, where $X$ is the percentage of blasted neck nodes 30 days after heading. For stem rust in wheat, Romig and Calpouzos (1970) found that

$$Y = -25.33 + \ln X \tag{9.3}$$

where $X$ is the disease severity when the developing caryopsis has reached three-quarters of its final size, was the best predictor of yield loss.

The critical-point and multiple-point approaches of equations (9.2) and (9.1) may be generalized and combined in a single equation

$$Y = \int w(t)X(t)\,\mathrm{d}t \tag{9.4}$$

where $w$ is a weighting function that depends on the variable $t$ which may denote chronological time or developmental time, and $X$ is disease severity at time $t$. Equation (9.4) is still a linear model, but this equation can be rewritten as

$$Y = \int w(t)f[X(t)]\,\mathrm{d}t \tag{9.5}$$

where $f$ denotes some function; this would then encompass nonlinear models such as equation (9.3). Effectively, equations (9.4) and (9.5) sum the area under the disease–progression curve.

Crop simulators (Chapter 8, pp. 164–173) are still at an early stage in their development, but a long-term prospect is to incorporate the effects of pests and diseases on the parameters or pools of a crop simulator. Modified yield predictions can then be directly obtained, and the effects of the pests and diseases on the physiological processes of the plant can then be understood and evaluated. We are not aware of any significant published work in this direction.

## Disease prediction and control

A disease is only able to progress if the conditions provided by the host plants and environment are favourable, and if some inoculum is present. The relevant components of the micro-environment largely depend upon weather, and weather cannot be forecast reliably for more than very short periods in advance. It is therefore usual to make predictions only after certain biological and meteorological conditions favourable to the disease have been fulfilled. If warranted, disease control measures may then be applied. In fact, only a few disease prediction systems take account of the amount of inoculum of disease present at a given stage in the plant's development (Eversmeyer and Burleigh, 1970); it is more common to assume that there is always enough inoculum present to initiate an epidemic, to monitor the weather or microclimate within the crop, and to base disease predictions upon these measurements.

Many methods have been developed to predict the occurrence of potato late blight. A simple system due to Hyre (1954) is described to illustrate the general principles.

A day is 'blight-favourable' if

$$\begin{aligned} &\text{the 5-day temperature average} < 25.5\,^{\circ}\mathrm{C} \\ &\text{the total rainfall for the last 10 days} \geq 3.0\ \mathrm{cm} \\ &\text{the minimum temperature on that day} \geq 7.2\,^{\circ}\mathrm{C} \end{aligned} \qquad (9.6)$$

When ten consecutive blight-favourable days are first recorded, late blight is forecast to appear 7 to 14 days later. Appropriate control measures are recommended. This system works well in the north-eastern United States, but less well in the mid-western States, where different relationships between rainfall, relative humidity and temperature prevail. For different climatic conditions, alternative systems may be needed. Krause, Massie and Hyre (1975) describe a combination of Hyre's system with one of these alternatives.

Multiple regression equations have also been used to predict disease-leaf rust of wheat (Eversmeyer and Burleigh, 1970), and stem rust of wheat (Eversmeyer, Burleigh and Roelfs, 1973). These authors define about 15 meteorological and biological parameters that predict the future rate of disease increase.

Automatic data collection, good communications, and the use of computers allow these schemes for disease prediction to make accurate, dependable and timely recommendations for disease control measures.

## Plant disease simulators

The methods of disease prediction described in the last section are essentially empirical, although elements of understanding and mechanism are sometimes present. The simulators of plant disease attempt to understand the system thoroughly, in terms of its parts and how they fit together (Waggoner, 1974). In *Figures 9.1* and *9.2* are shown the main features of the progress of a fungal disease epidemic. The disease microcycle of *Figure 9.1* is defined at the detailed level, largely that of microbiology, and using laboratory experiments. It is the activity of this disease microcycle, modified or stimulated by weather and host-plant status, that gives rise to the epidemic shown in *Figure 9.2*.

These are large models, and it is not possible to give a complete description of one of them here, any more than it was for the large crop models of Chapter 8. The best accounts are in some research reports, and the serious reader cannot do better than study EPIMAY, which simulates southern corn leaf blight (Waggoner, Horsfall and Lukens, 1972). However, the procedure is much the same as in the dynamic models discussed elsewhere in this book, and a brief outline is now given.

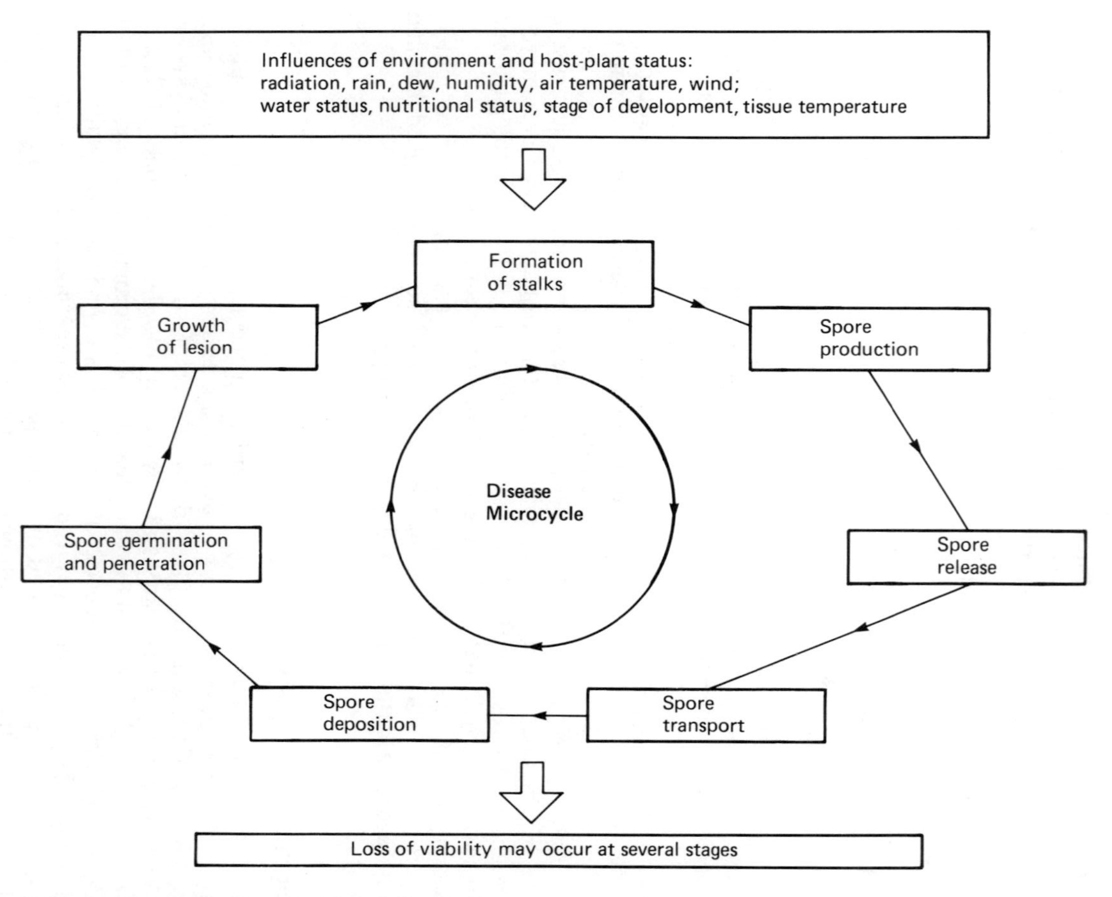


*Figure 9.1* Microcycle of a typical fungal disease

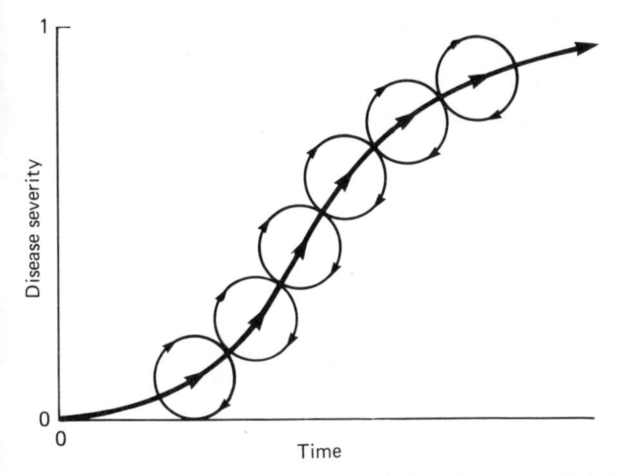


*Figure 9.2* Epidemic of a fungal disease in the field. The small circles represent turns of the disease microcycle shown in *Figure 9.1*

1. Choose state variables—quantities such as
   number of lesions, possibly with an age–size distribution;
   number of spores on stalks;
   number of spores on leaves;
   number of stalks, possibly with an age–size distribution.
2. Definite rates of processes that connect these state variables, and particularly how these depend on environmental and host-plant variables. Much of this is based on microbiological experimentation, and may be in terms of analytic functions or tables of numbers.
3. Specify initial conditions.
4. Read weather reports and update the state variables.

Validation of plant disease simulators presents similar problems to those encountered when validating most other large models—many parameters and often few data. The disease simulators are primarily research tools, aimed at increasing and organizing our understanding of the relevant processes. Practical disease prediction is still carried out by the empirical models described earlier.

## Pests, predators and parasites

Yield losses arising from pest damage are widespread and significant, and much effort has been expended in understanding the factors involved, and in controlling the pest populations so that the economic injury level of the pest is not exceeded. Classical biological control is concerned with the regulation of a pest population (pathogen, mite, insect, weed or mammal) by means of a natural enemy or enemies (pathogen, parasite or predator) (Caltagirone, 1981). More recently, the concept of integrated control, where both chemical and biological control measures are considered together, has become increasingly relevant, especially with the advent of powerful and discriminating chemicals (Stern *et al.*, 1959). Clearly, this is a most fruitful area for the mathematical modeller, where much has been done (Maynard Smith, 1974; Streifer, 1974; Conway, 1977), and much remains to be done. There is as yet no grand unified theory which can be applied to these aspects of agricultural practice. Practical problems are largely dealt with empirically, without recourse to mathematics or mathematical models.

In this section, a few simple models are first described, to illustrate some principles which are relevant to agriculture, and are not expounded elsewhere in this book. The ways in which the simple theory may be articulated and applied is outlined.

### Predator–prey model without age structure

Lotka (1925) and Volterra (1926) were the first to consider this topic, and there are many variations on the ways in which the dynamic equations for a simple predator–prey system can be formulated. In a given habitat, $y$ denotes the number of prey, and $x$ the number of predators. Following Volterra (1926):

$$\frac{dy}{dt} = \mu y(1 - y/y_m) - kxy \tag{9.7a}$$

$$\frac{dx}{dt} = k'xy - mx \tag{9.7b}$$

where $\mu$ [day$^{-1}$], $y_m$ [prey], $k$ [predator$^{-1}$ day$^{-1}$], $k'$ [prey$^{-1}$ day$^{-1}$] and $m$ [day$^{-1}$] are parameters. These deterministic equations contain many assumptions:

1. Large numbers are involved; the population of a species can be represented by a single variable; age and sex are ignored.
2. Effects are instantaneous, and there are no delays between (for example) egg laying and maturity, or food ingestion and conversion to a new predator.
3. In the absence of predators ($x = 0$), the prey ($y$) increase according to a logistic equation (p. 80) with an initial proportional growth rate ($(1/y)dy/dt$ at $y = 0$) of $\mu$, and a maximum prey number for the habitat of $y_m$.
4. The rate at which prey are eaten is proportional to the product of the prey and the predator populations with constant $k$.
5. $k'$ gives the rate of production of new predators, so that $k'/k$ is the number of predators produced per prey consumed.
6. A mortality rate $m$ is ascribed to the predators, for whom the prey are the only food source.

Steady-state solutions, if they exist, are obtained by putting $dy/dt$ and $dx/dt$ in equations (9.7) equal to zero:

$$\mu(1 - y/y_m) - kx = 0 \quad \text{and} \quad k'y - m = 0 \tag{9.8}$$

which gives

$$y = \frac{m}{k'} \quad \text{and} \quad x = \frac{\mu}{k}\left(1 - \frac{m}{k'y_m}\right) \tag{9.9}$$

Positive solutions are obtained only when

$$k'y_m > m \tag{9.10}$$

which states that the maximum prey population has to be sufficient to overcome the predator mortality rate. The steady-state solution S is illustrated in *Figure 9.3*, where the two straight lines of equations (9.8) are drawn. Above the line $y = m/k'$,

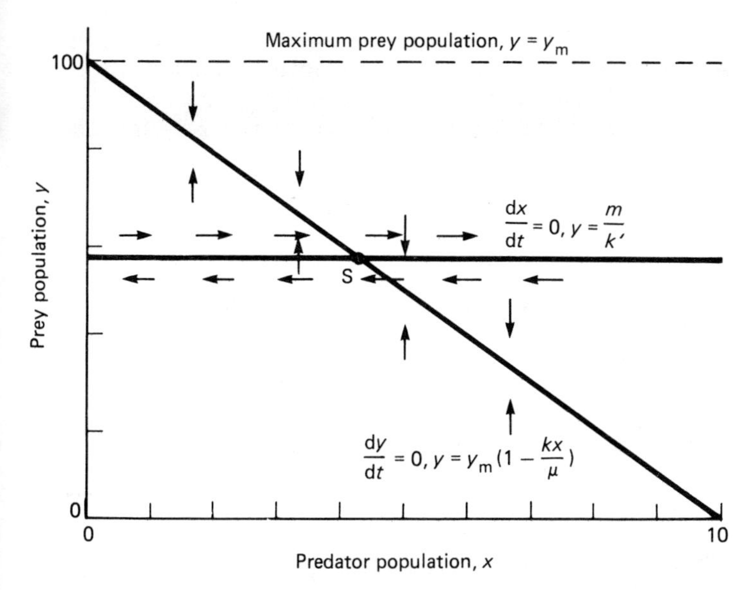


*Figure 9.3* Graphical representation of the steady-state solution, S, to Volterra's predator–prey model (equations (9.7)). The parameters of the model are $\mu = 0.5$ day$^{-1}$, $y_m = 100$ prey, $k = 0.05$ day$^{-1}$ predator$^{-1}$, $k' = 0.004$ day$^{-1}$ prey$^{-1}$ and $m = 0.22$ day$^{-1}$. See text for explanation

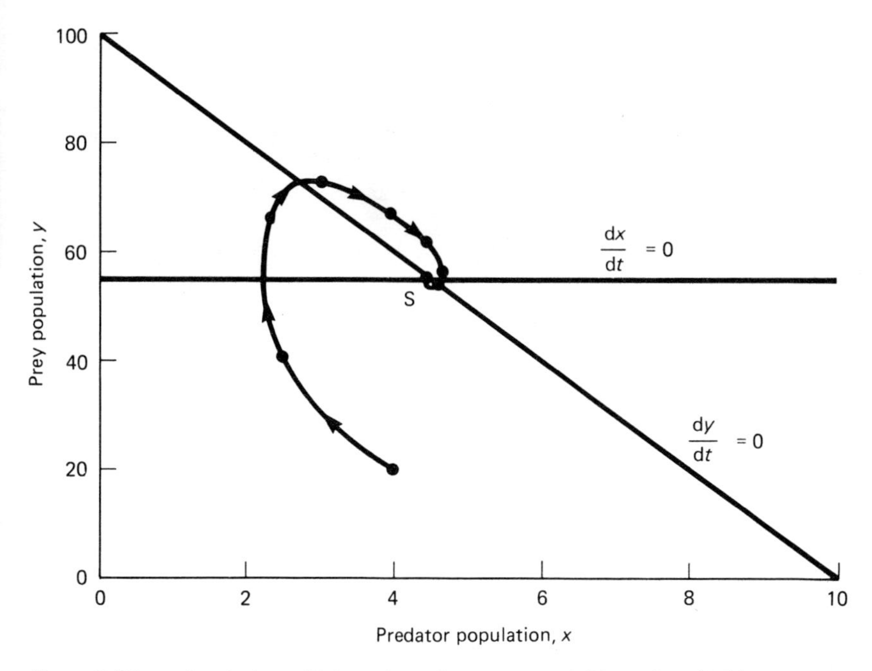


*Figure 9.4* Dynamic solution to Volterra's predator–prey model (equations (9.7)). Parameters are as in *Figure 9.3*. Integration is by Euler's method with a time step $\Delta t = 1$ day. The points (●) on the curve mark intervals of 4 days

$dx/dt$ is positive, and below this line $dx/dt$ is negative; this is denoted by the horizontal arrows. Above the inclined line, $dy/dt$ is negative, and below this line it is positive; this is similarly indicated with vertical arrows along the inclined line. It can be seen that the solution S will be approached along a clockwise spiral. This is shown in *Figure 9.4* where equations (9.7) are integrated numerically to give the dynamic behaviour of the system.

## MODIFICATIONS TO VOLTERRA'S EQUATIONS

There are many ways in which equations (9.7) can be modified to give greater realism or represent special situations.

*1. Autonomous growth of predators*

If the predators are able to use some other food supply as well as eating prey, then equation (9.7b) may be altered to

$$\frac{dx}{dt} = k'xy + \mu_x x(1 - x/x_m) \tag{9.11}$$

where it is assumed that the predators $x$ can grow according to a logistic equation with parameters $\mu_x$ and $x_m$.

*2. Limited predator intake of prey*

In equation (9.7b), the intake of prey per predator increases linearly with prey number $y$ without any limit. This is hardly realistic, and may be corrected by replacing equation (9.7b) by

$$\frac{dx}{dt} = k'x\left(\frac{y}{1 + ay}\right) - mx \tag{9.12}$$

where $a$ is a constant which causes the term to approach a maximum value of $k'x/a$ at high prey numbers.

*3. Cover for prey*

In some habitats, there may be cover for a limited number of prey where they are secure against predation. For this case, equations (9.7) may be written

$$\frac{dy}{dt} = \mu y(1 - y/y_m) - kx(y - y_o)H(y - y_o) \tag{9.13a}$$

$$\frac{dx}{dt} = k'x(y - y_o)H(y - y_o) - mx \tag{9.13b}$$

$y_o$ is the maximum number of prey that can use the cover. $H$ is the unit step function, which prevents the predator–prey interaction term going negative, defined by

$$H(y - y_o) = 1 \quad \text{for} \quad y - y_o \geqslant 0 \tag{9.14}$$
$$H(y - y_o) = 0 \quad \text{for} \quad y - y_o < 0$$

Steady-state solutions and dynamic behaviour for these modified predator–prey equations can be examined in the same way that equations (9.7) led to *Figures 9.3* and *9.4*.

**Effects of time delays**

The general effect of a time delay in a system is to promote instability. Familiar examples of this include the correcting of a skid in a car, which so easily makes the skid worse; our attempts to stabilize economic systems (at the micro or the macro level) can sometimes be counter-productive; and under a shower, one can easily be alternately frozen and scalded in one's efforts to adjust the water temperature.

Delayed responses in biology may arise in many ways. Two of these are

*1. Time required for developmental events*

For the growth of a biological population, $y$, one sometimes assumes the exponential growth equation

$$\frac{\mathrm{d}y}{\mathrm{d}t} = \mu\, y(t) \tag{9.15}$$

where $\mu$ is the proportional rate of growth, and the rate of growth in $y$ is related to the value of $y$ at time $t$. However, if it takes time $\tau$ for eggs to become adults (for instance), it is more correct to write

$$\frac{\mathrm{d}y}{\mathrm{d}t}(t) = \mu\, y(t - \tau) \tag{9.16}$$

where $y(t - \tau)$ is the (adult) population at time $t - \tau$.

*2. Delayed effects of environment*

The growth rate parameter $\mu$ of equation (9.15) may depend upon environment variables $E(t)$, which might include substrate supply. There may be delays in their effects upon the organism, in which case it is more accurate to write

$$\frac{\mathrm{d}y}{\mathrm{d}t}(t) = \mu[E(t - \tau)]y(t) \tag{9.17}$$

where $\mu$ is calculated from the values of the environmental variables $E$ time $\tau$ ago. For instance, a reduced food supply may have little effect on the rate of population growth for some considerable time.

AN EXAMPLE OF A SYSTEM WITH DELAYS

Let $Y$ denote an insect population which is nourished with a constant food supply $f$; the adults have a maintenance requirement $mY$, where $m$ is a constant; the surplus food $f - mY$ is converted into eggs with efficiency $\eta$, so that the rate of egg production is $\eta(f - mY)$; the eggs hatch into adults after time $\tau$; and finally, the adults suffer from a constant death rate $h$. Putting these assumptions together, the differential equation for $Y(t)$ at time $t$ is

$$\frac{\mathrm{d}Y}{\mathrm{d}t}(t) = \eta[f - m\, Y(t - \tau)] - h\, Y(t) \tag{9.18}$$

The steady-state solution occurs at $Y_s$ where

$$Y(t) = Y(t - \tau) = Y_s$$

and substituting into equation (9.18) with $\mathrm{d}Y/\mathrm{d}t = 0$, therefore

$$Y_s = \frac{\eta f}{\eta m + h} \tag{9.19}$$

In order to examine the stability of the system, small deviations $y$ from the steady state are considered. $y$ is defined by

$$Y = Y_s + y \tag{9.20}$$

and substituting equation (9.20) into equation (9.18), and using equation (9.19)

$$\frac{\mathrm{d}y}{\mathrm{d}t}(t) = -\eta m\, y(t-\tau) - h\, y(t) \tag{9.21}$$

This is written more simply as

$$\frac{\mathrm{d}y}{\mathrm{d}t}(t) = -a\, y(t-\tau) - h\, y(t) \tag{9.22}$$

With $h = 0$, it can be shown that this equation is stable if ($a \geqslant 0$)

$$a < \pi/2\tau \tag{9.23}$$

and is otherwise unstable (MacDonald, 1978). In *Figure 9.5*, stable and unstable solutions are shown. Increasing $h$ gives increasing stability for a given value of the parameter $a$.

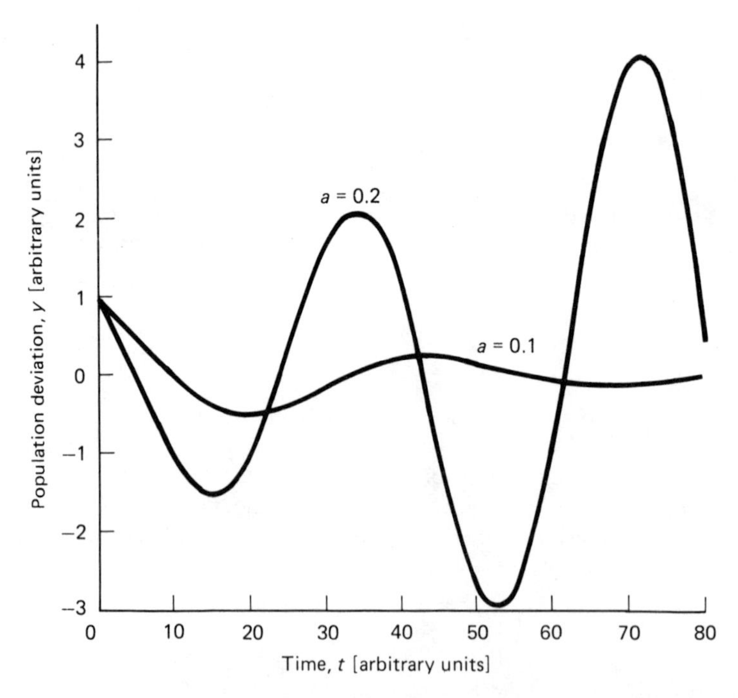


*Figure 9.5* Simple time delay and stability. Equation (9.22) is solved for $\tau = 10$, $h = 0$, $y(t \leqslant 0) = 1$ and two values of $a$, either side of $\pi/2\tau = 0.157$ (9.23). Integration is by Euler's method with $\Delta t = 1$

A SIMPLE EPIDEMIOLOGICAL MODEL

Another system that gives rise naturally to delays is an epidemic with an incubation period $\tau$. Let $Y_m$ be that total population and $Y$ the infected population, so that $(Y_m - Y)$ is the susceptible population. Due to the delay caused by the incubation period, the current probability of infection depends upon the infected population at time $t - \tau$. The relevant equation is

$$\frac{dY}{dt}(t) = \mu\, Y(t - \tau)\left(1 - \frac{Y(t)}{Y_m}\right) - c\, Y(t) \tag{9.24}$$

where $\mu$ is the probability per unit time of a susceptible organism becoming infected, and $c$ is a natural recovery or cure rate. The steady-state solution is

$$Y_s = Y_m\left(1 - \frac{c}{\mu}\right) \tag{9.25}$$

and writing

$$Y = Y_s + y \tag{9.26}$$

therefore

$$\frac{dy}{dt} = -\mu\, y(t) + c\, y(t - \tau) \tag{9.27}$$

## Models with age structure: the Leslie matrix approach

In population dynamics, the simplest modelling approach is in terms of total population number, $N$. However, total population models are unrealistically simple, and have been found to be generally too inaccurate. An age-specific distribution function $n(t, a)$ can be defined such that

$$n(t, a)\mathrm{d}a \tag{9.28}$$

is the number of organisms at time $t$ with ages lying between $a$ and $a + \mathrm{d}a$. The total population is obtained from

$$N(t) = \int_0^\infty n(t, a)\, \mathrm{d}a \tag{9.29}$$

Birth rate and death rate functions $B(t, a)$ and $D(t, a)$ can be defined as the rates of these processes at time $t$ for organisms of age $a$, so that

$$\text{birth rate} = \int_0^\infty B(t, a)\, n(t, a)\, \mathrm{d}a \tag{9.30}$$

$$\text{death rate} = \int_0^\infty D(t, a)\, n(t, a)\, \mathrm{d}a \tag{9.31}$$

Birth occurs into age $a = 0$ organisms, so that

$$n(t, 0) = \text{birth rate} \tag{9.32}$$

From a group of organisms of age $a$ at time $t$, organisms are lost by death and also by ageing. After time interval $\Delta t$, in the absence of death, organisms $n(t, a)\,\mathrm{d}a$ move into $n(t + \Delta t, a + \Delta a)\mathrm{d}a$, and it can be shown (Exercise 9.6) that this leads to

$$\frac{\partial n}{\partial t} + \frac{\partial n}{\partial a} = 0 \qquad (9.33a)$$

and in the presence of death

$$\frac{\partial n}{\partial t} + \frac{\partial n}{\partial a} = -D(t, a)\, n(t, a) \qquad (9.33b)$$

Even greater realism may be achieved by using an age–size-specific distribution function $n(t, a, w)$ where

$$n(t, a, w)\mathrm{d}a\,\mathrm{d}w \qquad (9.34)$$

is the number of organisms with ages lying between $a$ and $a + \mathrm{d}a$ and sizes between $w$ and $w + \mathrm{d}w$. Total population number $N$ is now obtained by a double integral

$$N = \int_0^\infty \int_0^\infty n(t, a, w)\, \mathrm{d}a\,\mathrm{d}w \qquad (9.35)$$

and equations analogous to equations (9.30)–(9.33) can be constructed. The approach can rapidly lead to very complicated equations (Streifer, 1974), and in solving the partial differential equations numerically, it is necessary to take finite intervals for $t$, $a$ and $w$ if it is included. The equations are then approximated by matrix equations which are identical to the matrix approach of Lewis (1942) and Leslie (1945).

## LESLIE MATRIX METHOD

In *Figure 9.6*, a continuous age-specific distribution is divided into four cohorts. Using the notation defined in *Figure 9.6*, a transition matrix can be written down:

$$\begin{pmatrix} n_1 \\ n_2 \\ n_3 \\ n_4 \end{pmatrix} (t = i + 1) = \begin{pmatrix} B_1 & B_2 & B_3 & B_4 \\ S_{12} & 0 & 0 & 0 \\ 0 & S_{23} & 0 & 0 \\ 0 & 0 & S_{34} & 0 \end{pmatrix} \begin{pmatrix} n_1 \\ n_2 \\ n_3 \\ n_4 \end{pmatrix} (t = i) \qquad (9.36)$$

The $B$'s are the birth rates of the four cohorts, and the $S$'s are survival rates—$S_{12}$ is the fraction of the first cohort which, after one time unit, survive and constitute the second cohort. Thus, equation (9.36) denotes the equations

$$\begin{aligned} n_1(i + 1) &= B_1 n_1(i) + B_2 n_2(i) + B_3 n_3(i) + B_4 n_4(i) \\ n_2(i + 1) &= S_{12} n_1(i) \\ n_3(i + 1) &= S_{23} n_2(i) \\ n_4(i + 1) &= S_{34} n_3(i) \end{aligned} \qquad (9.37)$$

The matrix approach has ben widely used in practical studies, and is discussed extensively by Williamson (1972) and many others.

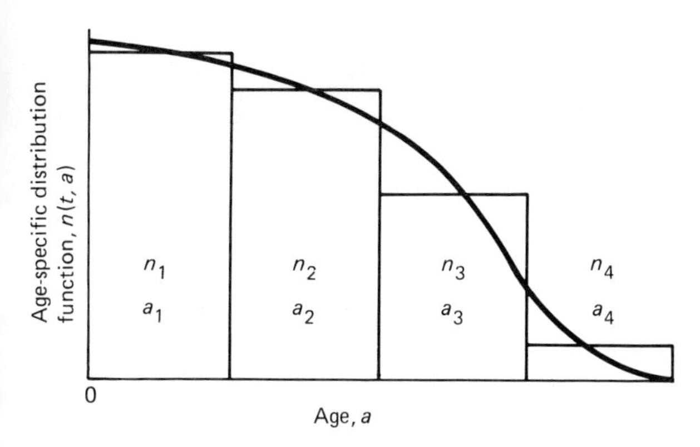


*Figure 9.6* The age-specific population distribution function shown by the continuous line is approximated by the four quartiles, $n_1$, $n_2$, $n_3$ and $n_4$

## An example of a biological control model

A much simplified scheme is shown in *Figure 9.7* to indicate a possible form for a biological control model. This is partially based on a study of the population dynamics of the aphid *Sitobion avenae* in winter wheat, and control of the aphid population by predation by hoverfly larvae (Rabbinge, Ankersmit and Pak, 1979).

The aphid larvae have four instars, $L_1$ to $L_4$. The developmental times of $L_1$, $L_2$ and $L_3$ are similar; rather longer is required in $L_4$. There are two forms of instar $L_4$, wingless and winged, and the relative proportions of each depends on factors such as temperature, crowding and plant status. The developmental rates, which determine the time spent in each instar, are also highly dependent on temperature (pp. 151–155).

Adult aphids are consumed by the larvae of the hoverfly, and it may be assumed that the mortality of hoverfly larvae is greatly reduced by an adequate supply of aphid prey. The rate of predation, for example, may be modelled using one of the relations described earlier in equations (9.7), (9.12) and (9.13). Many of the rates in *Figure 9.7*, mortality, reproduction, development and emigration, depend on weather as well as plant status.

It can be seen that models of these problems can quickly become highly complex, and they must be very carefully formulated if they are to contribute to understanding and problem solution in this area.

# Exercises

### Exercise 9.1

Write an outline computer program based on (9.6) and following sentence to give blight forecasts. The input data provided are daily values of mean temperature $\overline{T}$ [°C], minimum temperature $T_{min}$ [°C] and rainfall $h$ [mm].

### Exercise 9.2

Write a computer program to examine the dynamic behaviour of Volterra's equations (9.7), and verify *Figure 9.4*. A CSSL (Continuous System Simulation Language) computing package such as CSMP (p. 33) is ideal for this purpose.

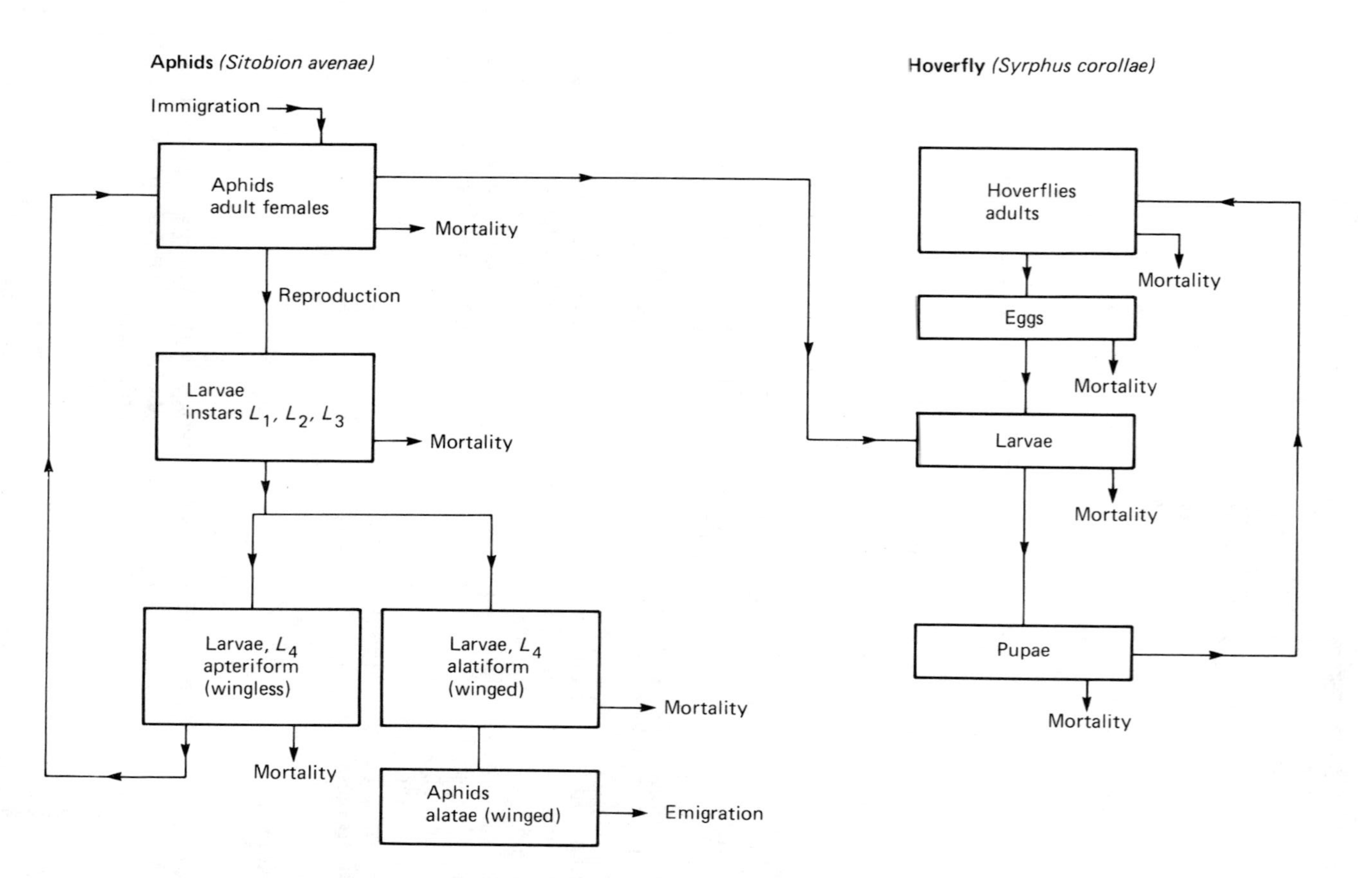


*Figure 9.7* Model for biological control of wheat aphid by predation by hoverfly larvae

**Exercise 9.3**

Modify your program of the last exercise to take account of the alterations to Volterra's equations (9.7) outlined in equations (9.11), (9.12) and (9.13). Check that your results are biologically plausible.

**Exercise 9.4**

Write a computer program to examine the behaviour of equation (9.22), which describes a system with a simple time delay; use your results to verify *Figure 9.5.*

**Exercise 9.5**

Use your program for the last exercise (equation (9.22)) to examine the epidemic model of equation (9.27).

**Exercise 9.6**

Satisfy yourself, graphically or otherwise, that equations (9.33) are valid.

Chapter 10

# Animal processes

## Introduction

In this chapter, we examine models of certain physiological processes in farm livestock, with particular reference to ruminants. Processes directly concerned with meat, milk and egg production, such as the growth of muscle tissue, milk synthesis in the udder, and the utilization of essential amino acids by the laying hen, are dealt with in a separate chapter on animal products (Chapter 11), as these topics have received considerable attention in their own right. The models considered in this chapter are grouped under two section headings—digestion and metabolism.

Much of the modelling work published in this area is empirical rather than mechanistic in nature, and is not easily summarized. It is worth noting the contrast between the state-of-the-art in animal process models and that in plant process models (Chapter 7); the latter area is comparatively highly developed with many more-mechanistic models. There is a pressing need for mechanistic models of animal processes, and it is possible that many of the important problems in animal production are not soluble by traditional empirical methods, and require more mechanistic approaches for their solution. A notable exception to the empirical approach is the excellent work of Baldwin and colleagues at the University of California, Davis, whose models are highly mechanistic and are pitched at a fairly detailed level of biochemistry. Their methodological approach is based on dynamic modelling of the kind described in Chapter 2. In this chapter, attention is focused on two dynamic models (one on ruminant digestion, the other on ruminant metabolism) drawn from our own work, which, in our view, illustrate the suitability of the dynamic modelling approach to the quantification of certain digestive and metabolic processes. The major differences between the two models presented here and those of Baldwin and his colleagues are differences of scale. Their models are large—they include many differential equations, variables and parameters, and consequently, they do not lend themselves to detailed exposition in a book of this nature.

## Digestion

The modelling of digestive processes in farm animals has centred on the rumen. Simple dynamic models of fibre digestion in the rumen have been developed by

Waldo, Smith and Cox (1972) and Mertens (1977), and of protein degradation by Ørskov and McDonald (1979) and McDonald (1981) (*see* Exercises 10.1 and 10.2). Significant attempts to model complete rumen function have been made by Black and colleagues in Australia (Black *et al.*, 1981; Beever, Black and Faichney, 1981; Faichney, Beever and Black, 1981) and by Baldwin and colleagues (Baldwin, Lucas and Cabrera, 1970; Reichl and Baldwin, 1975; Ulyatt, Baldwin and Koong, 1976; Baldwin, Koong and Ulyatt, 1977). Both are large models. The Black model is based on a number of empirically-derived relationships, whilst the Baldwin model is a much more mechanistic attempt. Here, a simpler rumen model, due to France, Thornley and Beever (1982), is described in some detail to illustrate the principles of dynamic modelling.

## A rumen model

A basic function of the rumen is to ferment the dietary carbohydrate and protein ingested by the ruminant. Hungate (1966) suggests that the mechanics of rumen processes are comparable to those in a well-stirred vat in which fresh feed and saliva mix with the fermenting mass, and fluid and feed residues leave in amounts similar to those entering. This analogy is adopted here, and thus the rumen is considered as a continually-mixing fermentation chamber, supplied with dietary and salivary inputs. Given those inputs, the objective of the model is to predict realistic outflows from the rumen, and to relate these to the rumen processes assumed in the model.

### INPUTS TO THE RUMEN

The dietary and salivary inputs, of which account is taken in the model, are shown in *Figures 10.1* and *10.2*. Carbohydrates are divided into four groups: non-rumen-degradable $\beta$-hexose, rumen-degradable $\beta$-hexose, $\alpha$-hexose, and water-soluble carbohydrates. The $\alpha$-hexose group contains all $\alpha$-linked polymers and pectin; the $\beta$-hexose group comprises cellulose, hemicellulose, and lignin; and the water-soluble carbohydrates consist of free sugars, fructosans, and short-chained non-glucose polymers. The nitrogenous material is divided into three groups: non-protein nitrogen, rumen-degradable protein, and non-rumen-degradable protein. The nucleic acids, if significant, are regarded as a small fraction of the rumen-degradable protein category. The non-protein nitrogen category comprises amino acids, ammonia, urea, amides and amines.

### STRUCTURE OF THE MODEL

This is shown schematically in *Figure 10.3*; the main symbols used are listed in *Table 10.1*. All the eight input categories of *Figures 10.1* and *10.2* are represented by compartments in *Figure 10.3*. The additional and ninth compartment of *Figure 10.3* is that for the microbial population of the rumen, which is considered in terms of total microbial mass with no distinction being made between bacteria and protozoa. Owing to the activities of the microbial population, it is possible for the quantities of nutrients flowing from the rumen to be quite different from those ingested. Microbial growth occurs, provided sufficient carbon and nitrogen substrates are available, and these substrates are incorporated directly into microbial protoplasm. If insufficient substrate is available for the maintenance of the

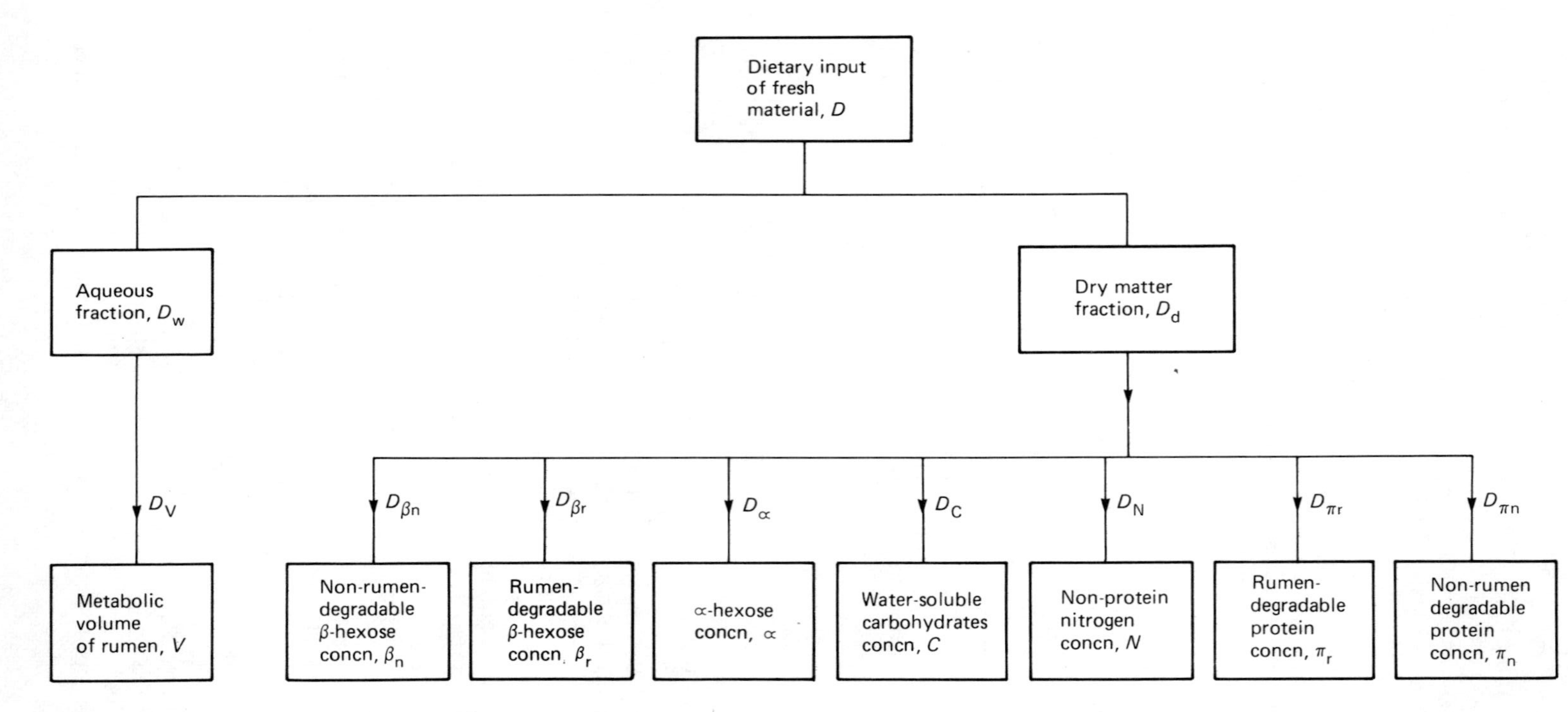


*Figure 10.1* Dietary inputs to the rumen. The components of the total dietary input $D$ are shown by $D_i$, where $i$ denotes the particular component. The eight boxes on the bottom row indicate compartments in the rumen of the animal

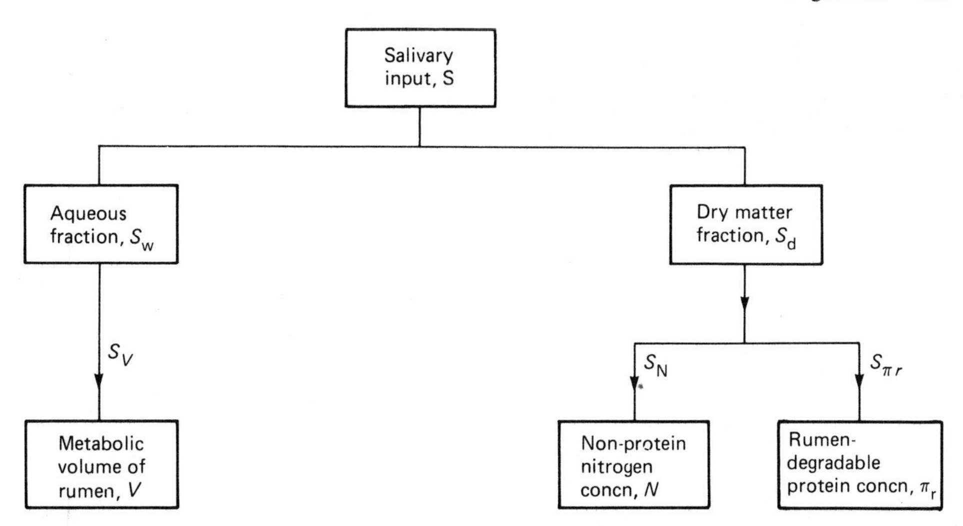


*Figure 10.2* Salivary inputs to the rumen. The components of the total salivary input $S$ are shown by $S_i$, where $i$ denotes the particular component. The three boxes on the bottom row indicate compartments in the rumen of the animal

microbial population, a portion of the microbial mass is catabolized and the substrate released can be utilized by the remaining microbes. It is assumed that the chemical composition of the rumen micro-organisms is constant.

## STATE VARIABLES, RUMEN VOLUME AND DILUTION RATE

Each of the nine compartments of *Figure 10.3* is described by a variable, $Y$ say, and for each of these state variables a differential equation of type $\mathrm{d}Y/\mathrm{d}t = \ldots$ is needed to define the static and dynamic properties of the system.

Let $V$ [m$^3$] be the metabolic volume of the rumen; it is envisaged that this is the effective volume within which the various reactions take place, and this is likely to be closely related to the aqueous volume of the rumen contents. If $v$ [m$^3$ day$^{-1}$] is the rate of outflow of digesta from the rumen to the omasum, then from *Figure 10.3*,

$$\frac{\mathrm{d}V}{\mathrm{d}t} = D_V + S_V - v \tag{10.1}$$

If $Y$ is a concentration, with units of kg m$^{-3}$, then the quantity of $Y$ in the rumen is $VY$ and the quantity of $Y$ leaving the rumen per unit time is $vY$. Thus, in the absence of any other factors (such as inputs or chemical reactions) affecting $Y$,

$$\frac{\mathrm{d}(VY)}{\mathrm{d}t} = -vY \tag{10.2}$$

On differentiating $VY$ with respect to time and dividing by $V$, equation (10.2) becomes

$$\frac{\mathrm{d}Y}{\mathrm{d}t} + \frac{Y}{V}\frac{\mathrm{d}V}{\mathrm{d}t} = \frac{-vY}{V} = -wY \tag{10.3}$$

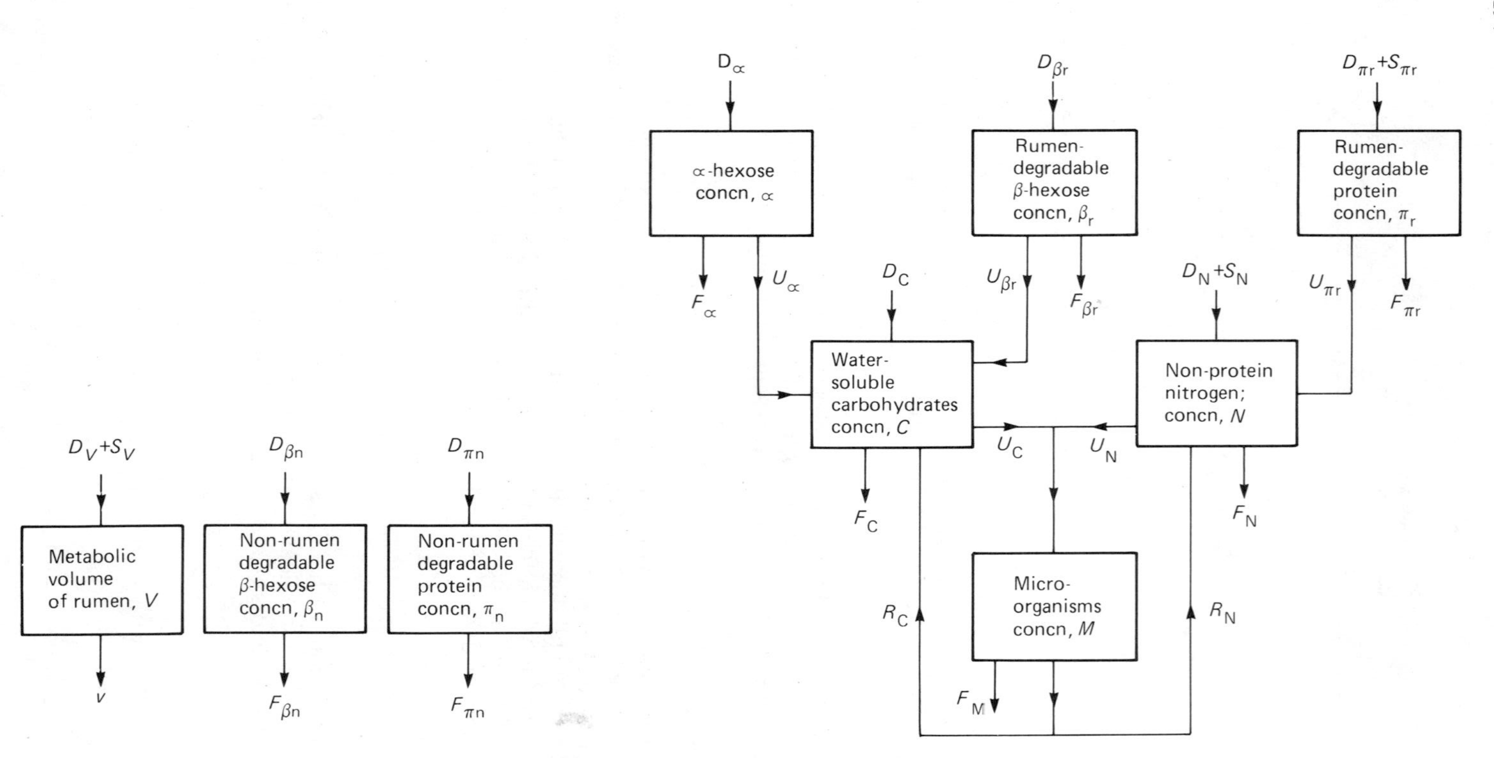


*Figure 10.3* Rumen model. The rumen is specified by the nine compartments (or state variables) shown. The lines and arrows indicate fluxes of material. The dietary and salivary inputs are denoted by $D_i$ (*Figure 10.1*) and $S_i$ (*Figure 10.2*). $F_i$ is the flow of material $i$ out of the rumen into the omasum. $U_i$ is the utilization rate of material $i$. $R_C$ and $R_N$ denote the rate of recycling of C and N from the breakdown of micro-organisms, M

**TABLE 10.1. Principal symbols used in the rumen model**

| | | |
|---|---|---|
| *Independent variable* | | |
| $t$ | Time | day |
| *State variables* | | |
| $C$ | Rumen concentration of water-soluble carbohydrates | kg $CH_2O$ $m^{-3}$ |
| $M$ | Rumen concentration of microbial dry matter | kg $m^{-3}$ |
| $N$ | Rumen concentration of non-protein nitrogen (NPN) | kg NPN $m^{-3}$ |
| $V$ | Metabolic volume of the rumen | $m^3$ |
| $\alpha$ | Rumen concentration of $\alpha$-hexose | kg $CH_2O$ $m^{-3}$ |
| $\beta_n$ | Rumen concentration of non-rumen-degradable $\beta$-hexose | kg $CH_2O$ $m^{-3}$ |
| $\beta_r$ | Rumen concentration of rumen-degradable $\beta$-hexose | kg $CH_2O$ $m^{-3}$ |
| $\pi_n$ | Rumen concentration of non-rumen-degradable protein | kg NPN units $m^{-3}$ |
| $\pi_r$ | Rumen concentration of rumen-degradable protein | kg NPN units $m^{-3}$ |
| *Driving variables* | | |
| $D_i$ | Dietary input of $i$ | kg or $m^3$ of $i$ per day |
| $S_i$ | Salivary input of $i$ | kg or $m^3$ of $i$ per day |
| *Parameters* | | |
| $f_C, f_N$ | Microbial composition parameters | kg $CH_2O$ (NPN units) per kg microbial dry matter |
| $k_C, k_N, k_{CN}$ | Parameters of the microbial growth equation (10.22) | |
| $k_{\alpha M}, k_{\beta r M}, k_{\pi r M}$ | Rate constants for the microbial degradation of $\alpha$-hexose, $\beta$-hexose, and protein | $m^3$ per kg microbial dry matter per day |
| $k_\mu$ | Constant determining dependence of microbial catabolism on microbial growth rate | days |
| $Y_M$ | Conversion efficiency of substrate carbohydrate into microbial carbohydrate | kg microbial $CH_2O$ per kg $CH_2O$ |
| $\lambda_m$ | Maximum specific rate of microbial catabolism | per day |
| $\mu_m$ | Maximum microbial specific growth rate | per day |

where the washout or dilution rate $w$ [per day] is defined by

$$w = \frac{v}{V} \tag{10.4}$$

$\alpha$-HEXOSE

For any substrate, conservation of matter gives the equation:

$$\text{rate of change of substrate pool} = \text{inflow} - \text{outflow} + \text{synthesis} - \text{utilization} \tag{10.5}$$

For $\alpha$-hexose with concentration $\alpha$ [kg m$^{-3}$], the rumen pool is $V\alpha$, inflow is the dietary input $D_\alpha$ (*Figures 10.1* and *10.3*), outflow is by washout to the omasum $v\alpha$, there is no synthesis of $\alpha$-hexose, and the utilization rate $U_\alpha$ is the breakdown of $\alpha$-hexose by microbial enzymes to simple sugars. Thus

$$\frac{\mathrm{d}(V\alpha)}{\mathrm{d}t} = D_\alpha - v\alpha - U_\alpha \tag{10.6}$$

On differentiating $V\alpha$ with respect to time, using equation (10.1), dividing by $V$ and rearranging, equation (10.6) becomes

$$\frac{\mathrm{d}\alpha}{\mathrm{d}t} = \frac{1}{V}(D_\alpha - U_\alpha - \alpha D_V - \alpha S_V) \tag{10.7}$$

The rate of utilization of $\alpha$-hexose, $U_\alpha$, must depend upon the concentration of substrate $\alpha$, the concentration of the microbial population, $M$ [kg m$^{-3}$], and the metabolic volume of the rumen, $V$. It is assumed that

$$U_\alpha = Vk_{\alpha \mathrm{M}}\,\alpha M \tag{10.8}$$

where $k_{\alpha \mathrm{M}}$ [m$^3$ (kg $M$)$^{-1}$ day$^{-1}$] is a constant of proportionality. Substituting equation (10.8) into equation (10.7) gives

$$\frac{\mathrm{d}\alpha}{\mathrm{d}t} = \frac{1}{V}(D_\alpha - \alpha D_V - \alpha S_V) - k_{\alpha \mathrm{M}}\,\alpha M \tag{10.9}$$

$\beta$-HEXOSE

Consider next the rumen-degradable $\beta$-hexose, and let the concentration of degradable $\beta$-hexose in the rumen at time $t$ be $\beta_\mathrm{r}$ [kg m$^{-3}$]. Following the same approach as for $\alpha$-hexose,

$$\frac{\mathrm{d}(V\beta_\mathrm{r})}{\mathrm{d}t} = D_{\beta\mathrm{r}} - v\beta_\mathrm{r} - U_{\beta\mathrm{r}} \tag{10.10}$$

where $D_{\beta\mathrm{r}}$ [kg day$^{-1}$] is the rate of ingestion of degradable $\beta$-hexose by the ruminant, the second term $v\beta_\mathrm{r}$ gives the outflow from the rumen, and $U_{\beta\mathrm{r}}$ [kg day$^{-1}$] is the rate of utilization of degradable $\beta$-hexose by the cellulolytic enzymes produced by the micro-organisms $M$. As in equation (10.8), it is assumed that $U_{\beta\mathrm{r}}$ is directly proportional to the product of substrate $\beta_\mathrm{r}$, micro-organisms $M$ and metabolic volume $V$, giving

$$U_{\beta\mathrm{r}} = Vk_{\beta\mathrm{rM}}\,\beta_\mathrm{r} M \tag{10.11}$$

where $k_{\beta rM}$ [$m^3 (kg\,M)^{-1} day^{-1}$] is the constant of proportionality. Substituting equation (10.11) into equation (10.10) and simplifying gives

$$\frac{d\beta_r}{dt} = \frac{1}{V}(D_{\beta r} - \beta_r D_V - \beta_r S_V) - k_{\beta rM} \beta_r M \tag{10.12}$$

For the non-rumen degradable $\beta$-hexose, with concentration $\beta_n$ [$kg\,m^{-3}$], the analogous equation to equation (10.12) is simplified because there is no utilization of this component of $\beta$-hexose in the rumen, and thus in this case, equation (10.12) is replaced by

$$\frac{d\beta_n}{dt} = \frac{1}{V}(D_{\beta n} - \beta_n D_V - \beta_n S_V) \tag{10.13}$$

PROTEIN

The non-rumen-degradable protein of concentration $\pi_n$ obeys a similar equation to the non-rumen-degradable $\beta$-hexose $\beta_n$ in equation (10.13), giving

$$\frac{d\pi_n}{dt} = \frac{1}{V}(D_{\pi n} - \pi_n D_V - \pi_n S_V) \tag{10.14}$$

The rumen-degradable protein of concentration $\pi_r$ is replenished both by inflows from the diet $D_{\pi r}$ and from the saliva $S_{\pi r}$, and it is utilized by the proteolytic enzymes produced by microbial activity at a rate $U_{\pi r}$. If $U_{\pi r}$ is the utilization rate of rumen-degradable protein, then equation (10.5) takes the form

$$\frac{d(V\pi_r)}{dt} = D_{\pi r} + S_{\pi r} - v\pi_r - U_{\pi r} \tag{10.15}$$

For the utilization rate $U_{\pi r}$, the same assumption is made as in equations (10.8) and (10.11), giving

$$U_{\pi r} = V k_{\pi rM} \pi_r M \tag{10.16}$$

where $k_{\pi rM}$ [$m^3 (kg\,M)^{-1} day^{-1}$] is a constant. Substituting equation (10.16) into equation (10.15) and simplifying gives

$$\frac{d\pi_r}{dt} = \frac{1}{V}(D_{\pi r} + S_{\pi r} - \pi_r D_V - \pi_r S_V) - k_{\pi rM} \pi_r M \tag{10.17}$$

WATER-SOLUBLE CARBOHYDRATES

Let the concentration of water-soluble carbohydrates in the rumen be $C$ [$kg\,m^{-3}$]. Inflow to this pool is from four sources: dietary intake $D_C$, degradation of $\alpha$-hexose by amylolytic enzymes, degradation of $\beta$-hexose by cellulolytic enzymes, and water-soluble carbohydrates released by microbial catabolism. Outflow is by washout to the omasum, $F_C$, and utilization is uptake by the microbial population for growth and maintenance.

It is assumed that both $\alpha$-hexose and $\beta$-hexose are converted into water-soluble carbohydrates without any change in mass, so that utilization of the polysaccharides is equal to production of water-soluble carbohydrates without any conversion factor. Equation (10.5) takes the form (*Figure 10.3*)

$$\frac{d(VC)}{dt} = D_C + U_\alpha + U_{\beta r} + R_C - vC - U_C \tag{10.18}$$

where $R_C$ [kg day$^{-1}$] is the rate of release of water-soluble carbohydrates from microbial catabolism, and $U_C$ [kg day$^{-1}$] is the rate of utilization of water-soluble carbohydrates by the microbial mass. Substitution by equations (10.1), (10.8) and (10.11), and simplification gives

$$\frac{dC}{dt} = \frac{1}{V}(D_C + R_C - U_C - CD_V - CS_V) + k_{\alpha M}\,\alpha M + k_{\beta r M}\,\beta_r\,M \tag{10.19}$$

NON-PROTEIN NITROGEN

The concentration of non-protein nitrogen in the rumen is denoted by $N$ [kg m$^{-3}$]. Inflow to this pool is from four sources: dietary intake $D_N$, salivary intake $S_N$, degradation of rumen-degradable protein by proteolytic enzymes $U_{\pi r}$, and non-protein nitrogen released by microbial catabolism $R_N$. Outflow is by washout to the omasum, and there is uptake $U_N$ by the microbes for maintenance and growth. Thus

$$\frac{d(VN)}{dt} = D_N + S_N + U_{\pi r} + R_N - vN - U_N \tag{10.20}$$

It is assumed here that the rumen-degradable protein is converted into non-protein nitrogen without any change in mass, so that no conversion factor is needed. Substitution with equation (10.16) and simplification gives

$$\frac{dN}{dt} = \frac{1}{V}(D_N + S_N + R_N - U_N - ND_V - NS_V) + k_{\pi r M}\,\pi_r\,M \tag{10.21}$$

MICROBIAL GROWTH

The microbial population is considered in terms of the concentration of microbial dry mass $M$ [kg m$^{-3}$]. The growth and maintenance of the microbial mass depends upon the availability of the substrates required for these processes. It is assumed here that growth depends only on the carbon and nitrogen substrates: water-soluble carbohydrates $C$, and non-protein nitrogen $N$. Other substrates, such as lipid and inorganic sulphur, are assumed to be present in non-limiting amounts and are therefore not considered. The specific growth rate $\mu$ [per day] of a system that depends on the availability of two substrates ($C$ and $N$) may often be usefully described by a rectangular hyperboloid, namely

$$\mu = \mu_m \frac{1}{1 + \dfrac{k_C}{C} + \dfrac{k_N}{N} + \dfrac{k_{CN}}{CN}} \tag{10.22}$$

where $\mu_m$ is the maximum (asymptotic) value of $\mu$ as $C$ and $N$ tend to infinity, and $k_C$, $k_N$ and $k_{CN}$ are constants. The growth rate of the microbial population, $g_M$ [kg day$^{-1}$] is then given by

$$g_M = \mu VM \tag{10.23}$$

MICROBIAL CATABOLISM

If insufficient substrates are available for the growth and maintenance of the microbial population, then some of the microbial mass is catabolized for this purpose. Microbial catabolism releases water-soluble carbohydrates and non-protein nitrogen to the respective substrate pools, where they can be utilized by the surviving microbes. Here, the specific rate of microbial catabolism, $\lambda$ [per day], is represented by a variable expression, so that $\lambda$ depends on microbial specific growth rate $\mu$, according to

$$\lambda = \lambda_m \frac{1}{1 + k_\mu \mu} \tag{10.24}$$

where $\lambda_m$ is the maximum rate of $\lambda$ obtained at $\mu = 0$, and $k_\mu$ is a constant. The death rate of the microbial population, $d_M$ [kg day$^{-1}$], is then given by

$$d_M = \lambda V M \tag{10.25}$$

Therefore, the instantaneous rate of change of the microbial mass is given by

$$\frac{d(VM)}{dt} = g_M - d_M - vM \tag{10.26}$$

where the last term is due to washout to the omasum. On substituting for $g_M$ and $d_M$ with equations (10.23) and (10.25), and simplifying, equation (10.26) becomes

$$\frac{dM}{dt} = \mu M - \lambda M - \frac{M}{V}(D_V + S_V) \tag{10.27}$$

MICROBIAL COMPOSITION AND SUBSTRATE UTILIZATION

It is assumed that the chemical composition of the rumen micro-organisms is constant. Let 1 kg of microbial dry mass contain carbon equivalent to $f_C$ kg of water-soluble carbohydrates and nitrogen equivalent to $f_N$ kg of non-protein nitrogen. The rate of release of water-soluble carbohydrates from microbial catabolism, $R_C$ [kg day$^{-1}$], is given by

$$R_C = f_C d_M \tag{10.28}$$

and the rate of release of non-protein nitrogen, $R_N$ [kg day$^{-1}$], by

$$R_N = f_N d_M \tag{10.29}$$

Let $U_C$ [kg day$^{-1}$] be the rate of utilization of water-soluble carbohydrates by the microbial population. For each unit of water-soluble carbohydrates utilized by the microbes, only some fraction of those carbon atoms, $Y_M$ say, will actually appear in the microbial mass. $Y_M$ is a conversion efficiency or yield factor, and it is less than unity because some of the utilized carbon atoms are converted to $CO_2$ and other products, providing energy for microbial biosynthesis. Therefore

$$U_C = \frac{1}{Y_M} f_C g_M \tag{10.30}$$

It is assumed that nitrogen from the non-protein nitrogen pool is utilized without loss, so that the utilization rate, $U_N$ [kg day$^{-1}$], is

$$U_N = f_N g_M \tag{10.31}$$

DIFFERENTIAL EQUATIONS OF THE MODEL

Here, a complete statement of the mathematical model is given in summary form. Of the nine state variables in *Figure 10.3*, the three on the left are given by simple expressions. The metabolic volume of the rumen is described by equation (10.1)

$$\frac{dV}{dt} = D_V + S_V - v \tag{10.32a}$$

and the non-rumen-degradable components of $\beta$-hexose and protein are described by equations (10.13) and (10.14)

$$\frac{d\beta_n}{dt} = \frac{1}{V}(D_{\beta n} - \beta_n D_V - \beta_n S_V) \tag{10.32b}$$

and

$$\frac{d\pi_n}{dt} = \frac{1}{V}(D_{\pi n} - \pi_n D_V - \pi_n S_V) \tag{10.32c}$$

For the remaining six state variables, the equations are more involved. Equation (10.9) gives

$$\frac{d\alpha}{dt} = \frac{1}{V}(D_\alpha - \alpha D_V - \alpha S_V) - k_{\alpha M}\, \alpha M \tag{10.32d}$$

From equations (10.12) and (10.17)

$$\frac{d\beta_r}{dt} = \frac{1}{V}(D_{\beta r} - \beta_r D_V - \beta_r S_V) - k_{\beta r M}\, \beta_r M \tag{10.32e}$$

and

$$\frac{d\pi_r}{dt} = \frac{1}{V}(D_{\pi r} + S_{\pi r} - \pi_r D_V - \pi_r S_V) - k_{\pi r M}\, \pi_r M \tag{10.32f}$$

Equations (10.28) and (10.30) with equations (10.23) and (10.25) are substituted into equation (10.19), to give

$$\frac{dC}{dt} = \frac{1}{V}(D_C - CD_V - CS_V) + \lambda f_C M - \frac{1}{Y_M}\mu f_C M + k_{\alpha M}\, \alpha M + k_{\beta r M}\, \beta_r M \tag{10.32g}$$

Similarly, equations (10.29) and (10.31) with equations (10.23) and (10.25) are substituted into equation (10.21), to give

$$\frac{dN}{dt} = \frac{1}{V}(D_N + S_N - ND_V - NS_V) + \lambda f_N M - \mu f_N M + k_{\pi r M}\, \pi_r M \tag{10.32h}$$

Finally, from equation (10.27)

$$\frac{dM}{dt} = \mu M - \lambda M - \frac{M}{V}(D_V + S_V) \tag{10.32i}$$

To evaluate these equations, equations (10.22) and (10.24) are used to provide values for $\mu$ and $\lambda$ by

$$\mu = \mu_m \frac{1}{1 + \frac{k_C}{C} + \frac{k_N}{N} + \frac{k_{CN}}{CN}} \tag{10.33a}$$

and

$$\lambda = \lambda_m \frac{1}{1 + k_\mu \mu} \tag{10.33b}$$

To solve the problem numerically requires specification of: the initial values at time $t = 0$ of the nine state variables ($V$, $\beta_n$, $\pi_n$, $\alpha$, $\beta_r$, $\pi_r$, $C$, $N$ and $M$); the 11 input functions of *Figure 10.1* ($D_V$, $D_{\beta n}$, $D_{\beta r}$, $D_\alpha$, $D_C$, $D_N$, $D_{\pi r}$, and $D_{\pi n}$) and of *Figure 10.2* ($S_V$, $S_N$, and $S_{\pi r}$); the output function $v$; and finally the 12 parameters ($k_{\alpha M}$, $k_{\beta r M}$, $k_{\pi r M}$, $\mu_m$, $k_N$, $k_C$, $k_{CN}$, $\lambda_m$, $k_\mu$, $f_C$, $f_N$ and $Y_M$).

### NUMERICAL ASSUMPTIONS AND RESULTS

These are described by France, Thornley and Beever (1982). The model was programmed in the modelling language CSMP (p. 33) and Euler's method used to integrate the differential equations (p. 21).

The rumen outflows predicted by the model were compared with experimental values for four different diets, and reasonable semi-quantitative agreement was obtained. Carbon and nitrogen balance sheets were constructed for each diet.

## Metabolism

Much of the modelling work undertaken in this area has been of an empirical, deterministic nature and has been conducted at a fairly general level, rather than at a biochemical level. These empirical models range from simple, static models, such as the energy and protein systems for ruminants adopted in the United Kingdom and the United States (Agricultural Research Council, 1980; National Research Council, 1970, 1971 and 1975), to more complicated dynamic models, such as Graham *et al.* (1976), Forbes (1977a, b), Geisler and Neal (1979), and Newton and Edelsten (1976). The model of Graham *et al.* (1976) predicts daily energy and nitrogen utilization in sheep of any age and physiological condition. Forbes (1977a, b) describes two models of voluntary food intake and energy balance, one for mature sheep and one for lactating dairy cows. Geisler and Neal (1979) give an account of a model for calculating the response of the pregnant ewe to different levels of energy nutrition and, in similar vein, Newton and Edelsten (1976) describe a model for predicting litter size, lamb birthweight, perinatal lamb mortality and ewe weight change during the latter part of pregnancy under different nutritional regimes.

The majority of mechanistic models of metabolism are attributable to Baldwin and his colleagues. Their work is biochemically orientated and an outline of some of it can be found in Smith, Baldwin and Sharp (1980) and Baldwin *et al.* (1981). Other mechanistic models reported in the literature include that of Schulz (1978), who describes a simulation of energy metabolism in the simple-stomached animal,

Keener (1979), who gives a simulation of energy utilization in beef cattle, and Koong, Falter and Lucas (1982), who describe a mathematical model for nitrogen and energy metabolism in cattle.

The remainder of this chapter is given over to an example from our own researches of a model that is biochemically based, but does not represent all known biochemical detail, and is aimed at understanding energy metabolism at the whole animal level (Gill *et al.*, 1984).

## A model for the efficiency of utilization of absorbed energy in ruminants

The objective of this model is to attempt to gain some understanding of why the efficiency of utilization of metabolizable energy for growth and fattening should vary so widely. Much research has been directed at this important and difficult problem.

### OVERALL STRUCTURE AND NOTATION

The scheme assumed is given in *Figure 10.4*; it is a compartmental model with pools and fluxes between the pools; only the principal fluxes are shown in *Figure 10.4*. Two-letter abbreviations are used to define the quantities in the model. These are given in *Table 10.2*. In *Table 10.3*, variables and parameters are given which

**TABLE 10.2. Symbols for abbreviations used in the metabolizable energy utilization model**

| *Substance, process* | *Symbol* | *Substance, process* | *Symbol* |
|---|---|---|---|
| Acetic acid | Ac | Glycogen | Gy |
| Acetyl co-enzyme A | Ca | Lipid | Li |
| Amino acids | Aa | Maintenance | Ma |
| ATP | At | NADPH | Np |
| Butyric acid | Bu | Oxygen | Ox |
| Carbon dioxide | Co | Propionic acid in: | |
| Degradation | Dg | adipose tissue | Pa |
| Fatty acids | Fa | liver | Pl |
| Glucose | Gu | Protein | Pt |
| | | Tri-glycerides | Tg |
| | | Urea | Ur |

describe properties of the model and which are subscripted with the symbols from *Table 10.2*. For example, $C_{Ac}$ is the concentration of acetic acid, $K_{AaGu}$ is a Michaelis–Menten constant for the amino acid to glucose reaction, $R_{FaTg,At}$ is the requirement of the fatty acid to tri-glyceride reaction for ATP. In *Table 10.4*, the principal transactions occurring in the model are listed, with reaction sites, principal and auxiliary substrates, inhibitors, and parameters required to specify the rate equations.

SI units are used throughout, with the exception of the SI units for concentration and amount of substance, which are inconsistent with the rest of the SI system. kg is used for mass, m for length, time is measured in days, amount of substance is measured in kg moles (1 kg mole of $^{12}C$ is 12 kg of $^{12}C$), and concentration in kg mole $m^{-3}$, which is identical to the familiar g mole $litre^{-1}$.

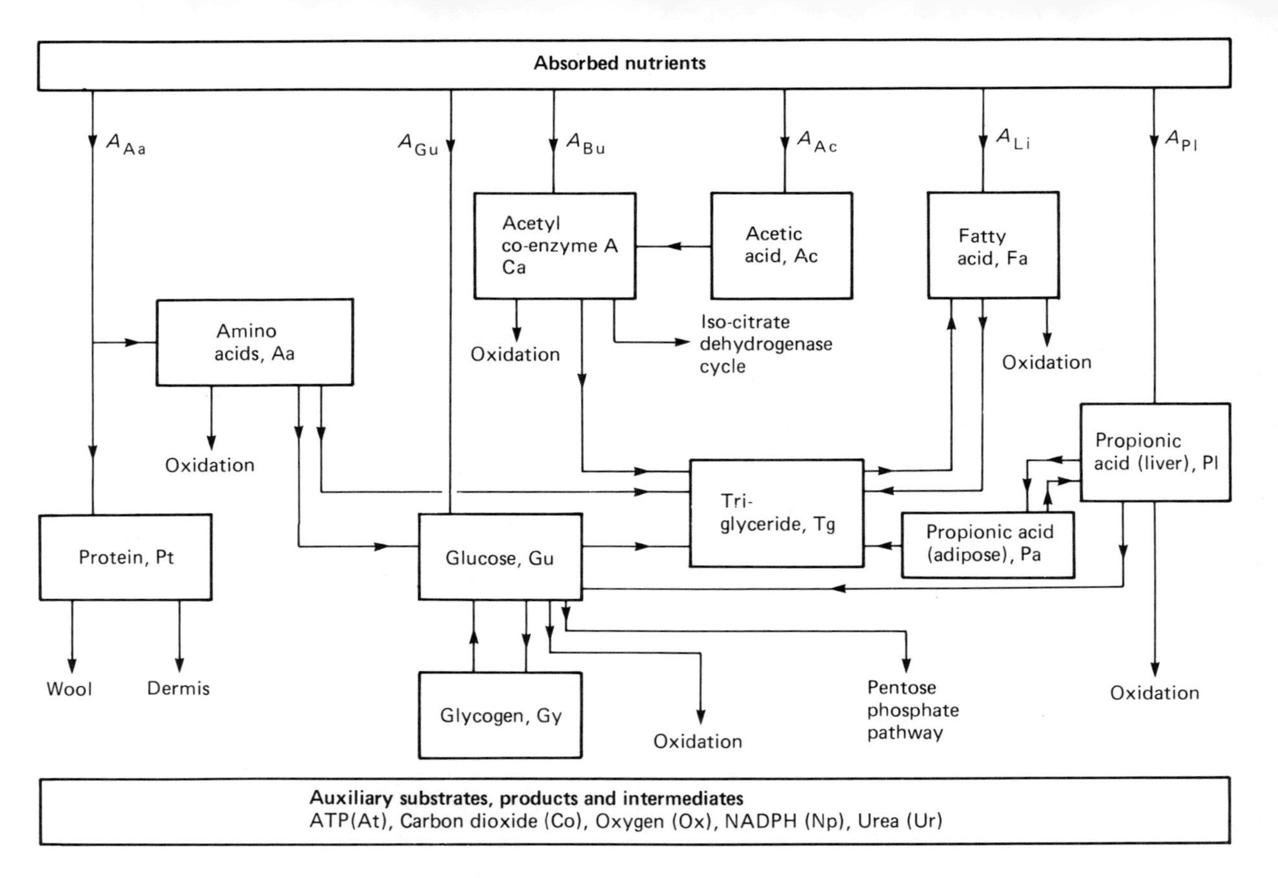


*Figure 10.4* Model for the efficiency of utilization of absorbed energy in ruminants. $A_i$ denotes the absorption per day of substance $i$ (*see Table 10.2*). The reactions shown are for the principal substrates and principal products only. The substances listed in the lower box are involved in many of these reactions. Maintenance is not shown above, but is a drain on the ATP (At) pool

**TABLE 10.3. General notation used in the efficiency of ME utilization model; subscripts take meanings from Table 10.2**

| Symbol | Meaning | Units |
|---|---|---|
| $A_i$ | Absorption rate of *i* | kg mole *i* per day |
| $C_i$ | Concentration of *i* | kg mole $m^{-3}$ |
| $J_{ij}$ | Inhibition constant of *ij* reaction with respect to *j* | kg mole $m^{-3}$ |
| $J_{ij,k}$ | Inhibition constant of *ij* reaction with respect to *k* | kg mole $m^{-3}$ |
| $K_{ij}$ | Michaelis–Menten constant of *ij* reaction with respect to *i* | kg mole $m^{-3}$ |
| $K_{ij,k}$ | Constant of *ij* reaction with respect to *k* | kg mole $m^{-3}$ |
| $M_{ij}$, $M_{ij,k}$ | Sigmoidicity constants for *ij* reaction | |
| $P_i$ | Rate of production of *i* | kg mole *i* per day |
| $P_{ij}$ | Rate of production of *j* by *ij* reaction | kg mole *j* per day |
| $P_{ij,k}$ | Rate of production of *k* by *ij* reaction | kg mole *k* per day |
| $Q_i$ | Quantity of *i* | kg mole of *i* |
| $R_{ij,k}$ | Requirement of *ij* reaction for *k* | kg mole of *k* required for each kg mole of *i* utilized in *ij* reaction |
| $U_{ij}$ | Rate of utilization of *i* by *ij* reaction | kg mole *i* per day |
| $U_{ij,k}$ | Rate of utilization of *k* by *ij* reaction | kg mole *k* per day |
| $V_{ij}$ | Maximum velocity of *ij* reaction | kg mole of *i* utilized in *ij* reaction per day per kg fresh tissue |
| $Y_{ij}$ | Yield of *j* from *ij* reaction | kg mole of *j* per kg mole of *i* utilized in *ij* reaction |
| $Y_{ij,k}$ | Yield of *k* from *ij* reaction | kg mole of *k* per kg mole of *i* utilized in the *ij* reaction |

**TABLE 10.4. Transactions occurring within the efficiency of ME utilization model. * denotes a sigmoidal response. See *Tables 10.2* and *10.3* for explanation of notation. Non-limiting substrates such as oxygen and carbon dioxide are ignored**

| *Transaction* | *Site* | *Substrates: principal; auxiliary* | *Inhibitors* | *Parameters* |
|---|---|---|---|---|
| AaCo | Whole-body | Aa; — | At | $V_{AaCo}, K_{AaCo}, J_{AaCo,At}$ |
| AaGu | Liver | Aa; — | Gu* | $V_{AaGu}, K_{AaGu}, J_{AaGu}, M_{AaGu}$ |
| AaPt | Empirical equation | | | |
| AaTg | Adipose + liver | Aa; At, Gu, Np | — | $V_{AaTg}, K_{AaTg}, K_{AaTg,At}, K_{AaTg,Gu}, K_{AaTg,Np}$ |
| AcCa | Whole-body | Ac; At | — | $V_{AcCa}, K_{AcCa}, K_{AcCa,At}$ |
| AtDg | Whole-body | At; — | — | $V_{AtDg}, K_{AtDg}, M_{AtDg}$ |
| AtMa | Empirical equation | | | |
| CaCo | Whole-body | Ca; — | At | $V_{CaCo}, K_{CaCo}, J_{CaCo,At}$ |
| CaNp | Adipose + liver | Ca; — | At, Np* | $V_{CaNp}, K_{CaNp}, J_{CaNp}, M_{CaNp}, J_{CaNp,At}$ |
| CaTg | Adipose + liver | Ca; At, Gu, Np | — | $V_{CaTg}, K_{CaTg}, K_{CaTg,At}, K_{CaTg,Gu}, K_{CaTg,Np}$ |
| FaCo | Whole-body | Fa; — | At | $V_{FaCo}, K_{FaCo}, J_{FaCo,At}$ |
| FaTg | Adipose + liver | Fa; At, Gu | — | $V_{FaTg}, K_{FaTg}, K_{FaTg,At}, K_{FaTg,Gu}$ |
| GuCo | Whole-body | Gu; — | At | $V_{GuCo}, K_{GuCo}, J_{GuCo,At}$ |
| GuGy | Liver + muscle | Gu; At | Gy | $V_{GuGy}, K_{GuGy}, K_{GuGy,At}, Q_{Gy,max}$ |
| GuNp | Adipose + liver | Gu; At | Np* | $V_{GuNp}, K_{GuNp}, K_{GuNp,At}, J_{GuNp}, M_{GuNp}$ |
| GuTg | Adipose + liver | Gu*; At, Np | — | $V_{GuTg}, K_{GuTg}, M_{GuTg}, K_{GuTg,At}, K_{GuTg,Np}$ |
| GyGu | Liver + muscle | Gy*; — | Gu* | $V_{GyGu}, K_{GyGu}, M_{GyGu}, J_{GyGu}, M_{GyGu,Gu}$ |
| PaPl | Transport in blood | Pa; — | — | $v_{bfr}$ |
| PaTg | Adipose | Pa; At, Gu, Np | — | $V_{PaTg}, K_{PaTg}, K_{PaTg,At}, K_{PaTg,Gu}, K_{PaTg,Np}$ |
| PlCo | Liver | Pl; — | At | $V_{PlCo}, K_{PlCo}, J_{PlCo,At}$ |
| PlGu | Liver | Pl; At | Gu* | $V_{PlGu}, K_{PlGu}, K_{PlGu,At}, J_{PlGu}, M_{PlGu}$ |
| PlPa | Transport in blood | Pl; — | — | *see* PaPl |
| TgFa | Adipose | Tg*; — | At* | $V_{TgFa}, K_{TgFa}, M_{TgFa}, J_{TgFa,At}, M_{TgFa,At}$ |

GENERAL REACTION KINETICS

Reaction rates are described using standard expressions from enzyme kinetics (Dixon and Webb, 1964; Mahler and Cordes, 1966). The utilization rate $U$ of substrate $s$ is

$$U = V\frac{s}{K + s} = V(1 + K/s)^{-1} \tag{10.34a}$$

where $V$ is the maximum velocity and $K$ is the Michaelis–Menten constant. For two substrates $s_1$ and $s_2$

$$U = V(1 + K_1/s_1 + K_2/s_2)^{-1} \tag{10.34b}$$

with an obvious extension to three or more substrates. Equation (10.34b) is preferred to the alternative form

$$U = V(1 + K_1/s_1)^{-1}(1 + K_2/s_2)^{-1} \tag{10.34c}$$

which weights the limitation produced by each substrate independently, and applies to substrates which can react in random order. Equation (10.34b) applies to ordered (and the so-called 'ping-pong') reaction mechanisms. For example, with $s_1 = K_1$ and $s_2 = K_2$, equation (10.34c) occurs at 0.25 of $V$ whereas equation (10.34b) gives 0.33 of $V$. With equation (10.34b), the substrate that is least limiting produces a much smaller effect on the reaction velocity than with (10.34c).

Inhibition by an inhibitor, $i$, is modelled using a multiplicative factor of

$$(1 + i/J)^{-1} \tag{10.34d}$$

where $J$ is an inhibition constant.

Equations (10.34a) and (10.34b) respond linearly to substrate at low substrate concentrations, and successive increments in substrate concentration give successively smaller increases in rate; there is no point of inflexion. For some reactions, for example, glucose to glycogen, at low glucose levels, the reaction should not proceed at an appreciable rate, whereas at above normal levels of glucose, the reaction should be very active. For these reactions, sigmoidal response equations are required, which give a 'switch-on' characteristic (Thornley, 1976, pp. 48–50; p. 229). Positive sigmoidal responses $y_+$ are described here by

$$y_+ = [1 + (K/s)^q]^{-1} \tag{10.34e}$$

where $q$ is a constant. $q = 1$ gives the familiar Michaelis–Menten equation (10.34a); $q = 2$ gives a weak sigmoidal response, and higher values of $q$ (4, 6, 8, etc.) give progressively sharper 'switch-on' characteristics. Sigmoidal inhibition ('switch-off') responses $y_-$ are similarly modelled with

$$y_- = [1 + (i/J)^q]^{-1} \tag{10.34f}$$

In equations (10.34e) and (10.34f), the quantity in square brackets varies between 0 and 1, and 1 and 0, it is 0.5 at $s = K$ or $i = J$, and the sharpness of the 'switch' depends on the magnitude of $q$.

WEIGHTS AND VOLUMES

The maximum reaction velocities (the $V$'s; *Tables 10.3* and *10.4*) are all defined with respect to fresh tissue weight. The weights of components of the animal (*see Table 10.5*) are given by

$$W_{adip} = 890Q_{Tg} \tag{10.35a}$$

where 890 is relative molecular mass (RMM) (i.e. molecular weight) of glycerol tristearate;

$$W_{body} = W_{adip} + (110Q_{Pt} + 0.2)/0.215 \tag{10.35b}$$

using the relationship between protein and fat-free empty body weight established for a wide range of treatments from birth to 120 kg for sheep (Searle and Griffiths, 1976), and 110 is the assumed mean RMM for the amino acids;

$$W_{live} = W_{body}/0.86 \tag{10.36}$$

from M. J. Gibb (personal communication);

$$W_{liver} = 0.0396W_{body}^{0.78} \tag{10.37}$$

from Baldwin and Black (1979, Table 1);

$$W_{met} = W_{body}^{0.75} \tag{10.38}$$

and

$$W_{mus} = 3.9 \times 110 \times Q_{Pt} \tag{10.39}$$

from Murray and Slezacek (1976).

The blood volume $v_{bld}$ is assumed to be given by

$$v_{bld} = 0.000\,0441W_{body}^{1.07} \tag{10.40}$$

from Baldwin and Black (1979, Table 1). The volume flow rate of blood $v_{bfr}$, which transports propionate between the liver and adipose tissue is given by

$$v_{bfr} = 0.0547W_{body} \tag{10.41a}$$

**TABLE 10.5. Efficiency of ME utilization model: definitions of specific symbols**

| Symbol | Definition | Units |
|---|---|---|
| $B_V$ | Biological value | |
| $E_{abs}$ | Absorbed energy | MJ $day^{-1}$ |
| $E_{dig}$ | Digestible energy | MJ $day^{-1}$ |
| $E_{ferm}$ | Heat of fermentation | MJ $day^{-1}$ |
| $E_{Ma}$ | Energy required for maintenance | MJ $day^{-1}$ |
| $E_{met}$ | Metabolizable energy | MJ $day^{-1}$ |
| $R_Q$ | Respiratory quotient | moles $CO_2$ (moles $O_2)^{-1}$ |
| $v_{adip}$ | Volume of adipose tissue | $m^3$ |
| $v_{bfr}$ | Volume flow rate of blood | $m^3$ $day^{-1}$ |
| $v_{bld}$ | Volume of blood | $m^3$ |
| $v_{liver}$ | Volume of liver | $m^3$ |
| $W_{adip}$ | Weight of adipose tissue | kg |
| $W_{body}$ | Empty body weight | kg |
| $W_{liver}$ | Weight of liver | kg |
| $W_{live}$ | Liveweight of animal | kg |
| $W_{met}$ | Metabolic weight of animal | $kg^{0.75}$ |
| $W_{mus}$ | Weight of muscle | kg |

from Pethick *et al.* (1981, p. 97). The volumes of adipose tissue and liver are assumed to be given by

$$v_{\text{adip}} = W_{\text{adip}}/1000 \quad \text{and} \quad v_{\text{liver}} = W_{\text{liver}}/1000 \tag{10.41b}$$

assuming that the tissue density is equal to that of water.

NUTRITIONAL INPUTS

Numerical values are ascribed to the six nutrient fluxes ($A_{\text{Aa}}$, $A_{\text{Gu}}$, $A_{\text{Bu}}$, $A_{\text{Ac}}$, $A_{\text{Li}}$, $A_{\text{Pl}}$) which are actually absorbed by the animal (*see Figure 10.4*). These inputs were held constant during each run of the computer program, and did not change with the progress of time, as they would in a fully dynamic model. The values used are summarized in *Table 10.6*, and are based on recent experimental work at the

**TABLE 10.6. Absorbed nutrient inputs for the efficiency of ME utilization model**

| | | [MJ day$^{-1}$] | $E_i$ [MJ kg$^{-1}$] | [kg day$^{-1}$] | $(RMM)_i$ | $A_i$ [kg moles day$^{-1}$] |
|---|---|---|---|---|---|---|
| Acetic acid | Ac | 4.34 | 14.6 | 0.297 | 60 | 0.004 952 |
| Amino acids | Aa | 3.06 | 23.0 | 0.133 | 110 | 0.001 209 |
| Butyric acid | Bu | 1.65 | 24.9 | 0.066 | 88 | 0.000 752 |
| Glucose | Gu | 0.45 | 16.0 | 0.028 | 180 | 0.000 156 |
| Lipid | Li(Tg) | 0.98 | 39.3 | 0.025 | 890 | 0.000 028 |
| Propionic acid | Pl | 2.45 | 20.8 | 0.118 | 74 | 0.001 592 |
| | Total | 12.93 | | | | |

Grassland Research Institute with growing lambs (D. J. Thomson, personal communication), adjusted for a 25 kg animal. This standard diet gives a total absorbed energy input of 12.93 MJ day$^{-1}$, of which 23.7% is derived from protein. The performance of the model is examined mostly in relation to variations from this standard diet.

Using the definitions in *Table 10.6*, the absorbed energy $E_{\text{abs}}$ is obtained by

$$E_{\text{abs}} = \sum_i E_i A_i (\text{RMM})_i \tag{10.42a}$$

The metabolizable energy $E_{\text{met}}$ is defined by

$$E_{\text{met}} = E_{\text{abs}} + E_{\text{ferm}} - E_{\text{Ur}} \tag{10.42b}$$

where the observed heat of fermentation $E_{\text{ferm}}$ calculated from Webster *et al.* (1975) and the observed urine energy $E_{\text{Ur}}$ (D. J. Thomson, personal communication) are

$$E_{\text{ferm}} = 0.93\ \text{MJ day}^{-1} \quad \text{and} \quad E_{\text{Ur}} = 1.26\ \text{MJ day}^{-1} \tag{10.42c}$$

Finally, the digestible energy $E_{\text{dig}}$ is obtained from

$$E_{\text{dig}} = E_{\text{met}}/0.81 \tag{10.42d}$$

from Ministry of Agriculture, Fisheries and Food (1975, p. 3, equation 1).

CONCENTRATIONS

With the exception of propionate, all the concentrations $C_i$ are obtained from the volume of blood $v_{bld}$ and the quantity $Q_i$ by means of

$$C_i = Q_i/v_{bld} \tag{10.43}$$

For example, for acetic acid

$$C_{Ac} = Q_{Ac}/v_{bld} \tag{10.44a}$$

The propionic acid concentrations in adipose and liver are obtained by

$$C_{Pa} = Q_{Pa}/v_{adip} \quad \text{and} \quad C_{Pl} = Q_{Pl}/v_{liver} \tag{10.44b}$$

STATE VARIABLES AND POOLS

The model has 12 state variables—the quantities Aa, Ac, At, Ca, Fa, Gu, Gy, Np, Pa, Pl, Pt and Tg—each of which is a representation of a metabolic pool. The description of these pools is similar in each case. Inputs and outputs are described separately. The rate of change of mass in the pool is then given by a differential equation which describes the difference between input and output over an interval of time:

$$\frac{d(\text{pool})}{dt} = \text{inputs} - \text{outputs}$$

The pools, together with carbon dioxide and oxygen relations, maintenance, and urea production, are described in alphabetical order.

ACETIC ACID POOL, Ac

*Inputs*
The only input to the pool is the absorbed nutrient input $A_{Ac}$.

*Outputs*
There is just the one reaction, with a flux of (*Table 10.4*)

$$U_{AcCa} = W_{body} V_{AcCa} (1 + K_{AcCa}/C_{Ac} + K_{AcCa,At}/C_{At})^{-1} \tag{10.45}$$

*Differential equation*
Summing the terms, the rate of change of the quantity of acetic acid $Q_{Ac}$ is given by

$$\frac{dQ_{Ac}}{dt} = A_{Ac} - U_{AcCa} \tag{10.46}$$

ACETYL CO-ENZYME A POOL, Ca

*Inputs*
Ca is produced at rate $P_{AcCa}$ from acetic acid (Ac), given by

$$P_{AcCa} = Y_{AcCa}\, U_{AcCa} \tag{10.47a}$$

where $U_{AcCa}$ is defined by equation (10.45). There is a second input as it is assumed that the nutrient input of butyric acid ($A_{Bu}$) is converted immediately into Ca; this is written

$$P_{BuCa} = Y_{BuCa} A_{Bu} \tag{10.47b}$$

*Outputs*

There are three outputs (*Figure 10.4*); from *Table 10.4* these are

$$U_{CaCo} = W_{body} V_{CaCo} (1 + K_{CaCo}/C_{Ca})^{-1} (1 + C_{At}/J_{CaCo,At})^{-1} \tag{10.48a}$$

$$U_{CaNp} = (W_{adip} + W_{liver}) V_{CaNp} (1 + K_{CaNp}/C_{Ca})^{-1} [1 + (C_{Np}/J_{CaNp})^{M_{CaNp}}]^{-1} \times (1 + C_{At}/J_{CaNp,At})^{-1} \tag{10.48b}$$

$$U_{CaTg} = (W_{adip} + W_{liver}) V_{CaTg} (1 + K_{CaTg}/C_{Ca} + K_{CaTg,At}/C_{At} + K_{CaTg,Gu}/C_{Gu} + K_{CaTg,Np}/C_{Np})^{-1} \tag{10.48c}$$

*Differential equation*

Summing the terms, this is

$$\frac{dQ_{Ca}}{dt} = P_{AcCa} + P_{BuCa} - U_{CaCo} - U_{CaNp} - U_{CaTg} \tag{10.49}$$

AMINO ACID POOL, Aa

*Inputs*

The only inputs to this catabolic pool are derived from the absorbed nutrient input $A_{Aa}$. This is assigned a biological value $B_V$, so that amino acids which are in excess of the limiting amino acid for protein synthesis, $x_1$, described by

$$x_1 = (1 - B_V)A_{Aa} \tag{10.50a}$$

are input to the amino acid pool. $P_{Pt,max}$ is defined as the maximum possible rate of protein accretion at a particular body-weight and energy intake (equation (10.83)). $P_{Pt,max}$ includes the synthesis of wool and dermis. Since the maximum rate of protein accretion allowed by diet input is $B_V A_{Aa}$, then the actual rate of protein accretion $P_{AaPt}$ is the smaller of $P_{Pt,max}$ and the dietary limit $B_V A_{Aa}$, which is denoted by

$$U_{AaPt} = P_{AaPt} = \text{minimum}\,(P_{Pt,max}, B_V A_{Aa}) \tag{10.50b}$$

There is thus a possible second input to the amino acid pool representing amino acids in excess of those of correct $B_V$ required for protein deposition, $x_2$, of

$$x_2 = B_V A_{Aa} - P_{AaPt} \tag{10.50c}$$

*Outputs*

There are three outputs from the amino acid pool (*Figure 10.4* and *Table 10.4*):

$$U_{AaCo} = W_{body} V_{AaCo} (1 + K_{AaCo}/C_{Aa})^{-1} (1 + C_{At}/J_{AaCo,At})^{-1} \tag{10.51a}$$

$$U_{AaGu} = W_{liver} V_{AaGu} (1 + K_{AaGu}/C_{Aa})^{-1} [1 + (C_{Gu}/J_{AaGu})^{M_{AaGu}}]^{-1} \tag{10.51b}$$

$$U_{AaTg} = (W_{adip} + W_{liver}) V_{AaTg} (1 + K_{AaTg}/C_{Aa} + K_{AaTg,At}/C_{At} + K_{AaTg,Gu}/C_{Gu} + K_{AaTg,Np}/C_{Np})^{-1} \tag{10.51c}$$

*Differential equation*
From equations (10.50) and (10.51), therefore

$$\frac{dQ_{Aa}}{dt} = A_{Aa} - P_{AaPt} - U_{AaCo} - U_{AaGu} - U_{AaTg} \tag{10.52}$$

ATP POOL, At

Inputs and outputs from this pool arise mainly from stoichiometric descriptions of the ATP required to drive other fluxes in the model, or as yields of ATP as a result of other fluxes.

*Inputs*

$$P_{AaCo,At} = Y_{AaCo,At}\, U_{AaCo} \tag{10.53a}$$

$$P_{AaGu,At} = Y_{AaGu,At}\, U_{AaGu} \tag{10.53b}$$

$$P_{AaTg,At} = Y_{AaTg,At}\, U_{AaTg} \tag{10.53c}$$

$$P_{BuCa,At} = Y_{BuCa,At}\, A_{Bu} \tag{10.53d}$$

$$P_{CaCo,At} = Y_{CaCo,At}\, U_{CaCo} \tag{10.53e}$$

$$P_{CaNp,At} = Y_{CaNp,At}\, U_{CaNp} \tag{10.53f}$$

$$P_{FaCo,At} = Y_{FaCo,At}\, U_{FaCo} \tag{10.53g}$$

$$P_{GuCo,At} = Y_{GuCo,At}\, U_{GuCo} \tag{10.53h}$$

$$P_{GuTg,At} = Y_{GuTg,At}\, U_{GuTg} \tag{10.53i}$$

$$P_{PlCo,At} = Y_{PlCo,At}\, U_{PlCo} \tag{10.53j}$$

*Outputs*

$$U_{AaPt,At} = R_{AaPt,At}\, U_{AaPt} \tag{10.54a}$$

$$U_{AcCa,At} = R_{AcCa,At}\, U_{AcCa} \tag{10.54b}$$

$$U_{CaTg,At} = R_{CaTg,At}\, U_{CaTg} \tag{10.54c}$$

$$U_{FaTg,At} = R_{FaTg,At}\, U_{FaTg} \tag{10.54d}$$

$$U_{GuGy,At} = R_{GuGy,At}\, U_{GuGy} \tag{10.54e}$$

$$U_{GuNp,At} = R_{GuNp,At}\, U_{GuNp} \tag{10.54f}$$

$U_{Ma,At}$ [*see* maintenance section and equation (10.73b)] (10.54g)

$$U_{PlGu,At} = R_{PlGu,At}\, U_{PlGu} \tag{10.54h}$$

$$U_{PaTg,At} = R_{PaTg,At}\, U_{PaTg} \tag{10.54i}$$

There is an additional energy cost associated with nitrogen excretion in urea (Ur); it is assumed that this cost is met from the ATP pool, according to

$$U_{Ur,At} = R_{Ur,At}\frac{dQ_{Ur}}{dt} \tag{10.54j}$$

In some early runs of the model, it was found that the ATP pool would sometimes increase without limit. This arises because not all futile cycles have been considered. It was found necessary to introduce an ATP degrader into the equations to put an upper limit on ATP. This is achieved by means of

$$U_{AtDg} = V_{AtDg}\,[1 + (K_{AtDg}/C_{At})^{M_{AtDg}}]^{-1} \qquad (10.54k)$$

*Differential equation*

$$\frac{dQ_{At}}{dt} = \text{inputs} - \text{outputs} \qquad (10.55)$$

CARBON DIOXIDE (Co), OXYGEN (Ox) AND RESPIRATORY QUOTIENT

*Carbon dioxide production*

$$P_{AaCo} = Y_{AaCo}\,U_{AaCo} \qquad (10.56a)$$

$$P_{AaGu,Co} = Y_{AaGu,Co}\,U_{AaGu} \qquad (10.56b)$$

$$P_{AaTg,Co} = Y_{AaTg,Co}\,U_{AaTg} \qquad (10.56c)$$

$$P_{CaCo} = Y_{CaCo}\,U_{CaCo} \qquad (10.56d)$$

$$P_{CaNp,Co} = Y_{CaNp,Co}\,U_{CaNp} \qquad (10.56e)$$

$$P_{FaCo} = Y_{FaCo}\,U_{FaCo} \qquad (10.56f)$$

$$P_{GuCo} = Y_{GuCo}\,U_{GuCo} \qquad (10.56g)$$

$$P_{GuNp,Co} = Y_{GuNp,Co}\,U_{GuNp} \qquad (10.56h)$$

$$P_{GuTg,Co} = Y_{GuTg,Co}\,U_{GuTg} \qquad (10.56i)$$

$$P_{PlCo} = Y_{PlCo}\,U_{PlCo} \qquad (10.56j)$$

*Carbon dioxide requirement*

$$U_{PaTg,Co} = R_{PaTg,Co}\,U_{PaTg} \qquad (10.57)$$

*Net carbon production*
This is given by

$$\frac{dQ_{Co}}{dt} = \text{production terms} - U_{PaTg,Co} \qquad (10.58)$$

*Oxygen requirement*

$$U_{AaCo,Ox} = R_{AaCo,Ox}\,U_{AaCo} \qquad (10.59a)$$

$$U_{AaGu,Ox} = R_{AaGu,Ox}\,U_{AaGu} \qquad (10.59b)$$

$$U_{AaTg,Ox} = R_{AaTg,Ox}\,U_{AaTg} \qquad (10.59c)$$

$$U_{CaCo,Ox} = R_{CaCo,Ox}\,U_{CaCo} \qquad (10.59d)$$

$$U_{CaNp,Ox} = R_{CaNp,Ox}\,U_{CaNp} \qquad (10.59e)$$

$$U_{FaCo,Ox} = R_{FaCo,Ox}\,U_{FaCo} \qquad (10.59f)$$

$$U_{\mathrm{GuCo,Ox}} = R_{\mathrm{GuCo,Ox}}\, U_{\mathrm{GuCo}} \tag{10.59g}$$

$$U_{\mathrm{GuNp,Ox}} = R_{\mathrm{GuNp,Ox}}\, U_{\mathrm{GuNp}} \tag{10.59h}$$

$$U_{\mathrm{GuTg,Ox}} = R_{\mathrm{GuTg,Ox}}\, U_{\mathrm{GuTg}} \tag{10.59i}$$

$$U_{\mathrm{PlCo,Ox}} = R_{\mathrm{PlCo,Ox}}\, U_{\mathrm{PlCo}} \tag{10.59j}$$

$$U_{\mathrm{PlGu,Ox}} = R_{\mathrm{PlGu,Ox}}\, U_{\mathrm{PlGu}} \tag{10.59k}$$

*Net oxygen requirement*

$$\frac{\mathrm{d}Q_{\mathrm{Ox}}}{\mathrm{d}t} = \Sigma \text{ requirement terms} \tag{10.60}$$

*Respiratory quotient, $R_{\mathrm{Q}}$*
This is defined by

$$R_{\mathrm{Q}} = \frac{\mathrm{d}Q_{\mathrm{Co}}}{\mathrm{d}t} \bigg/ \frac{\mathrm{d}Q_{\mathrm{Ox}}}{\mathrm{d}t} \tag{10.61}$$

FATTY ACID POOL, Fa

*Inputs*

$$P_{\mathrm{LiFa}} = Y_{\mathrm{LiFa}}\, A_{\mathrm{Li}} \tag{10.62a}$$

$$P_{\mathrm{TgFa}} = Y_{\mathrm{TgFa}}\, U_{\mathrm{TgFa}} \tag{10.62b}$$

*Outputs*

$$U_{\mathrm{FaCo}} = W_{\mathrm{body}}\, V_{\mathrm{FaCo}}\, (1 + K_{\mathrm{FaCo}}/C_{\mathrm{Fa}})^{-1}\, (1 + C_{\mathrm{At}}/J_{\mathrm{FaCo,At}})^{-1} \tag{10.63a}$$

$$U_{\mathrm{FaTg}} = (W_{\mathrm{adip}} + W_{\mathrm{liver}})\, V_{\mathrm{FaTg}}\, (1 + K_{\mathrm{FaTg}}/C_{\mathrm{Fa}} + K_{\mathrm{FaTg,At}}/C_{\mathrm{At}} + K_{\mathrm{FaTg,Gu}}/C_{\mathrm{Gu}})^{-1} \tag{10.63b}$$

*Differential equation*

$$\frac{\mathrm{d}Q_{\mathrm{Fa}}}{\mathrm{d}t} = P_{\mathrm{LiFa}} + P_{\mathrm{TgFa}} - U_{\mathrm{FaCo}} - U_{\mathrm{FaTg}} \tag{10.64}$$

GLUCOSE POOL, Gu

*Inputs*

$$P_{\mathrm{AaGu}} = Y_{\mathrm{AaGu}}\, U_{\mathrm{AaGu}} \tag{10.65a}$$

$$P_{\mathrm{GyGu}} = Y_{\mathrm{GyGu}}\, U_{\mathrm{GyGu}} \tag{10.65b}$$

$$P_{\mathrm{LiGu}} = Y_{\mathrm{LiGu}}\, A_{\mathrm{Li}} \tag{10.65c}$$

$$P_{\mathrm{PlGu}} = Y_{\mathrm{PlGu}}\, U_{\mathrm{PlGu}} \tag{10.65d}$$

$$P_{\mathrm{TgFa,Gu}} = Y_{\mathrm{TgFa,Gu}}\, U_{\mathrm{TgFa}} \tag{10.65e}$$

*Outputs*

$$U_{GuCo} = W_{body} V_{GuCo} (1 + K_{GuCo}/C_{Gu})^{-1} (1 + C_{At}/J_{GuCo,At})^{-1} \tag{10.66a}$$

For glycogen synthesis from glucose, it is assumed that not more than a certain quantity of glycogen, $Q_{Gy,max}$, can be laid down, and glucose utilization for glycogen synthesis is limited by an extra term, according to

$$U_{GuGy} = (W_{liver} + W_{mus}) V_{GuGy} (1 + K_{GuGy}/C_{Gu} + K_{GuGy,At}/C_{At})^{-1} \times (1 - Q_{Gy}/Q_{Gy,max}) \tag{10.66b}$$

where $Q_{Gy,max}$ is obtained from equation (10.67) below.

$$U_{GuNp} = (W_{adip} + W_{liver}) V_{GuNp} (1 + K_{GuNp}/C_{Gu} + K_{GuNp,At}/C_{At})^{-1} \times [1 + (C_{Np}/J_{GuNp})^{M_{GuNp}}]^{-1} \tag{10.66c}$$

$$U_{GuTg} = (W_{adip} + W_{liver}) V_{GuTg} [1 + (K_{GuTg}/C_{Gu})^{M_{GuTg}}]^{-1} (1 + K_{GuTg,Np}/C_{Np})^{-1} \tag{10.66d}$$

$$U_{AaTg,Gu} = R_{AaTg,Gu} U_{AaTg} \tag{10.66e}$$

$$U_{CaTg,Gu} = R_{CaTg,Gu} U_{CaTg} \tag{10.66f}$$

$$U_{FaTg,Gu} = R_{FaTg,Gu} U_{FaTg} \tag{10.66g}$$

$$U_{PaTg,Gu} = R_{PaTg,Gu} U_{PaTg} \tag{10.66h}$$

The auxiliary equation required to evaluate $Q_{Gy,max}$ for equation (10.66b) is

$$Q_{Gy,max} = (0.02W_{mus} + 0.06W_{liver})/180 \tag{10.67}$$

where it is assumed that at most muscle contains 2% glycogen (McVeigh and Tarrant, 1982) and liver 6% glycogen (Leng and Annison, 1963).

*Differential equation*

$$\frac{dQ_{Gu}}{dt} = \text{inputs} - \text{outputs} \tag{10.68}$$

GLYCOGEN POOL, Gy

*Input*

$$P_{GuGy} = Y_{GuGy} U_{GuGy} \tag{10.69}$$

*Output*

$$U_{GyGu} = (W_{liver} + W_{mus}) V_{GyGu} [1 + (K_{GyGu}/C_{Gy})^{M_{GyGu}}]^{-1} \times [1 + (C_{Gu}/J_{GyGu})^{M_{GyGu,Gu}}]^{-1} \tag{10.70}$$

*Differential equation*

$$\frac{dQ_{Gy}}{dt} = P_{GuGy} - U_{GyGu} \tag{10.71}$$

*Maintenance*

From equations (10.35) the rate of body weight gain is

$$\frac{dW_{body}}{dt} = \frac{110}{0.215}\frac{dQ_{Pt}}{dt} + 890\frac{dQ_{Tg}}{dt} \tag{10.72}$$

The energy required for maintenance $E_{Ma}$ [MJ day$^{-1}$] is (Graham *et al.*, 1976, p. 124)

$$E_{Ma} = 0.257W_{met}\exp[-0.000\,219(t-98)] + 2.8\frac{dW_{body}}{dt} + 0.046E_{dig} \tag{10.73a}$$

where $W_{met}$ and $E_{dig}$ are given by equations (10.38) and (10.42d). This energy is provided at the expense of the ATP pool, giving an ATP utilization rate for maintenance of

$$U_{Ma,At} = E_{Ma}/E_{At} \tag{10.73b}$$

where $E_{At}$ is the energy content of ATP and is given by

$$E_{At} = 77.3\ \text{MJ (kg mole)}^{-1} \tag{10.73c}$$

## NADPH POOL, Np

*Inputs*

$$P_{AaGu,Np} = Y_{AaGu,Np}\,U_{AaGu} \tag{10.74a}$$

$$P_{CaNp} = Y_{CaNp}\,U_{CaNp} \tag{10.74b}$$

$$P_{GuNp} = Y_{GuNp}\,U_{GuNp} \tag{10.74c}$$

*Outputs*

$$U_{AaTg,Np} = R_{AaTg,Np}\,U_{AaTg} \tag{10.75a}$$

$$U_{CaTg,Np} = R_{CaTg,Np}\,U_{CaTg} \tag{10.75b}$$

$$U_{GuTg,Np} = R_{GuTg,Np}\,U_{GuTg} \tag{10.75c}$$

$$U_{PaTg,Np} = R_{PaTg,Np}\,U_{PaTg} \tag{10.75d}$$

*Differential equation*

$$\frac{dQ_{Np}}{dt} = \text{inputs} - \text{outputs} \tag{10.76}$$

## PROPIONATE POOLS, Pa AND Pl

There are assumed to be two significant propionate pools in the animal, in the adipose tissue and in the liver, connected by the blood flow rate (equation (10.41a)).

For the adipose tissue, the only input term is

$$P_{PlPa} = v_{bfr}\,C_{Pl} \tag{10.77}$$

and the two output terms are

$$U_{PaPl} = v_{bfr}\,C_{Pa} \tag{10.78a}$$

$$U_{PaTg} = W_{adip}\,V_{PaTg}\,(1 + K_{PaTg}/C_{Pa} + K_{PaTg,At}/C_{At} + K_{PaTg,Gu}/C_{Gu} + K_{PaTg,Np}/C_{Np})^{-1} \tag{10.78b}$$

The differential equation for Pa is

$$\frac{dQ_{Pa}}{dt} = v_{bfr}\,(C_{Pl} - C_{Pa}) - U_{PaTg} \tag{10.79}$$

For the liver pool, there are two inputs: the absorbed nutrient input $A_{Pl}$ and

$$U_{PaPl} = v_{bfr}\,C_{Pa} \tag{10.80}$$

There are three outputs:

$$U_{PlPa} = v_{bfr}\,C_{Pl} \tag{10.81a}$$

$$U_{PlCo} = W_{liver}\,V_{PlCo}\,(1 + K_{PlCo}/C_{Pl})^{-1}\,(1 + C_{At}/J_{PlCo,At})^{-1} \tag{10.81b}$$

$$U_{PlGu} = W_{liver}\,V_{PlGu}\,(1 + K_{PlGu}/C_{Pl} + K_{PlGu,At}/C_{At})^{-1} \times [1 + (C_{Gu}/J_{PlGu})^{M_{PlGu}}]^{-1} \tag{10.81c}$$

The differential equation for Pl is

$$\frac{dQ_{Pl}}{dt} = A_{Pl} + v_{bfr}\,(C_{Pa} - C_{Pl}) - U_{PlCo} - U_{PlGu} \tag{10.82}$$

PROTEIN POOL, Pt

The maximum rate of protein synthesis $P_{Pt,max}$ is assumed to be described by the empirical expression (Black and Griffiths, 1975)

$$P_{Pt,max} = 0.001 \times \frac{6.25}{110}\,(2.017E_{met} - 0.401W_{met} - 0.106E_{met}\,W_{met} + 0.024W_{met}^{2}) \tag{10.83}$$

where metabolizable energy intake $E_{met}$ and the metabolic weight $W_{met}$ are given by equations (10.42b) and (10.38). Equation (10.83) allows the computation of $U_{AaPt}$ in equation (10.50b). The production of protein is given by

$$P_{AaPt} = Y_{AaPt}\,U_{AaPt} \tag{10.84}$$

The protein laid down in wool and dermis must be subtracted from this. That laid down in wool is

$$\frac{dQ_{Pt,wool}}{dt} = \frac{6.25}{110} \times 0.0016 \tag{10.85a}$$

from Black and Griffiths (1975, p. 404). That laid down in dermis is

$$\frac{dQ_{Pt,derm}}{dt} = \frac{6.25}{110} \times 0.000018W_{Met} \tag{10.85b}$$

from Agricultural Research Council (1980, p. 133). The differential equation for $Q_{Pt}$ is therefore

$$\frac{dQ_{Pt}}{dt} = P_{AaPt} - \frac{dQ_{Pt,wool}}{dt} - \frac{dQ_{Pt,derm}}{dt} \tag{10.85c}$$

TRI-GLYCERIDE POOL, Tg

*Inputs*

$$P_{AaTg} = Y_{AaTg}\, U_{AaTg} \tag{10.86a}$$

$$P_{CaTg} = Y_{CaTg}\, U_{CaTg} \tag{10.86b}$$

$$P_{FaTg} = Y_{FaTg}\, U_{FaTg} \tag{10.86c}$$

$$P_{GuTg} = Y_{GuTg}\, U_{GuTg} \tag{10.86d}$$

$$P_{PaTg} = Y_{PaTg}\, U_{PaTg} \tag{10.86e}$$

*Output*

$$U_{TgFa} = W_{adip}\, V_{TgFa} [1 + (K_{TgFa}/C_{Tg})^{M_{TgFa}}]^{-1} \times [1 + (C_{At}/J_{TgFa,At})^{M_{TgFa,At}}]^{-1} \tag{10.87}$$

*Differential equation*

$$\frac{dQ_{Tg}}{dt} = \text{inputs} - \text{output} \tag{10.88}$$

UREA (Ur) PRODUCTION

The transactions producing urea are

$$P_{AaCo,Ur} = Y_{AaCo,Ur}\, U_{AaCo} \tag{10.89a}$$

$$P_{AaGu,Ur} = Y_{AaGu,Ur}\, U_{AaGu} \tag{10.89b}$$

$$P_{AaTg,Ur} = Y_{AaTg,Ur}\, U_{AaTg} \tag{10.89c}$$

The rate of production of urea is the sum of these terms:

$$\frac{dQ_{Ur}}{dt} = P_{AaCo,Ur} + P_{AaGu,Ur} + P_{AaTg,Ur} \tag{10.90}$$

The energy associated with this urea excretion is

$$E_{Ur} = 22.6 \times 0.47 \times 60 \frac{dQ_{Ur}}{dt} \tag{10.91}$$

where urea has an energy content of 22.6 MJ (kg nitrogen)$^{-1}$, a RMM of 60 and is 47% nitrogen. Note that although equation (10.91) predicts the urea energy, an observed urine energy is used in equations (10.42b), (10.42c) and (10.42d) to calculate metabolizable energy and digestible energy.

EFFICIENCIES AND HEAT PRODUCTION

The energy retained by the animal, $E_{ret}$, is obtained by

$$E_{ret} = (\mathrm{RMM})_{Aa} E_{Aa} P_{AaPt} + (\mathrm{RMM})_{Tg} E_{Tg} \frac{dQ_{Tg}}{dt} + (\mathrm{RMM})_{Gu} E_{Gu} \frac{dQ_{Gy}}{dt} \tag{10.92a}$$

where *Table 10.6* and equations (10.84), (10.88) and (10.71) are used.

The total efficiency $k_{total}$ is defined by

$$k_{total} = (E_{ret} + E_{Ma})/E_{abs} \tag{10.92b}$$

using equations (10.92a), (10.73a) and (10.42a). The gross growth efficiency $k_{growth}$ is given by

$$k_{growth} = E_{ret}/E_{abs} \tag{10.92c}$$

Heat production can be calculated in two ways, which gives a useful check on the formulation of the model. Brouwer (1965) gives the following formula for the rate of heat production, $H_1$, in MJ per day:

$$H_1 = 22.4\left(16.18\frac{dQ_{Ox}}{dt} + 5.02\frac{dQ_{Co}}{dt}\right) - 28 \times 5.99 \times \frac{dQ_{Ur}}{dt} \tag{10.93a}$$

with equations (10.60), (10.58) and (10.90). Heat production $H_2$ can also be estimated from the total energy balance by means of

$$H_2 = E_{abs} - E_{ret} - E_{Ur} \tag{10.93b}$$

using equations (10.42a), (10.92a) and (10.91). The heat increment $H_I$ is obtained from

$$H_I = (H_2 - E_{Ma})/E_{abs} \tag{10.93c}$$

SUMMARY OF MODEL

The model has 12 state variables—the quantities of Aa, Ac, At, Ca, Fa, Gu, Gy, Np, Pa, Pl, Pt and Tg. There are 12 corresponding differential equations, which can be integrated for given initial conditions and parameter values. All other quantities of interest that lie within the scope of the model can be calculated from the state variables and their derivatives. The model was programmed in CSMP (p. 33). Further details and results are given by Gill *et al.* (1984).

## Exercises

### Exercise 10.1

Assume that a quantity of feed protein entering the rumen, $\pi(0)$ [kg], can be divided into three parts: (1) an instantly degradable component, $\pi_1(0)$, (2) a potentially degradable component, $\pi_2(0)$, and (3) an undegradable component, $\pi_3(0)$, that is

$$\pi(0) = \pi_1(0) + \pi_2(0) + \pi_3(0)$$

If $c$ [per day] is the specific rate of degradation within the rumen and $k$ [per day] is the specific rate of passage out of the rumen, then by applying the principles of dynamic modelling show that the fraction of the feed protein which is degraded in the rumen, $f_{deg}$, is given by the identity

$$f_{deg} = \left[\pi_1(0) + \frac{c}{c+k}\pi_2(0)\right] / \pi(0)$$

**Exercise 10.2**

If the situation described in Exercise 10.1 is further complicated by a time lag of $t_0$ [days] between ingestion and commencement of degradation of the potentially degradable portion, show that $f_{deg}$ is now given by

$$f_{deg} = \left[\pi_1(0) + \frac{c}{c+k}\mathrm{e}^{-kt_0}\pi_2(0)\right] / \pi(0)$$

Chapter 11

# Animal products

## Introduction

In this chapter, we continue the examination of physiological processes in farm animals started in Chapter 10, by considering processes directly concerned with livestock and poultry production. These processes are considered under three headings—milk, meat and egg production. Again, much of the published modelling work in this area is empirical in nature and therefore, as an attempt to balance empiricism and mechanism, a mechanistic model of the cow's mammary gland drawn from our own research is presented in some detail.

## Milk production by the dairy cow

Fresh milk provides us perhaps with a better balance of nutrients than any other single food: it is also the starting material for a range of equally nutritious dairy products, such as butter, cheese, yogurt, etc. More mathematical approaches can help the dairy farmer who is attempting to achieve more controlled, continuous, and economic means of production.

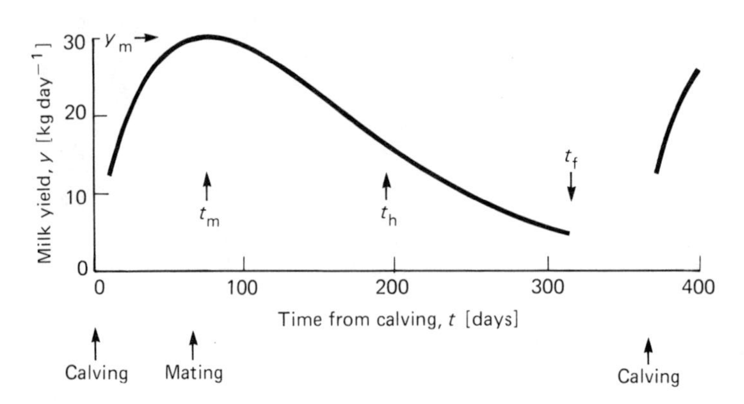


*Figure 11.1* An idealized lactation curve for a dairy cow, with one year between calvings

Milk is synthesized by specialized cells in the mammary glands of the animal from simple nutrients carried by the blood. The lactation curve of the dairy cow increases rapidly from around calving to a peak a few weeks later, followed by a gradual decline until the cow is dried off about ten months after calving, giving a dry period of about 45 days. This is shown in an idealized form in *Figure 11.1*, where any possible seasonal effects have been ignored. The changes in daily milk yield as lactation progresses result from changes in cell numbers and in cell activity in the mammary glands. Attempts to describe this curve mathematically have aimed mostly at the prediction of quantities such as milk yield, feed requirements and cash flow, and have not been much concerned with obtaining a quantitative understanding of the lactation process.

## Review of lactation models

An early attempt to model the lactation curve was due to Gaines (1927), who proposed

$$y_i = k_1 \mathrm{e}^{-k_2 i}$$

where $y_i$ is the daily yield averaged over the month $i$, and $k_1$ and $k_2$ are parameters. $k_1$ is the initial yield (when $i = 0$), and $k_2$ is the rate of decline per month. However, Gaines' model takes no account of the initial rise to a maximum, which is of interest and importance.

Possibly the first attempt to represent the entire lactation curve was that of Vujičić and Bačić (1961), who modified Gaines' model to give an equation equivalent to

$$y_i = k_1 i \mathrm{e}^{-k_2 i} \tag{11.1}$$

where $y_i$ is the yield in the $i$th period, and $k_1$ and $k_2$ are parameters. Unfortunately, this equation does not fit the lactation data satisfactorily.

A major advance on the model of equation (11.1) was made by Wood (1967), who proposed using a gamma curve:

$$y_i = k_1 i^{k_2} \mathrm{e}^{-k_3 i} \tag{11.2}$$

where now $y_i$ is the average daily milk yield [kg day$^{-1}$] in the $i$th week of lactation, $k_1$ is a scale parameter, and $k_2$ and $k_3$ determine the shape of the lactation curve ($k_1$, $k_2$, $k_3 > 0$). The model is not readily given a realistic mechanistic representation, so the parameters cannot easily be associated with particular biological processes; however, Wood (1979) has attempted a speculative interpretation. A modified account of the approach pioneered by Wood is now given.

The continuous equivalent of equation (11.2) is

$$y(t) = at^b \mathrm{e}^{-ct} \tag{11.3}$$

where $y(t)$ is the rate of milk production at time $t$ [kg day$^{-1}$], $t$ is in days, $a$, $b$ and $c$ are parameters; $a$ and $c$ have units of kg day$^{-(b+1)}$ and day$^{-1}$ respectively. $Y_i$ is defined as the animal's milk production over the time interval $t_{i-1}$ to $t_i$, so that

$$Y_i = \int_{t_{i-1}}^{t_i} y(t)\,\mathrm{d}t \tag{11.4}$$

Substituting for $y(t)$ with equation (11.3), therefore

$$Y_i = \int_{t_{i-1}}^{t_i} at^b e^{-ct}\, dt$$

$$= a\left[\int_0^{t_i} t^b e^{-ct}\, dt - \int_0^{t_{i-1}} t^b e^{-ct}\, dt\right] \tag{11.5}$$

The incomplete gamma function, $\gamma(q,x)$, is defined by (Abramowitz and Stegun, 1965, p. 260)

$$\gamma(q,x) = \int_0^x z^{q-1}\, e^{-z}\, dz \tag{11.6}$$

where $z$ is a dummy variable of integration, and $q$ and $x$ are the arguments of the function. By substituting $ct = z$ and $b = q - 1$ in equation (11.5), and using equation (11.6), equation (11.5) becomes

$$Y_i = \frac{a}{c^{b+1}}\,[\gamma(b + 1, ct_i) - \gamma(b + 1, ct_{i-1})] \tag{11.7}$$

Thus, given values of $a$, $b$ and $c$, the milk production over any portion of the cow's lactation can be evaluated from equation (11.7). The appropriate values of the incomplete gamma function can either be computed directly, for example by using the algorithm of Bhattacharjee (1970), or taken from tables (Pearson, 1922). Values of the incomplete gamma function, $\gamma(q,x)$, for a range of $q$ and $x$ that are useful in this application, are given in *Table 11.1*.

Equation (11.7) is an exact expression for the milk yield over the time interval $t_{i-1}$ to $t_i$. For many purposes, an adequate approximation to $Y_i$ is (from equation (11.4))

$$Y_i \approx y(\bar{t}_i)\Delta t \tag{11.8}$$

where $\Delta t$ is the length of the time interval and $\bar{t}_i$ is the value of the time variable $t$ at the middle of the $i$th interval, so that

$$\bar{t}_i = \frac{1}{2}(t_{i-1} + t_i)$$

Substituting equation (11.3) into equation (11.8), therefore

$$Y = a\bar{t}_i^b \exp(-c\bar{t}_i)\, \Delta t \tag{11.9}$$

The parameters $a$, $b$ and $c$ may be determined by a nonlinear least-squares or maximum-likelihood fitting procedure (*see* Glossary), or more simply by linear regression (*see* Glossary). Taking logarithms of both sides of equation (11.9) gives

$$\ln Y_i = \ln(a\Delta t) + b \ln \bar{t}_i - c\bar{t}_i$$

Thus a regression of $\ln Y_i$ on $\ln \bar{t}_i$ and $\bar{t}_i$ gives estimates of $a$, $b$ and $c$.

Wood (1969) used this method to fit curves to 859 British Friesian cow lactations, accounting for 82.3% of the variation in the logarithm of milk yield. However, Cobby and Le Du (1978) have shown that this method of fitting equations (11.2) or (11.9) to lactation data can result in a poor fit. Better results were obtained by using

**TABLE 11.1. Some values of the incomplete gamma function, $\gamma(q, x) = \int_0^x z^{q-1}\, e^{-z}\, dz$, in the range $1.1 \leqslant q \leqslant 1.4$, $0 \leqslant x \leqslant 3$**

| $x$ | $\gamma(1.10, x)$ | $\gamma(1.15, x)$ | $\gamma(1.20, x)$ | $\gamma(1.25, x)$ | $\gamma(1.30, x)$ | $\gamma(1.35, x)$ | $\gamma(1.40, x)$ |
|---|---|---|---|---|---|---|---|
| 0.00 | 0.000 000 | 0.000 000 | 0.000 000 | 0.000 000 | 0.000 000 | 0.000 000 | 0.000 000 |
| 0.05 | 0.032 820 | 0.027 011 | 0.022 273 | 0.018 398 | 0.015 222 | 0.012 614 | 0.010 467 |
| 0.10 | 0.068 554 | 0.058 377 | 0.049 808 | 0.042 572 | 0.036 448 | 0.031 252 | 0.026 835 |
| 0.15 | 0.104 370 | 0.090 646 | 0.078 881 | 0.068 768 | 0.060 051 | 0.052 520 | 0.045 999 |
| 0.20 | 0.139 618 | 0.122 946 | 0.108 480 | 0.095 892 | 0.084 908 | 0.075 300 | 0.066 876 |
| 0.25 | 0.174 005 | 0.154 858 | 0.138 096 | 0.123 376 | 0.110 414 | 0.098 971 | 0.088 844 |
| 0.30 | 0.207 381 | 0.186 144 | 0.167 424 | 0.150 869 | 0.136 187 | 0.123 131 | 0.111 493 |
| 0.35 | 0.239 665 | 0.216 662 | 0.196 272 | 0.178 139 | 0.161 966 | 0.147 500 | 0.134 529 |
| 0.40 | 0.270 818 | 0.246 323 | 0.224 512 | 0.205 026 | 0.187 565 | 0.171 873 | 0.157 734 |
| 0.45 | 0.300 826 | 0.275 072 | 0.252 056 | 0.231 416 | 0.212 848 | 0.196 096 | 0.180 943 |
| 0.50 | 0.329 690 | 0.302 881 | 0.278 848 | 0.257 228 | 0.237 716 | 0.220 055 | 0.204 025 |
| 0.55 | 0.357 423 | 0.329 733 | 0.304 848 | 0.282 403 | 0.262 092 | 0.243 658 | 0.226 879 |
| 0.60 | 0.384 044 | 0.355 628 | 0.330 035 | 0.306 902 | 0.285 922 | 0.266 837 | 0.249 425 |
| 0.65 | 0.409 580 | 0.380 569 | 0.354 397 | 0.330 697 | 0.309 164 | 0.289 538 | 0.271 599 |
| 0.70 | 0.434 058 | 0.404 570 | 0.377 930 | 0.353 772 | 0.331 790 | 0.311 723 | 0.293 352 |
| 0.75 | 0.457 509 | 0.427 647 | 0.400 639 | 0.376 118 | 0.353 778 | 0.333 360 | 0.314 644 |
| 0.80 | 0.479 965 | 0.449 819 | 0.422 529 | 0.397 730 | 0.375 117 | 0.354 429 | 0.335 445 |
| 0.85 | 0.501 462 | 0.471 108 | 0.443 614 | 0.418 613 | 0.395 800 | 0.374 914 | 0.355 733 |
| 0.90 | 0.522 029 | 0.491 539 | 0.463 908 | 0.438 773 | 0.415 825 | 0.394 805 | 0.375 492 |
| 0.95 | 0.541 703 | 0.511 137 | 0.483 429 | 0.458 217 | 0.435 193 | 0.414 098 | 0.394 709 |
| 1.00 | 0.560 516 | 0.529 925 | 0.502 194 | 0.476 958 | 0.453 910 | 0.432 790 | 0.413 378 |
| 1.05 | 0.578 501 | 0.547 932 | 0.520 222 | 0.495 009 | 0.471 983 | 0.450 885 | 0.431 496 |
| 1.10 | 0.595 691 | 0.565 184 | 0.537 537 | 0.512 386 | 0.489 422 | 0.468 389 | 0.449 061 |
| 1.15 | 0.612 118 | 0.581 707 | 0.554 157 | 0.529 104 | 0.506 240 | 0.485 304 | 0.466 078 |
| 1.20 | 0.627 812 | 0.597 529 | 0.570 106 | 0.545 181 | 0.522 446 | 0.501 642 | 0.482 547 |
| 1.25 | 0.642 802 | 0.612 671 | 0.585 406 | 0.560 636 | 0.538 057 | 0.517 412 | 0.498 478 |
| 1.30 | 0.657 119 | 0.627 163 | 0.600 073 | 0.575 485 | 0.553 086 | 0.532 624 | 0.513 876 |
| 1.35 | 0.670 791 | 0.641 027 | 0.614 133 | 0.589 744 | 0.567 549 | 0.547 290 | 0.528 750 |
| 1.40 | 0.683 842 | 0.654 288 | 0.627 608 | 0.603 435 | 0.581 461 | 0.561 426 | 0.543 109 |
| 1.45 | 0.696 301 | 0.666 971 | 0.640 517 | 0.616 574 | 0.594 834 | 0.575 040 | 0.556 967 |
| 1.50 | 0.708 193 | 0.679 097 | 0.652 881 | 0.629 181 | 0.607 690 | 0.588 144 | 0.570 328 |
| 1.55 | 0.719 544 | 0.690 690 | 0.664 720 | 0.641 273 | 0.620 038 | 0.600 759 | 0.583 212 |
| 1.60 | 0.730 377 | 0.701 771 | 0.676 056 | 0.652 868 | 0.631 900 | 0.612 892 | 0.595 625 |
| 1.65 | 0.740 713 | 0.712 361 | 0.686 905 | 0.663 985 | 0.643 290 | 0.624 563 | 0.607 582 |
| 1.70 | 0.750 575 | 0.722 480 | 0.697 290 | 0.674 640 | 0.654 224 | 0.635 781 | 0.619 094 |
| 1.75 | 0.759 984 | 0.732 149 | 0.707 226 | 0.684 851 | 0.664 716 | 0.646 564 | 0.630 174 |
| 1.80 | 0.768 960 | 0.741 386 | 0.716 732 | 0.694 633 | 0.674 783 | 0.656 923 | 0.640 835 |
| 1.85 | 0.777 521 | 0.750 208 | 0.725 823 | 0.704 002 | 0.684 439 | 0.666 874 | 0.651 090 |
| 1.90 | 0.785 688 | 0.758 635 | 0.734 519 | 0.712 976 | 0.693 698 | 0.676 429 | 0.660 950 |
| 1.95 | 0.793 474 | 0.766 683 | 0.742 834 | 0.721 568 | 0.702 575 | 0.685 603 | 0.670 428 |
| 2.00 | 0.800 901 | 0.774 367 | 0.750 784 | 0.729 793 | 0.711 086 | 0.694 407 | 0.679 538 |
| 2.05 | 0.807 985 | 0.781 704 | 0.758 385 | 0.737 666 | 0.719 241 | 0.702 857 | 0.688 290 |
| 2.10 | 0.814 739 | 0.788 709 | 0.765 650 | 0.745 201 | 0.727 057 | 0.710 961 | 0.696 697 |
| 2.15 | 0.821 179 | 0.795 395 | 0.772 595 | 0.752 412 | 0.734 544 | 0.718 737 | 0.704 771 |
| 2.20 | 0.827 319 | 0.801 778 | 0.779 230 | 0.759 310 | 0.741 717 | 0.726 193 | 0.712 522 |
| 2.25 | 0.833 174 | 0.807 872 | 0.785 572 | 0.765 911 | 0.748 587 | 0.733 342 | 0.719 962 |
| 2.30 | 0.838 754 | 0.813 688 | 0.791 631 | 0.772 224 | 0.755 164 | 0.740 196 | 0.727 104 |
| 2.35 | 0.844 074 | 0.819 237 | 0.797 419 | 0.778 262 | 0.761 462 | 0.746 765 | 0.733 957 |
| 2.40 | 0.849 145 | 0.824 532 | 0.802 949 | 0.784 036 | 0.767 490 | 0.753 061 | 0.740 530 |
| 2.45 | 0.853 980 | 0.829 586 | 0.808 231 | 0.789 557 | 0.773 262 | 0.759 094 | 0.746 836 |
| 2.50 | 0.858 588 | 0.834 407 | 0.813 276 | 0.794 836 | 0.778 786 | 0.764 874 | 0.752 884 |
| 2.55 | 0.862 980 | 0.839 007 | 0.818 094 | 0.799 882 | 0.784 072 | 0.770 409 | 0.758 682 |
| 2.60 | 0.867 166 | 0.843 395 | 0.822 695 | 0.804 706 | 0.789 129 | 0.775 712 | 0.764 241 |
| 2.65 | 0.871 155 | 0.847 582 | 0.827 089 | 0.809 317 | 0.793 967 | 0.780 789 | 0.769 570 |
| 2.70 | 0.874 956 | 0.851 575 | 0.831 284 | 0.813 724 | 0.798 597 | 0.785 652 | 0.774 678 |
| 2.75 | 0.878 579 | 0.855 385 | 0.835 289 | 0.817 934 | 0.803 024 | 0.790 306 | 0.779 572 |
| 2.80 | 0.882 033 | 0.859 018 | 0.839 113 | 0.821 959 | 0.807 258 | 0.794 763 | 0.784 262 |
| 2.85 | 0.885 323 | 0.862 484 | 0.842 763 | 0.825 803 | 0.811 308 | 0.799 029 | 0.788 754 |
| 2.90 | 0.888 458 | 0.865 789 | 0.846 247 | 0.829 477 | 0.815 181 | 0.803 111 | 0.793 058 |
| 2.95 | 0.891 445 | 0.868 941 | 0.849 573 | 0.832 986 | 0.818 884 | 0.807 018 | 0.797 180 |
| 3.00 | 0.894 292 | 0.871 947 | 0.852 748 | 0.836 338 | 0.822 424 | 0.810 757 | 0.801 128 |

a weighted linear regression (*see* Glossary) of $\ln Y_i$ on $\ln \bar{t}_i$ and $\bar{t}_i$. Guest (1961, p. 334) showed that the appropriate weights are proportional to $Y_i^2$, the square of the corresponding milk yield.

Useful approximations to $a$, $b$ and $c$ can be obtained from easily recognized features of the lactation (*Figure 11.1*), namely, the length of the lactation, $t_f$, total milk yield, $Y$ [kg], the time to maximum yield $t_m$, and the average relative rate of decline after peak yield, $\bar{r}$. Taking equation (11.3)

$$y(t) = at^b e^{-ct}$$

and differentiating with respect to time $t$, therefore

$$\frac{dy}{dt} = at^{b-1}(b - ct)e^{-ct} = (b - ct)\frac{y}{t} \qquad (11.10)$$

The maximum value $y$, $y_m$, occurs when $dy/dt = 0$, hence at time $t_m$ where

$$t_m = b/c \qquad y_m = a(b/c)^b e^{-b} \qquad (11.11)$$

The relative rate of change of $y$ is defined by

$$r = \frac{1}{y}\frac{dy}{dt}$$

which with equation (11.10) gives

$$r = \frac{b}{t} - c \qquad (11.12)$$

At the time $t_h$ halfway between the time of maximum yield and the end of lactation, $t_h$ is given by

$$t_h = \frac{1}{2}(t_m + t_f)$$

and substituting this into equation (11.12), and using equation (11.11) to substitute for $b$, therefore

$$r(t_h) = \frac{2b}{t_m + t_f} - c = \frac{2ct_m}{t_m + t_f} - c$$

$$= c\,\frac{t_m - t_f}{t_m + t_f} \qquad (11.13)$$

According to Wood (1979), the relative rate of decline at a point halfway between peak yield and the end of lactation, $r(t_h)$, is an acceptable approximation to the average rate of decline after peak yield, $\bar{r}$, which is easily estimated. The total yield $Y$ is given by taking $t_{i-1} = 0$ and $t_i = t_f$ in equation (11.7), and since $\gamma(b + 1, 0) = 0$, thus

$$Y = \frac{a}{c^{b+1}}\,\gamma(b + 1, ct_f) \qquad (11.14)$$

Summarizing this approximate way of estimating the parameters of the gamma curve, equation (11.13) gives (with Wood's approximation)

$$c = \bar{r}\,\frac{t_m + t_f}{t_m - t_f} \tag{11.15}$$

From equation (11.11)

$$b = ct_m \tag{11.16}$$

The parameter $a$ can be obtained either from the maximum milk yield using equation (11.11)

$$a = y_m\,(c/b)^b\,e^b \tag{11.17}$$

or alternatively from the total yield $Y$ using equation (11.14) and the *Table 11.1* of the incomplete gamma function, giving

$$a = \frac{Yc^{b+1}}{\gamma(b + 1, ct_f)} \tag{11.18}$$

Thus, given estimates of the length of the lactation $t_f$, total milk yield $Y$, the maximum daily yield $y_m$, the time to maximum yield $t_m$, and the average relative rate of decline of yield $\bar{r}$, equations (11.15), (11.16) and (11.17) or (11.18) enable $c$, $b$ and $a$ to be calculated. The whole of the lactation curve can then be evaluated from equation (11.3), which can be useful when simulating milk production, for instance when calculating feeds.

Other models have been discussed in the literature which describe the entire lactation of a dairy cow as well, if not better than Wood's model (equations (11.2) and (11.3)). For example, Nelder (1966) proposed the inverse polynomial

$$y_i = i(k_1 + k_2 i + k_3 i^2)^{-1}$$

where $y_i$ is the average daily milk yield in the $i$th week, and $k_1$, $k_2$ and $k_3$ are parameters.

Cobby and Le Du (1978) proposed

$$y_i = k_1(1 - e^{-k_2 i}) - k_3 i$$

with similar terminology. Dhanoa (1981) suggested a reparametrization of Wood's model (equation (11.2)):

$$y_i = k_1\,i^{k_2 k_3}\,e^{-k_3 i}$$

where the parameter $k_2$ now represents the time to peak yield ($t_m$). Dhanoa and Le Du (1982) put forward the partial adjustment model

$$y_i = k_1\lambda - k_2\lambda i + (1 - \lambda)y_{i-1}$$

where $y_i$ and $y_{i-1}$ are the average daily milk yields in the current ($i$th) and preceding ($i-1$ th) weeks, $k_1$ and $k_2$ are parameters, and $\lambda$ is an adjustment parameter.

Except for the last of these four models, the parameters of the equations are more difficult to determine, as the models do not lend themselves to ordinary linear regression and hence to routine statistical analysis. Further, there are no simple approximate methods of estimating all the parameters from the essential features of the lactation ($t_f$, $Y$, $y_m$, $t_m$ and $\bar{r}$).

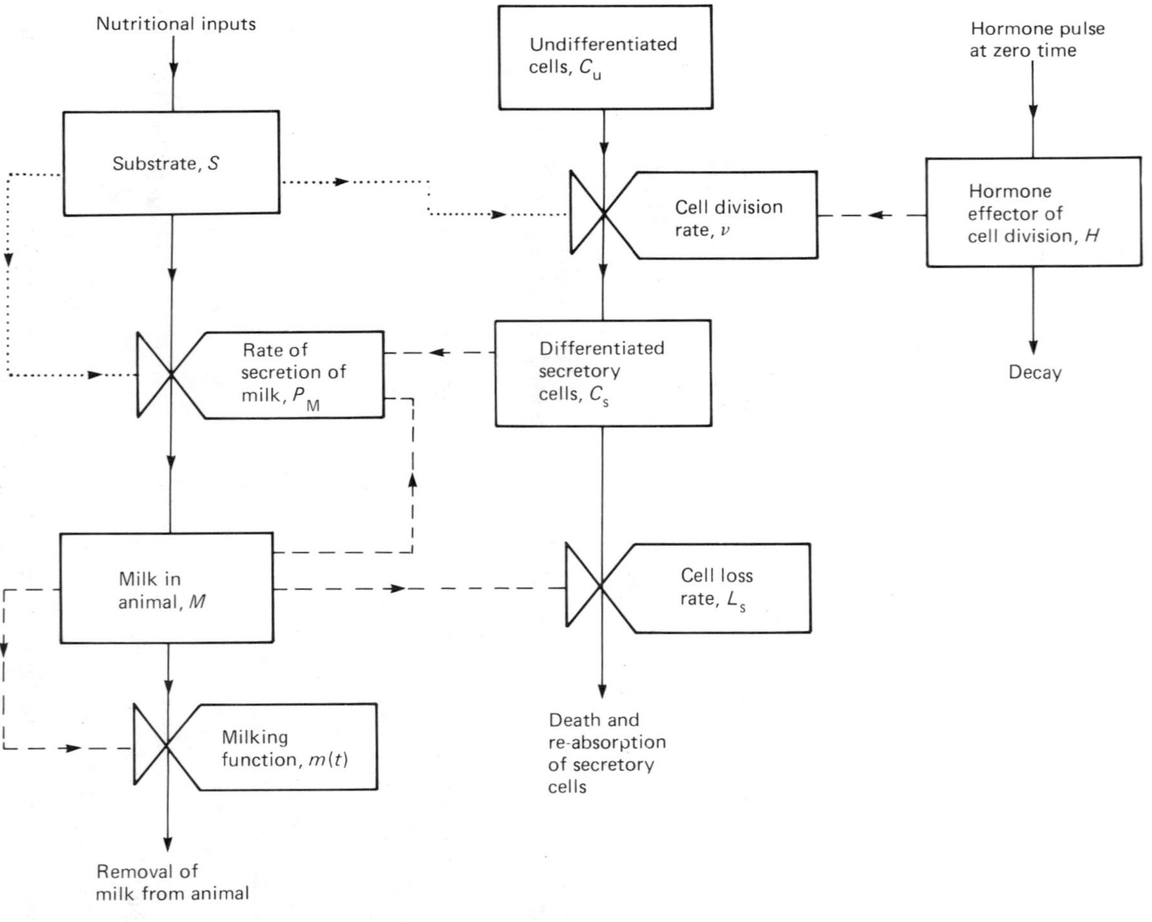


*Figure 11.2* Model of mammary gland. State variables are shown in the boxes; the valve symbols denote processes of transformation or transport. The dashed lines indicate where variables are assumed to affect the rates of processes

Little work has been done on modelling milk production in livestock other than dairy cows, although Torres-Hernandez and Hohenboken (1980) have found Wood's model suitable for describing milk production in lactating ewes.

## Mechanistic model for lactation

The models of the lactation curve in cattle that have been described so far have been primarily empirical (although see Wood, 1979). A more mechanistic model (due to Neal and Thornley, 1983), which utilizes the principles of dynamic modelling discussed in Chapter 2, is now described in detail. The principle objective is to predict the time course of milk production over a single lactation, by means of a simple model of the mammary gland of the dairy cow. The model is connected to the animal in two points: nutritional inputs and a hormonal input. It is based on the number and activity of secretory cells. Only those biological components are included which appear to dominate the system and determine the main features of the typical lactation curve.

The overall scheme is shown in *Figure 11.2* and the principal symbols are listed in *Table 11.2*. It is assumed that the supply of metabolites by the blood for milk synthesis and cell growth in the mammary gland is non-limiting. The mammary gland itself is represented by undifferentiated cells of number $C_u$, cells of number $C_s$ which have differentiated and are adapted to secrete milk, and a storage compartment representing the ducts, alveoli and gland cistern containing $M$ kg

**TABLE 11.2. Principal symbols used in the mammary gland model**

| *Independent variable* | | *Unit* |
|---|---|---|
| $t$ | Time | days |
| *State variables* | | |
| $C_s$ | Number of secretory cells | |
| $H$ | Hormone effector of cell division | kg m$^{-3}$ |
| $M$ | Quantity of milk in animal | kg |
| $\bar{M}$ | Time average of $M$ | kg |
| *Other variables and quantities* | | |
| $m(t)$ | Milking function, a forced variable | kg day$^{-1}$ |
| $Y$ | Total milk yield over lactation | kg |
| *Parameters* | | |
| $C_u$ | Number of undifferentiated cells | |
| $K_H$ | Cell-division rate Michaelis–Menten constant | kg hormone m$^{-3}$ |
| $K_M$ | Milk removal constant | kg |
| $K_R$ | Milk secretion rate Michaelis–Menten constant | kg |
| $k_H$ | Hormone decay rate | per day |
| $k_M$ | Milk secretion constant | kg cell$^{-1}$ day$^{-1}$ |
| $k_r$ | Milk averaging constant | per day |
| $k_s$ | Basal cell degradation rate | per day |
| $k_{sM}$ | Milk-induced cell degradation rate constant | per day |
| $M_h$ | Parameter of equation (11.22) | kg |
| $M_m$ | Milk capacity of animal | kg |
| $q$ | Parameter of equation (11.22) | kg |
| $\nu_m$ | Cell division rate | per day |
| $r_c, r_m$ | Parameters of milking function $m(t)$ | kg day$^{-1}$ |
| $t_1, \ldots$ | Parameters of milking function $m(t)$ | day |

milk. It is postulated that a hormone $H$ controls the rate at which the undifferentiated cells $C_u$ divide to give active secretory cells. The level of milk in the animal, $M$, which if it is high may indicate incomplete removal of milk from the mammary gland, may set up immediate direct and long-term indirect responses mediated by hormones, biochemistry and physical pressure. The time variable $t$ is measured in days, and the period of time under consideration is from parturition, throughout a whole lactation, until the cow is dried off.

The concentration of metabolites in the blood is denoted by a single variable, 'substrate', with concentration $S$ [kg m$^{-3}$].

### HORMONE, $H$

In mammals, onset of lactation is caused by a change of hormone levels at or around parturition. In cattle, several hormones are involved and their concentrations in the blood decrease as lactation proceeds (Mepham, 1976). In the model, these hormones are represented by a concentration of a single hormone, $H$, with units of kg m$^{-3}$. It is assumed that a single pulse of hormone is produced at time $t$ = 0 in response to parturition, and this decays exponentially with rate constant $k_H$ [per day], to give the differential equation

$$\frac{dH}{dt} = -k_H H \quad \text{with} \quad H = H_0 \quad \text{at } t = 0 \tag{11.19}$$

Integrating equation (11.19), therefore

$$H = H_0 e^{-k_H t} \tag{11.20}$$

### DIVISION OF UNDIFFERENTIATED CELLS, $C_u$

The progeny of these cell divisions may be undifferentiated, or may differentiate into specialized cells. It is assumed that the rates of division and differentiation are such that the number of undifferentiated cells does not change, to give

$$\frac{dC_u}{dt} = 0 \quad \text{and} \quad C_u = \text{constant}$$

With binary division, this may be achieved if, for each division producing two cells, one of these cells is committed to differentiation, and the other cell remains undifferentiated. As shown by the dashed line in *Figure 11.2*, the cell division rate $v$ [divisions day$^{-1}$], is influenced by the hormone level $H$, and it is assumed that

$$v = v_m \left( \frac{H}{K_H + H} \right) C_u \tag{11.21}$$

where $v_m$ defines the maximum response, and $K_H$ [kg m$^{-3}$] is a Michaelis–Menten constant giving half-maximal hormone response.

### PRODUCTION AND LOSS OF SECRETORY CELLS, $C_s$

The number of secretory cells continues to increase in early lactation (Mumford, 1964). Here it is assumed that secretory cells are produced at a rate $P_s$ equal to the cell division rate given by equation (11.21), so that

$$P_s = v = v_m \left( \frac{H}{K_H + H} \right) C_u$$

Although differentiating cells will often divide slowly, it is assumed here that the differentiated secretory cells are produced fully mature and no further growth or division occurs.

It is assumed that the cells die after a certain length of time with a specific degradation rate $k_s$ [per day]. However, this degradation rate will be increased if the milk is not removed from the animal over a period of time, due to biochemical and physical pressure effects (Mepham, 1976). Thus, a high value of milk in the animal, $M$, will lead to a higher rate of degradation. It is assumed that it is an average over a recent time interval that causes this effect, and this average is denoted by $\bar{M}$, which is derived from $M$ by equation (11.31) below. The basal degradation rate $k_s$ is supplemented by a term dependent on $\bar{M}$/, to give a total rate of loss of secretory cells $L_s$ given by

$$L_s = -\left\{ k_s + k_{sM} \left[ \frac{(\bar{M}/M_h)^q}{1 + (\bar{M}/M_h)^q} \right] \right\} C_s \tag{11.22}$$

where $k_{sM}$ is a constant [per day] giving the asymptotic value of the second term, $M_h$ [kg] is a parameter giving the half-response point (when $\bar{M} = M_h$, the quantity in square brackets equals 0.5), and $q$ is a dimensionless parameter which determines the steepness of the response. The specific cell loss rate (equal to the quantity in braces) is shown in *Figure 11.3* as a function of $\bar{M}$.

The differential equation for the number of secretory cells $C_s$ is therefore

$$\frac{dC_s}{dt} = P_s - L_s = v_m \left( \frac{H}{K_H + H} \right) C_u - \left\{ k_s + k_{sM} \left[ \frac{(\bar{M}/M_h)^q}{1 + (\bar{M}/M_h)^q} \right] \right\} C_s \tag{11.23}$$

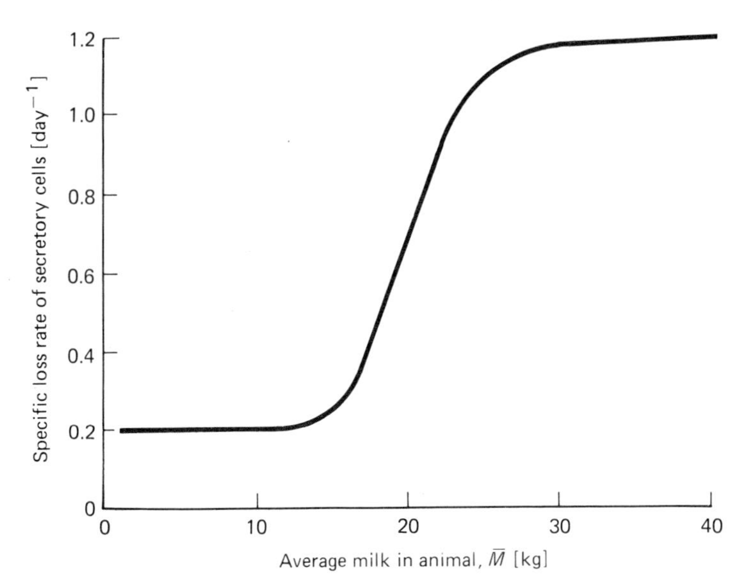


*Figure 11.3* Specific loss rate of secretory cells (the quantity in braces in equation (11.22)) plotted against the averaged amount of milk in the animal, $\bar{M}$ (equation (11.31)). Parameters are: $k_s = 0.2$ per day, $k_{sM} = 1.0$ per day, $M_h = 20$ kg, and $q = 10$

SECRETION AND REMOVAL OF MILK, $M$

It is assumed that the rate of secretion of milk, $P_M$ [kg day$^{-1}$], is proportional to the number of secretory cells $C_s$. According to Knight (1982), secretory cells have a maximum rate of secretion determined by maximum cell size, although here cell size distribution is ignored. The secretory rate is inhibited if the amount of milk in the animal, $M$, approaches the maximum capacity of the animal, $M_m$ [kg] (Mepham, 1976). Thus, in the model, it is assumed that

$$P_M = k_M C_s (M_m - M)/(M_m - M + K_R) \tag{11.24}$$

where $k_M$ is a constant [kg cell$^{-1}$ day$^{-1}$] and $K_R$ is a Michaelis–Menten constant [kg]. $M = M_m - K_R$ gives the half maximal response. In equation (11.24), hormones do not play a direct role in regulating the activity of the secreting cells. A high level of milk in the animal, $M$, directly inhibits milk secretion in equation (11.24), but it should be noted that these high levels also increase the rate of loss of secretory cells, reducing the number of secretory cells (equation (11.23)) and hence milk secretion. Any further reduction, for instance due to the animal becoming pregnant during lactation, is not included.

The present model allows only a very simple representation of the process of milk ejection and removal. Let $R_M$ [kg day$^{-1}$] be the rate of removal of milk from the animal. A milking function $m(t)$ [kg day$^{-1}$] is defined as the potential rate of removal of milk at time $t$ when the quantity of milk is non-limiting; the function $m(t)$ is technically known as a driving function—it is set by the environment, that is, the demands of the suckling calf or of the machine in the milking parlour. If milk in the animal is non-limiting, then $R_M = m(t)$. However, as the amount of milk in the cow falls, the actual rate of removal $R_M$ must fall increasingly below the demand function $m(t)$. It is assumed that the relation between $R_M$ and $m(t)$ is given by

$$R_M = \left(\frac{M}{K_M + M}\right) m(t) \tag{11.25}$$

where $K_M$ is a constant [kg].

A calf removes increasing amounts of milk during the first week or two, and thereafter, its appetite and the cow's milk production are generally in balance (Roy, 1980). It feeds usually between six and ten times a day, taking small amounts. In the model, it is assumed that a continuous low rate is a reasonable approximation for removal of milk by a calf, with a milking function $m_c$(t) given by

$$m_c(\mathrm{t}) = r_c \tag{11.26}$$

where $r_c$ [kg day$^{-1}$] is a constant.

The milking function for machine milking, $m_m(t)$, is approximated by a pulse of constant height lasting for several minutes, possibly repeated two or three times a day. $t^*$ is defined as the decimal part of the time variable $t$, by

$$t^* = \text{decimal part}(t) \tag{11.27}$$

The pulse function $\pi$ is defined by

$$\begin{aligned} \pi(t^*; t_1, t_2) &= 1 \quad \text{for} \quad t_1 < t^* \leqslant t_2 \\ &= 0 \text{ otherwise} \end{aligned} \tag{11.28}$$

With two milking periods a day, the machine milking function $m_m(t)$ is given by

$$m_m(t) = r_m[\pi(t^*; t_1, t_2) + \pi(t^*; t_3, t_4)] \tag{11.29}$$

where $t_1$ and $t_2$ denote the beginning and end of the first milking period, and $t_3$ and $t_4$ those of the second. For example, if the animal is milked at 6 am and 6 pm for periods of 0.01 day (about 15 minutes), $t_1 = 0.25$ day, $t_2 = 0.26$ day, $t_3 = 0.75$ day and $t_4 = 0.76$ day. $r_m$ is a constant [kg day$^{-1}$].

The rate of change of the variable $M$ is given by combining equations (11.24) and (11.25), to give

$$\frac{dM}{dt} = P_M - R_M = k_M C_s \left(\frac{M_m - M}{M_m - M + K_R}\right) - \left(\frac{M}{K_M + M}\right) m(t) \tag{11.30}$$

AVERAGED AMOUNT OF MILK IN ANIMAL, $\overline{M}$

Although the amount of milk in the animal, $M$, appears in equation (11.24), affecting directly and immediately the rate of secretion of milk, it is the length of time that $M$ has a high value that will have an effect on the rate of loss of secretory cells (Mepham, 1976). Indeed, in equation (11.22) for the loss rate of these cells, the variable $\overline{M}$ is introduced as a quantity derived from $M$ by averaging, which introduces some delay. $\overline{M}(t)$ is defined as an average with an exponential weighting factor, to give

$$\overline{M}(t) = k_r \int_{-\infty}^{t} M(\tau)\exp[-k_r(t - \tau)]\, d\tau \tag{11.31}$$

where $k_r$ is a constant [per day] and $\tau$ is a dummy time variable of integration. This equation produces the result that $\overline{M}(t)$ is approximately the average of $M(t)$ values over a time interval of $1/k_r$ days.

Computationally, it is more convenient to use a differential equation than an integral equation. It can be shown that equation (11.31) is equivalent to

$$\frac{d\overline{M}}{dt} = k_r(M - \overline{M}) \tag{11.32}$$

SUBSTRATE, $S$

Since the supply of substrate is assumed to be non-limiting, this implies that the substrate concentration $S$ is maintained at a constant or high value. The variable $S$ then no longer appears in the model. In *Figure 11.2*, broken lines are shown connecting $S$ to the cell division rate (equation (11.21)) and the rate of secretion of milk (equation (11.24)). This is to indicate that the substrate level $S$ might reasonably be expected to influence these two processes; in a model where the nutritional inputs are limiting or are variable, it may be important to write the substrate concentration $S$ explicitly into equations (11.21) and (11.24).

DIFFERENTIAL EQUATIONS OF THE MODEL

In this section, the mathematical definition of the model is summarized. The four significant state variables are $H$, $C_s$, $M$ and $\overline{M}$ ($S$ and $C_u$ are constant). From equations (11.19) and (11.20)

$$\frac{dH}{dt} = -k_H H \quad \text{and} \quad H = H_0 e^{-k_H t} \tag{11.33a}$$

From equation (11.23)

$$\frac{dC_s}{dt} = v_m \left(\frac{H}{K_H + H}\right) C_u - \left\{ k_s + k_{sM} \left[\frac{(\overline{M}/M_h)^q}{1 + (\overline{M}/M_h)^q}\right] \right\} C_s \tag{11.33b}$$

From equation (11.30)

$$\frac{dM}{dt} = k_M C_s \left(\frac{M_m - M}{M_m - M + K_R}\right) - \left(\frac{M}{K_M + M}\right) m(t) \tag{11.33c}$$

From equation (11.32)

$$\frac{d\overline{M}}{dt} = k_r (M - \overline{M}) \tag{11.33d}$$

To complete the numerical definition of the model, four initial values (at time $t = 0$) of $H$, $C_s$, $M$ and $\overline{M}$ are required. The 13 parameters $k_H$, $v_m$, $K_H$, $C_u$, $k_s$, $k_{sM}$, $M_h$, $q$, $k_M$, $M_m$, $K_R$, $K_M$ and $k_r$ must be specified. In addition, the milking function $m(t)$ must be defined, either with equation (11.26) for a suckling calf

$$m(t) = r_c \tag{11.33e}$$

which involves one extra parameter $r_c$, or with equation (11.29) for machine milking with say two milking periods per day

$$m(t) = r_m [\pi(t^*; t_1, t_2) + \pi(t^*; t_3, t_4)] \tag{11.33f}$$

which involves five extra parameters $r_m$, $t_1$, $t_2$, $t_3$ and $t_4$. $t^*$ and the pulse function $\pi$ are defined in equations (11.27) and (11.28):

$$t^* = \text{decimal part}(t) \tag{11.33g}$$

and

$$\begin{aligned} \pi(t^*; t_1, t_2) &= 1 \quad \text{for} \quad t_1 < t^* \leqslant t_2 \\ &= 0 \text{ otherwise} \end{aligned} \tag{11.33h}$$

Integration of equations (11.33a–d) gives the time course of the lactation, with the rate of removal of milk from the animal being given by equation (11.25)

$$R_M(t) = \left(\frac{M}{K_M + M}\right) m(t) \tag{11.33i}$$

and the total milk yield $Y$ [kg] over the lactation of length $t_L$ [days] is

$$Y = \int_0^{t_L} R_M(t)\, dt \tag{11.33j}$$

NUMERICAL ASSUMPTIONS AND RESULTS

The numerical assumptions are fully described in Neal and Thornley (1983). The model was programmed in the simulation language CSMP (p. 33) and the differential equations were solved numerically using Euler's method (p. 21).

The model was tested against experiment and is able to predict a range of experimental data; full details are given by Neal and Thornley (1983).

# Meat production

Meat is the term used to describe the edible flesh and organs of animals and birds acceptable for human consumption. The scientific definition of meat is the *post mortem* aspect of the 300 or so anatomically distinct muscles of the body, together with the connective tissue in which the muscle fibres are deposited and such intermuscular fat as cannot be trimmed off without breaking up the muscle as a whole (Lawrie, 1975). Intramuscular fat, being physically inseparable, is automatically included. Meat production is therefore inextricably linked with the growth of animals and, in particular, the nature and growth of the muscle tissue. Growth in animals, according to Fowler (1968), has two aspects. The first is measured as an increase in weight per unit time, and the second involves changes in form and composition resulting from differential growth of the component parts of the body. The use of mathematical functions to describe growth is discussed in Chapter 5; the application of these methods to describe animal growth has been considerable, and much of the pioneering work in this field was due to Brody (1945). Growth equations have generally been used to describe carcass quantity, and much less attention has been paid to quality and composition.

## Growth equations

The curve of bodyweight growth with time is normally sigmoidal in shape in both animals (Fowler, 1980) and birds (Wilson, 1980). Brody (1945) considered the growth curve as consisting of two distinct phases—the self-accelerating or auto-catalytic phase, when growth is unrestricted by environment, and the self-inhibiting or nutrient-limited phase, when growth is restricted by other factors. Brody assumed that the growth in the first phase is a function of the growth already made, and used a simple exponential function to describe this first phase (equation (5.9) *et seq.*):

$$\frac{\mathrm{d}W}{\mathrm{d}t} = \mu W \qquad (11.34)$$

where $W$ is the bodyweight at time $t$, and $\mu$ is the specific growth rate constant. Integrating equation (11.34) gives

$$W = W_0 \mathrm{e}^{\mu t} \qquad (11.35)$$

where $W_0$ is the value of $W$ at time $t = 0$. In the self-inhibiting phase, Brody assumed that the growth rate is a function, not of the growth already made, but of the growth yet to be made to reach maturity; he used a monomolecular growth function (equation (5.15)) to describe this phase:

$$\frac{\mathrm{d}W}{\mathrm{d}t} = k(W_\mathrm{f} - W) \qquad (11.36)$$

where $W$ and $t$ are defined as above, $W_\mathrm{f}$ is the final or mature weight, and $k$ is a constant describing the growth rate in relation to the growth yet to be made. Integrating equation (11.36) gives

$$W = W_\mathrm{f} - B\mathrm{e}^{-kt} \qquad (11.37)$$

where $B$ is a constant. Brody used equations (11.35) and (11.37) for comparing different species and different breeds of the same species. Note that it is possible to interpret the limited growth of the monomolecular equation (11.37) as arising from the exhaustion of a limited supply of a nutrient, or being due to an approach to a predetermined final size.

Brody's original model has now been superseded by the use of a Gompertz function (equation (5.29)) to describe animal growth with time:

$$\frac{dW}{dt} = \mu_0 W e^{-Dt} \tag{11.38}$$

where $W$ and $t$ are as previously defined, and $\mu_0$ and $D$ are constants. Many applications of this equation have been reported. For example, Laird, Tyler and Barton (1965) and Laird (1965) used this function to predict organ weights within a species from early embryonic life to maturity with considerable accuracy. The Agricultural Research Council (1980) used a Gompertz equation to describe pregnancy in cattle and sheep:

$$\frac{dE}{dt} = aE e^{-bt}$$

where $t$ = time [days] since conception, $E$ is the energy stored [MJ] in the gravid uterus at time $t$, $a = 0.0201$ for cattle or $a = 0.0737$ for sheep, and $b = 0.000\,0576$ for cattle or $b = 0.006\,43$ for sheep. (These values of $a$ and $b$ relate to a calf birthweight of 40 kg and a lamb birthweight of 4 kg; the energy retentions are in proportion for other birthweights.) Wilson (1980) showed that growth in birds for meat production is also well described by a Gompertz curve. An advantage of the Gompertz function of equation (11.38) is that one can attempt to separate out the effects of nutrition (this may be viewed as modifying $\mu_0$), of developmental rate $D$, and of the initial weight of the animal or tissue ($W = W_0$ at $t = 0$).

Other more complex dynamic growth models have been described in the literature. For example, Baldwin and Black (1979), in modelling the effects of nutritional and physiological status on the growth of animal tissues, used a series of differential equations to describe the growth rate of nine tissues, organs and organ groups.

## Allometry

A growth model that has been widely used in animal production is the allometric equation (equation (5.66)):

$$P = aQ^b$$

where $P$ and $Q$ are variables, and $a$ and $b$ are constants. Brody (1945) observed that, during the growth of a body, its surface area tends to increase more rapidly than its linear size, and its weight or volume tends to increase more rapidly than its surface area. In other words, linear size, surface area, and weight or volume are all allometrically related. Brody then went on to use the allometric equation to estimate the weight of cattle from their girth, the nutritive condition of cattle from height at the withers, the amount of wool or feathers from the bodyweight, and organ weight from body weight. Following Brody's early work, the allometric equation has been extensively used in animal production modelling. For example, Barton and Laird (1969) and Mukhoty and Berg (1971) used it to relate organ or tissue weight to bodyweight. Adolph (1949) and Munro (1969) applied the relationship to describe a number of physiological and metabolic functions. The Agricultural Research Council (1980) used the allometric equation to calculate the fasting metabolism of cattle and sheep, $F$ [MJ day$^{-1}$], from their liveweight, $W$ [kg]:

$$F = 0.503W^{0.67} \quad \text{for cattle}$$

and

$$F = 0.208W^{0.75} \quad \text{for sheep}$$

Whittemore and Fawcett (1974, 1976) used allometric equations, together with other empirically derived relationships, to model pig production. Roux (1976) incorporated both the allometric equation and the Gompertz function in a model for describing whole-animal growth with time, and the allometric equation described the growth of one animal component relative to another.

### Carcass composition

Another important aspect of meat production is the assessment of carcass composition. Carcasses are sold to butchers on the basis of weight, shape, and some estimate of the proportions of lean and fat they contain. Linear regression models (*see* Glossary) are often used to evaluate carcass composition. For example, Williams *et al.* (1974) used linear regression to determine the composition of beef carcasses from observations on intact and quartered sides and partial dissection data. Harries, Williams and Pomeroy (1975) applied the method to predict an index of retail value from side weight and the proportional area of muscle (determined from photographs of the cut surface after quartering). Harries, Williams and Pomeroy (1976) used linear regression models for comparing the retail value of beef carcasses from different groups of animals.

## Egg production

The major constituent of an egg, apart from water, is protein. A typical chicken's egg weighs about 58 g and contains just under 7 g of protein. Of this about 42% is yolk protein, which is synthesized in the liver; about 54% is egg white protein, which is mainly synthesized in the magnum region of the oviduct; and the remaining 4% or so is contained in the shell and associated membranes, which originate from the isthmus and uterus regions of the oviduct (Svensson, 1964; Fisher, 1980). Egg production in laying birds is a reproductive activity and represents a high level of material turnover (Gilbert, 1971). The dietary supply of essential amino acids is a major determinant of egg production, and is the factor most amenable to short-term manipulation. In practical terms, the nutritional objective is normally to supply sufficient dietary protein to meet the needs for maximum productivity.

Combs (1960) was one of the first to quantify the dietary requirement of a laying hen for an essential amino acid when he derived the following equation, using linear regression, for describing the methionine requirement:

$$A = 5.0E + 50.0W + 6.2\Delta W$$

where $A$ is methionine intake [mg day$^{-1}$], $E$ egg production [g bird$^{-1}$ day$^{-1}$], $W$ bodyweight [kg], and $\Delta W$ bodyweight gain [g day$^{-1}$]. Similar equations for other amino acids in both laying hens and growing chicks were subsequently derived (Combs, 1968; Thomas, Twining and Bossard, 1977; and others). In each case, regression techniques were applied to performance data of either individual birds or groups of birds, for which it could be assumed that the supply of amino acid in question was the factor limiting production.

A more mechanistic approach was adopted by Hurwitz and Bornstein (1973), who proposed two models for the estimation of the essential amino acid requirements of the laying hen. Amino acid requirements were partitioned into three components, namely maintenance, growth and egg production. Both models assumed the same requirements for maintenance and growth, and it was assumed that yolk synthesis was a continuous process in which the supply of amino acids for yolk protein synthesis was derived directly from the diet. The phasic nature of egg white protein synthesis was the basis of both models. Model I assumed that egg white and shell membrane (membrane plus shell) proteins were synthesized from tissue protein at the time of secretion, and the model was stated thus:

$$A = [k_m W + 0.3k_t + 121rE_w(0.44k_y + 0.56 \times 2k_t)]/0.85$$

where the variable $A$ [mg day$^{-1}$] denotes the dietary amino acid requirement of the laying hen, $W$ [kg] the hen's bodyweight, $r$ [eggs bird$^{-1}$ day$^{-1}$] the rate of egg production, and $E_w$ [g] the egg weight. The parameter $k_m$ [mg kg$^{-1}$ day$^{-1}$] is the experimentally derived amino acid requirement per unit of bodyweight for maintenance, $k_t$ [mg g$^{-1}$] the amino acid content of tissue protein, and $k_y$ [mg g$^{-1}$] the amino acid content of yolk protein. The constants in Model I were based on the following assumptions:

1. a hen makes a constant weight gain between the onset of egg production and the end of the laying period in which tissue protein is laid down at the rate of 0.3 g day$^{-1}$;
2. egg protein accounts for 12.1% of egg weight;
3. the proportion of egg protein in egg yolk is 44% and in egg white and shell membrane 56%;
4. the amount of sulphur amino acids in egg white protein is twice as high as that in tissue protein; and
5. the efficiency of absorption of dietary amino acids is 85%.

Model II, on the other hand, assumed that all oviduct products were synthesized continuously from dietary amino acids, except for the ovoglycoproteins and shell membrane protein, which were synthesized from tissue protein during secretion. Model II was stated thus:

$$A = [k_m W + 0.3k_t + 121rE_w(0.44k_y + 0.42k_o + 0.1 \times 2.2k_t + 0.04 \times 4k_t)]/0.85$$

where parameter $k_o$ [mg g$^{-1}$] is the amino acid content of ovoglycoprotein and the other parameters and variables are as before. The constants in Model II were based on assumptions (1), (2) and (5) plus the following two assumptions:

6. the proportions of the protein components of the egg are—egg yolk 44%, ovalbumins 42%, ovoglycoprotein 10%, and shell membrane protein 4%; and
7. the amounts of sulphur amino acids in ovoglycoprotein and shell membrane protein are 2.2 and 4 times respectively that in tissue protein.

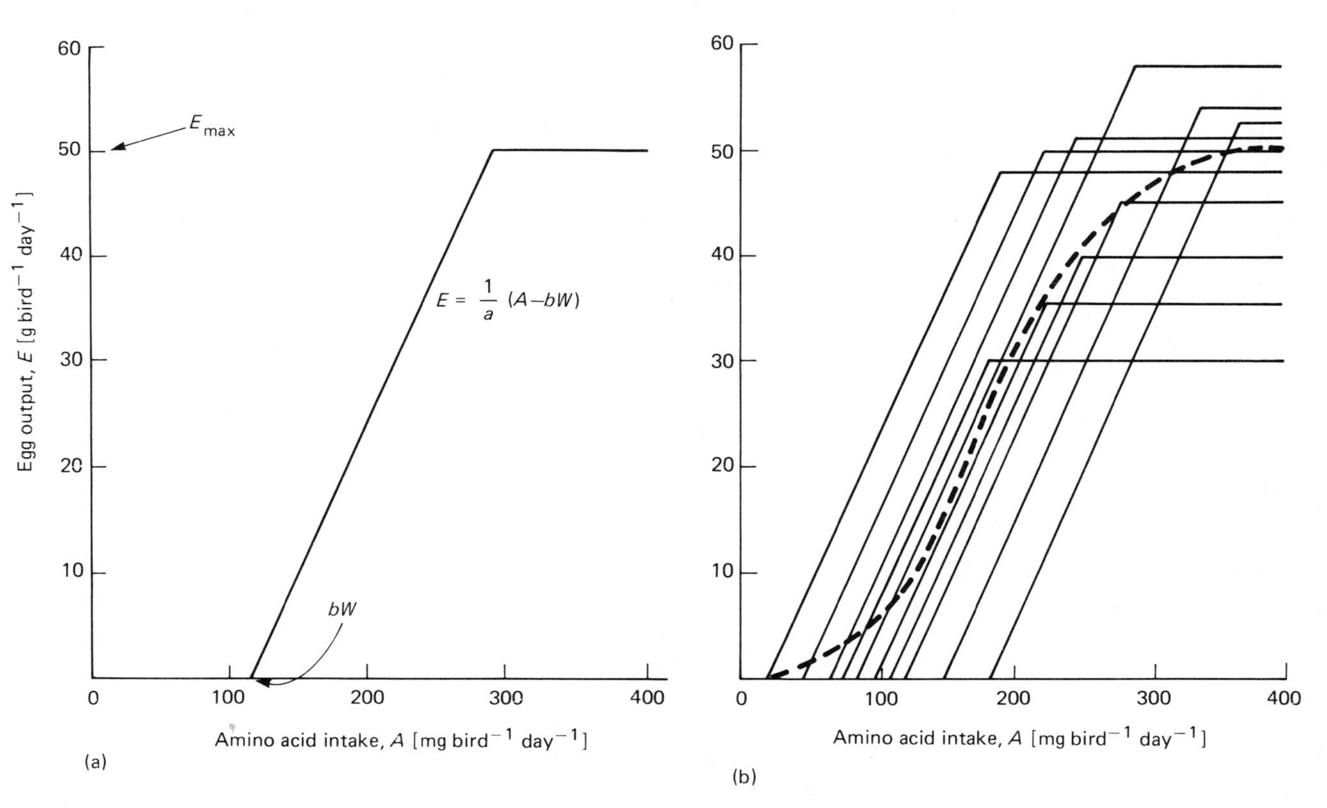


*Figure 11.4* A model for the response of laying hens to amino acid intake: (a) the response of an individual bird, (b) individual (——) and mean (– – –) responses for a small group of birds. (After Fisher *et al.*, 1973)

When compared with existing literature values for amino acid requirements, Hurwitz and Bornstein showed that Model II gave closer estimates of requirements than Model I. In subsequent experimental verification of the models, Hurwitz and Bornstein (1977) have shown that Model II provided a more accurate estimate of the requirements of essential amino acids (with the possible exception of methionine) than did Model I. Diets based on Model II consistently produced an egg mass close to the target, while diets based on Model I allowed hens to exceed the target output, showing that Model I provided an excess of all essential amino acids. Smith (1978a), in an extension of Hurwitz and Bornstein's work, suggested a series of refinements and then, using a similar approach, went on to develop two alternative models for estimating the amino acid requirement (Smith, 1978b).

Fisher, Morris and Jennings (1973) describe a model which determines the response of a group of laying hens to different levels of amino acid intake. The model is based on the assumption of simple linear relationships between amino acid intake, $A$ [mg bird$^{-1}$ day$^{-1}$], and egg production, $E$ [g bird$^{-1}$ day$^{-1}$], and maintenance requirement [mg bird$^{-1}$ day$^{-1}$] for individual birds. It is assumed that each individual bird has a characteristic maximum level of egg output $E_{max}$ and that, for each bird

$$A = aE + bW \quad \text{when} \quad E < E_{max}$$

where $W$ [kg] is the weight of the bird, and $a$ [mg g$^{-1}$] and $b$ [mg kg$^{-1}$ bird$^{-1}$ day$^{-1}$] are the quantities of amino acid associated with a unit of $E$ and a unit of $W$ respectively. It also assumes that $E = 0$ when $A < bW$, thus excluding negative egg production. These relationships are illustrated in *Figure 11.4a*. When a group of birds is considered, the response is the average of the responses for individual birds. This is illustrated for a small group in *Figure 11.4b*; the parameters $a$ and $b$ are assumed to be invariate. The exact equation for describing the group response curve, together with methods for estimating the parameters and their associated statistical properties, are given by Curnow (1973). As well as Curnow's exact approach, Fisher, Morris and Jennings (1973) discuss Monte Carlo simulation methods (*see* Glossary) for determining the curve.

The model has potentially wider application and can be used in any situation where a smooth response curve is required for an animal or plant population, in which an individual animal or plant starts to respond to a stimulus only when that stimulus exceeds a certain level $X_0$, and where the response is then linear until a plateau is reached, after which the response stays constant at $Y_1$. The values of $X_0$ and $Y_1$ may be different for different individuals in the population.

Attention has also been given to developing econometric forecasting models of national and regional egg production which predict trends in the major variables affecting egg markets—chick placings, flock size, prices and so on. Such models are generally based on time-series analysis (*see* Glossary) or regression procedures, and examples include Kulshreshta (1971), Roy and Johnson (1973), and Hallam (1981).

## Exercises

### Exercise 11.1

A Holstein Friesian cow is expected to produce milk according to the pattern of *Figure 11.1* (p. 220), with a length of lactation of 44 weeks, a total milk yield of 6500 kg, time to peak yield of 8 weeks and an average relative rate of decline after the peak of 3% a week. Evaluate the expected milk production in each four-week (28 day) period of the lactation.

### Exercise 11.2

Show that

$$\int_0^t e^{-a\exp(-bx)} e^{-cx}\, dx = \frac{1}{ba^{c/b}}\left[\gamma(c/b, a) - \gamma(c/b, ae^{-bt})\right]$$

and evaluate the integral for $t = \infty$, $a = 2$, $b = 1$, $c = 1.2$.

### Exercise 11.3

If a hen's egg production is described by the differential equation

$$\frac{dE}{dt} = k_1 A - k_2 W^{0.67}$$

where $E$ is egg production [kg], $A$ is amino acid intake [g day$^{-1}$], $W$ is the liveweight of the bird [kg], and $t$ is time [days], what are the units of the parameters $k_1$ and $k_2$?

Chapter 12

# Farm planning and control: I

## Introduction

In this and the next chapter, an account is given of models designed for application in farm planning and control. Farm planning and control are generally concerned with the organization of a farm's resources—such as land, labour, machinery, buildings and capital, and the management of particular enterprises undertaken on the farm—for example, a dairy herd, sugar beet production, or a sheep flock (*see* Barnard and Nix, 1973). The material presented is split into four parts. Part one covers resource allocation models and parts two to four cover management models of whole (and components of) enterprises. The enterprise management models are considered under the headings: crop production, grazing and conservation, and livestock production. Resource allocation, crop production, and grazing and conservation are dealt with in this chapter; livestock production is dealt with in Chapter 13.

Much of the subject matter of these two chapters falls within the realm of *agricultural systems*, a topic concerned with the analysis of production systems, enterprises and farming systems (Spedding, 1979). Computer simulation is the tool of the agricultural systems analyst (Dent and Blackie, 1979) and there are numerous such modelling applications reported in the literature, of which a wide cross-section can be found in Dent and Anderson (1971) and Dalton (1975). Systems simulation reports can be lengthy and generally lack a mathematical statement of the model. For these reasons, they are generally difficult to present in summary form and so relevant applications are merely cited here; attention is concentrated on models that can be stated concisely in mathematical form and illustrate the application of a particular method.

## Resource allocation

Resource allocation problems arise whenever there is a number of activities to be performed and there are limitations on either the resources available or on how they can be utilized. In such situations, the objective is to allocate the available resources to the activities in an optimal way. Many resource allocation problems encountered in practice are amenable to solution by mathematical programming

(Chapter 3) and allied techniques, and a general introduction to such problems can be found in most standard operational research or mathematical programming textbooks (e.g. Taha, 1971).

Resource allocation models developed at the farm management level fall into two classes—those which recognize risk and uncertainty, and those which ignore it. Risk and uncertainty in farm planning have been a subject of considerable academic interest amongst agricultural economists in recent years, and many elegant models have been constructed using decision theory, mathematical programming techniques or mixtures of both. Such models tend to pivot around a utility function constructed to describe the preference of a farmer or group of similar farmers. The construction of utility functions is often beset with difficulties, owing to the practical problems of quantifying risk and uncertainty. As a result, these models have found but limited application in agriculture and agricultural research generally. They are considered to be outside the scope of this book, and thus are omitted from our discussion here. The interested reader is referred to Hardaker (1979) for a review of this class of model, and to Roumasset, Boussard and Singh (1979) for a more detailed discourse on risk and uncertainty in agriculture. Here, we concern ourselves with deterministic resource allocation problems where things are known with certainty and the farmer's utility function can be represented as straightforward profit maximization or cost minimization. To illustrate this type of allocation problem, we describe a much reduced version of a model developed by Audsley, Dumont and Boyce (1978) which demonstrates the application of linear programming, which is essentially a static technique, to problems having a dynamic component. The problem considered here (adapted from Audsley, 1979) is concerned with allocating land to particular crops and determining a farmer's labour and machinery needs for his arable enterprise and though hypothetical, it serves to illustrate the amenability of many farm planning and control problems to solution by mathematical programming.

The farmer grows two crops $X$ and $Y$ on 160 areal units of land. He wishes to allocate land, labour and machinery in such a way as to maximize his profits from these crops. Each crop needs three operations—ploughing, drilling and harvesting. Four periods of the year are set aside for performing these tasks and the time available in each period, the timing of each operation, together with the time required to perform it, are shown below.

| *Period* | *Time available* [hours] | *Time required* [hour (unit area)$^{-1}$] | | | | | |
|---|---|---|---|---|---|---|---|
| | | *Crop X* | | | *Crop Y* | | |
| | | *Plough* | *Drill* | *Harvest* | *Plough* | *Drill* | *Harvest* |
| 1 | 5 | 0.5 | — | — | — | — | — |
| 2 | 10 | — | 0.3 | — | 0.5 | — | — |
| 3 | 15 | — | — | 0.75 | — | 0.6 | — |
| 4 | 20 | — | — | 0.75 | — | — | 2.0 |

Crop $X$ may be harvested in either period 3 or 4, but there is a timeliness cost of 0.2 units (unit area)$^{-1}$ associated with crop $X$ harvested in period 4. Let $x_3$ and $x_4$ denote those areas allocated to crop $X$ that are harvested in period 3 and 4 respectively, and let $y$ be the land allocated to crop $Y$. The areas allocated to each crop are limited by the total land area available, therefore

$$x_3 + x_4 + y \leqslant 160 \tag{12.1}$$

The same type of self-propelled harvester can be used on each crop. Let $\delta_H$ be the number of harvesters required to harvest the crops. $\delta_H$ can only assume non-negative, integer values (0, 1, 2, 3, etc.). We thus have the constraints

$$\text{period 3: } 0.75x_3 \leqslant 15\delta_H \tag{12.2}$$

$$\text{period 4: } 0.75x_4 + 2.0y \leqslant 20\delta_H \tag{12.3}$$

A tractor is needed for ploughing and drilling; the harvesting operation also requires a tractor in addition to a harvester. Let $\delta_T$ be a non-negative integer variable denoting the number of tractors needed, therefore,

$$\text{period 1: } 0.5(x_3 + x_4) \leqslant 5\delta_T \tag{12.4}$$

$$\text{period 2: } 0.3(x_3 + x_4) + 0.5y \leqslant 10\delta_T \tag{12.5}$$

$$\text{period 3: } 0.75x_3 + 0.6y \leqslant 15\delta_T \tag{12.6}$$

$$\text{period 4: } 0.75x_4 + 2.0y \leqslant 20\delta_T \tag{12.7}$$

The labour requirements are as follows. Ploughing and drilling both require a man to drive the tractor. To harvest crop $X$, two men are needed to operate the harvester and one to drive a tractor; crop $Y$, on the other hand, requires one man to drive the harvester and another the tractor. Again, let $\delta_M$ be an integer variable denoting the total number of men required, then crop production is restricted by available man-hours in the following way:

$$\text{period 1: } 0.5(x_3 + x_4) \leqslant 5\delta_M \tag{12.8}$$

$$\text{period 2: } 0.3(x_3 + x_4) + 0.5y \leqslant 10\delta_M \tag{12.9}$$

$$\text{period 3: } 3 \times 0.75x_3 + 0.6y \leqslant 15\delta_M \tag{12.10}$$

$$\text{period 4: } 3 \times 0.75x_4 + 2 \times 2.0y \leqslant 20\delta_M \tag{12.11}$$

Crop $Y$ can only be grown on the same piece of land one year in two and crop $X$ two years in three. The farmer intends to grow both crops every year and therefore has to include this crop rotation in his planning. Such a rotation is allowed for by introducing the constraints:

$$y \leqslant \frac{160}{2} \tag{12.12a}$$

and

$$x_3 + x_4 \leqslant \frac{2}{3} \times 160 \tag{12.12b}$$

The income received for crop $X$ and $Y$ is 2 and 4 units (unit area)$^{-1}$ respectively. The respective annual costs of labour, harvesters and tractors are 15 units man$^{-1}$, 3 units machine$^{-1}$ and 1.5 units machine$^{-1}$. The farmer's profit function, $Z$ [units year$^{-1}$], is therefore:

$$Z = 2x_3 + (2 - 0.2)x_4 + 4y - 15\delta_M - 3\delta_H - 1.5\delta_T \tag{12.13}$$

which he wishes to maximize.

To summarize, this problem has been formulated as a mixed-integer linear programme (p. 58) in six decision variables ($x_3$, $x_4$, $y$, $\delta_H$, $\delta_T$ and $\delta_M$) and 13 constraints (equations (12.1)–(12.12)), namely

$$\text{maximize} \quad Z = 2x_3 + 1.8x_4 + 4y - 15\delta_M - 3\delta_H - 1.5\delta_T$$

subject to

$$
\begin{array}{rrrrrrll}
x_3 + & x_4 + & y & & & \leqslant & 160 & \text{[land constraint]} \\
0.75x_3 & & & -15\delta_H & & \leqslant & 0 & \left.\right\} \text{[harvester constraints]} \\
 & 0.75x_4 + & 2.0y & -20\delta_H & & \leqslant & 0 & \\
0.5x_3 + & 0.5x_4 & & & -5\delta_T & \leqslant & 0 & \\
0.3x_3 + & 0.3x_4 + & 0.5y & & -10\delta_T & \leqslant & 0 & \left.\right\} \text{[tractor constraints]} \\
0.75x_3 & & +0.6y & & -15\delta_T & \leqslant & 0 & \\
 & 0.75x_4 + & 2.0y & & -20\delta_T & \leqslant & 0 & \\
0.5x_3 + & 0.5x_4 & & & -5\delta_M & \leqslant & 0 & \\
0.3x_3 + & 0.3x_4 + & 0.5y & & -10\delta_M & \leqslant & 0 & \left.\right\} \text{[labour constraints]} \\
2.25x_3 & & +0.6y & & -15\delta_M & \leqslant & 0 & \\
 & 2.25x_4 + & 4y & & -20\delta_M & \leqslant & 0 & \\
 & & 2y & & & \leqslant & 160 & \left.\right\} \text{[rotational constraints]} \\
3x_3 + & 3x_4 & & & & \leqslant & 320 & \\
 & & & x_3, x_4, y, \delta_H, \delta_T, \delta_M & & \geqslant & 0 & \text{[non-negativity conditions]} \\
 & & & \delta_H, \delta_T, \delta_M & & = & \text{an integer} & \text{[integer conditions]}
\end{array}
$$

The optimal solution to the problem is $Z = 204.0$, $x_3 = 80.0$, $x_4 = 0.0$, $y = 80.0$, $\delta_H = 8$, $\delta_T = 8$ and $\delta_M = 16$.

## Crop production

Fewer models have been developed to aid the decision-making process in crop production than in other areas of farm enterprise management, and most of the accounts published have been systems simulations. Here, in order to illustrate a more mathematical approach, we focus on an application of dynamic programming (p. 63) to wheat production in the Great Plains of the USA, reported by Burt and Allison (1963).

Rainfall during the growing season and soil moisture at planting time are critical to the subsequent yield of wheat. Rainfall is outside the farmer's control but soil moisture at planting time can, to an extent, be influenced by fallowing, as cropping a piece of land the previous year serves to reduce its soil moisture level at planting. Typically, the wheat farmer aims to maximize his total return over a number of years from an area of land, and thus it may be optimal for him to fallow it some of the time. Essentially, the farmer's problem is a multi-stage decision process in which he has to decide at the beginning of each year whether to plant or to fallow. The model of Burt and Allison (1963) computes the optimal strategy the farmer should pursue in order to maximize the total returns over his planning horizon.

The basic details of the model are as follows. The stage of this multi-decision process is defined for a time interval of one year, the beginning of which is wheat planting time. The $n$th stage denotes the $n$th year from the *end* of the farmer's planning horizon ($n = 1, 2, \ldots, N$). Soil moisture at the beginning of a stage is the state variable, whose value falls into one of $M$ discrete classes of soil moisture level. During the passage of a year, soil moisture may change from one soil moisture level to another; in other words, as the process goes from the $n$th to the $(n-1)$th stage, there is movement from the $i$th to the $j$th state. We are looking for a management policy that will specify whether to plant wheat or fallow, given a specific state and stage, which maximizes expected return during the entire planning horizon of the farmer. A policy for a given year is defined by a decision, plant or fallow, for each soil moisture level. Burt and Allison (1963) define the following recurrence relation for the problem:

$$f_n(i) = \max_k \left[R_i^{(k)} + \beta \sum_{j=1}^{M} p_{ij}^{(k)} f_{n-1}(j)\right] \qquad (n = 1, 2, \ldots, N;\ i = 1, 2, \ldots, M) \tag{12.14}$$

where

$f_n(i)$ = the discounted expected return from an $n$-stage decision process under an optimal policy, given a soil moisture level of $i$

$k$ = the decision variable, where $k = W$ and $k = F$ indicate decisions to plant wheat and to fallow respectively

$\beta$ = a discount factor which is the reciprocal of one plus the relevant interest rate

$R_i^{(k)}$ = the expected immediate returns under decision $k$ given a soil moisture level of $i$. Under the decision to fallow (i.e. $k = F$), the returns are the negative of fallow costs and are the same regardless of soil moisture

$p_{ij}^{(k)}$ = the probability of moving from the $i$th to the $j$th state under decision $k$

$f_0(i)$ is assumed to be a constant for all $i$ and is arbitrarily set equal to zero. The recurrence relation (equation (12.14)) calculates $f_n$ from $f_{n-1}$, where $n$ is measured from the end of the farmer's planning horizon, and this computational procedure is therefore an example of dynamic programming of the *backward* type (p. 66). It can be shown that the recurrence relation (equation (12.14)) leads to a constant decision made for large $N$.

Application of equation (12.14) enables optimal management policy to be determined, which specifies whether to plant wheat or fallow given a specific state and stage, so that the farmer's expected return over his entire planning horizon is maximized. Computational details of applying the algorithm are given by Burt and Allison (1963). The authors use their model to demonstrate that an optimal policy based on soil moisture at wheat planting time gives a considerably better return than either a policy of continuous wheat or alternating fallow and wheat.

## Grazing and conservation

### Grazing

Grasslands and grazing lands account for nearly a quarter of the world's land surface and are mostly exploited as a basis for grazing systems. It is hardly surprising, therefore, that several attempts have been made to develop grazing models to aid the management process. Most of the models reported have been systems simulations, usually programmed in FORTRAN and of a semi-empirical nature. Typical of these are three recent attempts by Christian *et al.* (1978), by Sibbald, Maxwell and Eadie (1979), and by France, Brockington and Newton (1981). Christian *et al.*'s model simulates a farm producing prime lambs on improved pasture in temperate regions of Australia. Sibbald, Maxwell and Eadie's model is concerned with investigating alternative management strategies for improved systems of Scottish hill sheep production. France *et al.* describe the simulation of an English lowland sheep flock grazing a perennial ryegrass sward; the model is constructed to compare continuous and rotational grazing systems, and to investigate the effects on stocking rate of different levels of nitrogen fertilizer application. Such models, whilst often of limited predictive value and explanatory power, can provide a sound logical framework for reviewing the information available on the various components of this difficult subject area, thus highlighting the most urgent requirements for further applied research. In this section, we focus on two simple grazing models, one due to Morley (1968), and the other by Noy-Meir (1976, 1978), which illustrate some of the principles of dynamic mathematical modelling.

Morley (1968) uses the pasture growth curves of Brougham (1956) to examine variations in systems of rotational grazing. Brougham fitted the logistic growth equation (cf. equation (5.22))

$$W = a/(1 + be^{-ct})$$

where $W$ is herbage mass at time $t$ and $a$, $b$ and $c$ are parameters, to three sets of data describing the regrowth of a sward after grazing. Three curves were thus obtained, each with a unique set of parameter values. Morley analysed the effect of varying the interval between grazings on production of plant material by examining 1000 regrowth curves, generated by varying $a$, $b$ and $c$ over a range of values suggested by Brougham's analysis. For each curve, the average growth rate ($W/t$) was considered and the optimum period between grazings, $t_{max}$, defined as the time taken to reach the *maximum* average growth rate. $t_{max}$ is therefore given by

$$\frac{d(W/t)}{dt} = 0$$

that is

$$be^{-ct}(ct - 1) = 1$$

$t_{max}$ was calculated for each curve generated, along with the times at which 90% of the maximum average growth rate was achieved, and the range between these times about the optimum. The optimal interval between grazings was found to range between 5 and 12 weeks, and Morley showed that only two grazing intervals during a year are sufficient to maintain pasture growth rates within 90% of optimum.

The optimum number of subdivisions was determined in the following way. The amount of herbage available for consumption, $C$ [kg animal$^{-1}$ day$^{-1}$], was calculated as the sum of two terms—one representing the amount present when the animals are introduced, the other representing that which grows during the grazing period. Morley used the following expression to calculate $C$:

$$C = W(t = t^*)/(psd) + kg/(ps) \qquad (12.15)$$

where

$W(t = t^*)$ = the amount of herbage present [kg ha$^{-1}$] when grazing of the subdivision recommences after an interval between grazings of $t^*$ days
$p$ = the number of subdivisions [ha$^{-1}$] in the rotational system
$s$ = the number of grazing animals
$d$ = the length of the grazing period [days]
$g$ = the average rate of herbage growth immediately prior to the start of the current grazing period [kg ha$^{-1}$ day$^{-1}$]
$k$ = the ratio of growth during grazing to $g$ ($0 \leqslant k \leqslant 1$)

A relationship between $C$ and $p$ was derived by considering $R$, the ratio of $C$ to $C(k = 1)$. From equation (12.15),

$$R = [W(t^*) + kdg]/[W(t^*) + dg] \qquad (12.16)$$

If the interval between grazing periods is optimal, then $t^* = t_{max}$, $g = W(t_{max})/t_{max}$, and the optimal length of the grazing period is given by

$$d = t_{max}/(p - 1)$$

Substituting these values for $t^*$, $g$ and $d$, equation (12.16) becomes

$$R = (p + k - 1)/p \qquad (12.17)$$

$R$, as given by equation (12.17), is interpreted as the actual herbage available for consumption expressed as a fraction of its theoretical maximum. This is plotted

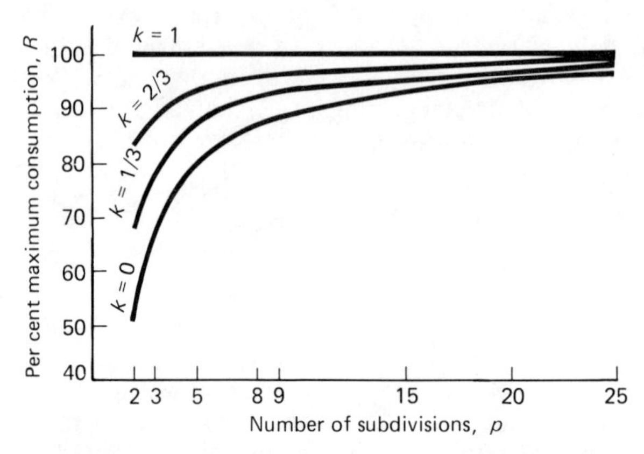


*Figure 12.1* Effect of number of subdivisions on herbage available for consumption for various values of $k$. (After Morley, 1968)

against $p$ for different values of $k$ in *Figure 12.1*. *Figure 12.1* suggests that there is little advantage to be gained by increasing the number of subdivisions beyond nine, and if $k \geqslant 0.5$ even five subdivisions achieve 90% of the theoretical maximum.

Noy-Meir (1976) gives an interesting account of the use of a simple mathematical model, representing both herbage growth and animal consumption as functions of green biomass $V$ [kg m$^{-2}$], to examine the effects of continuous and rotational grazing on steady-state productivity of a pasture. The model has just a single state variable (namely $V$) and a first-order differential equation is used to define the dynamic and static properties of the pasture system:

$$\frac{dV}{dt} = G - C \tag{12.18}$$

where $G$ is the herbage growth rate [kg m$^{-2}$ day$^{-1}$], and $C$ is the rate of consumption of green biomass by the animals [kg m$^{-2}$ day$^{-1}$]. A logistic function of herbage biomass is used to represent growth:

$$G = gV\left(1 - \frac{V}{V_m}\right) \tag{12.19}$$

where $g$ is the maximum specific growth rate [day$^{-1}$] and $V_m$ is the maximum herbage biomass [kg m$^{-2}$]. The rate of consumption by an animal, $C$ [kg animal$^{-1}$ day$^{-1}$], is given by a rectangular hyperbolic function of herbage biomass above an ungrazeable residual:

$$c = c_m \frac{V - V_r}{(V - V_r) + (V_k - V_r)} \quad \text{if} \quad V > V_r \tag{12.20}$$

$$= 0 \text{ otherwise}$$

where $c_m$ is the maximum (satiated) consumption rate [kg animal$^{-1}$ day$^{-1}$], $V_r$ is the ungrazeable residual herbage biomass [kg m$^{-2}$], and $V_k$ is a Michaelis–Menten constant giving herbage biomass at which consumption is half of satiation. The rate of consumption of green biomass $C$ [kg day$^{-1}$] is given by

$$C = cH \tag{12.21}$$

where $H$ is the stocking density [animals m$^{-2}$]. On using equations (12.19), (12.20) and (12.21) to substitute in (12.18), equation (12.18) becomes

$$\frac{dV}{dt} = gV\left(1 - \frac{V}{V_m}\right) - c_m \frac{(V - V_r)}{(V - V_r) + (V_k - V_r)} H \tag{12.22}$$

The stability properties of the system under continuous grazing are examined by setting the equilibrium condition:

$$\frac{dV}{dt} = 0$$

and considering values of $H$ and $V$ at which this equilibrium is possible. Setting equation (12.22) to zero, putting $H = H^*$ and $V = V^*$, and rearranging gives

$$H^* = \frac{gV^*(V_m - V^*)(V^* + V_k - 2V_r)}{c_m V_m (V^* - V_r)} \tag{12.23}$$

The solution of equation (12.23) for $V^*$ and the solution of $dH^*/dV^* = 0$ for maxima and minima involve solving higher-order polynomials. However, the solutions are simpler when there is no ungrazeable residual (i.e. $V_r = 0$):

$$H^* = \frac{g}{c_m V_m} (V_m - V^*)(V^* + V_k) \tag{12.24}$$

and

$$\frac{dH^*}{dV^*} = \frac{g}{c_m V_m}(-2V^* + V_m - V_k)$$

$$= 0 \quad \text{when} \quad V^* = \frac{V_m - V_k}{2}$$

Equation (12.24) is a quadratic in $V^*$:

$$V^{*2} - (V_m - V_k)V^* + V_m\left(\frac{c_m}{g} H^* - V_k\right) = 0$$

whose solution is

$$V^* = \frac{1}{2}\{(V_m - V_k) \pm \sqrt{[(V_m - V_k)^2 - 4V_m (c_m/g)H^* - V_k)]}\} \qquad (12.25)$$

Taking $V_m \geqslant V_k$ (a grazing system in which $V_k > V_m$, i.e. consumption is less than half satiated at maximum possible herbage biomass, is considered of no practical interest), the following is deduced from an inspection of equation (12.25):

1. The system has only a single stable equilibrium if

   $$H^* < \frac{gV_k}{c_m}$$

   Noy-Meir defines $H_s = gV_k/c_m$ as the 'safe carrying capacity'—the stocking density below which there is no danger of the system crashing to extinction.
2. The system has no stable equilibrium and the vegetation becomes extinct if

   $$H^* > \frac{g}{4c_m V_m}[(V_m - V_k)^2 + 4V_k V_m]$$

   $H_x = g(V_m + V_k)^2/(4c_m V_m)$ is defined as the 'maximal carrying capacity'—the stocking density above which a crash to extinction is certain.
3. When the stocking density is between $H_s$ and $H_x$, there are two equilibria. The lower one is a threshold point; if $V$ is below this point, the pasture goes to extinction.

Noy-Meir (1976) uses the model in the form of equation (12.22) to conduct a series of computer simulation experiments with rotational grazing. The rotational scheme is defined by two parameters: $n$, the number of paddocks (degree of subdivision), and $t_r$, the length [days] of the grazing cycle. The stocking density $H$ is set equal to $n\bar{H}$ (where $\bar{H}$ is the average total stocking rate) for the first $t_r/n$ days (the grazing period) in each cycle of $t_r$ days, and $H$ is set to zero for the remaining $t_r(n - 1)/n$ days (the rest period). The model was programmed in CSMP and contains eight parameters: $g$, $V_m$, $c_m$, $V_r$, $V_k$, $\bar{H}$, $n$ and $t_r$. Equation (12.22) was solved numerically using Euler's method. A full account of the simulation experiments is given in Noy-Meir (1976).

In a subsequent paper, Noy-Meir (1978) uses a modified version of the model to analyse herbage production in a continuously-grazed, seasonal pasture. The modification consists of expressing both the rates of herbage growth and consumption of green biomass in terms of two linear segments. This permits $dV/dt$ to be integrated analytically.

## Conservation

The development of models to aid the management of conservation systems would appear from the literature to have commanded less attention than that of grazing. The published work, however, shows a diversity of approach that encompasses simulation, linear programming and nonlinear regression methods. For example, Parke, Dumont and Boyce (1978) describe a simulation of a single-cut conservation system primarily designed to investigate the effect of machine performance on the nutrient content of conserved forage. The effects of other aspects such as crop growth characteristics, climatic differences and management policy can also be assessed. Simulation is also used by Corrall, Neal and Wilkinson (1982) to study the economic impact of management decisions in the production and use of silage in a dairy enterprise. Dumont and Boyce (1976) use linear programming to provide information on the economic feasibility of on-farm production and use of leaf protein from grass. Edelsten and Corrall (1979) describe regression models constructed to predict the yields and digestibilities of herbage cut in different sequences of harvests. In this section, however, we focus on two conservation models which are more relevant to the theme of the book and more easily summarized—one by Audsley (1974), the other by Richards and Hobson (1979).

### A LINEAR-PROGRAMMING MODEL OF A GRASS DRYING ENTERPRISE

Audsley (1974) describes a linear-programming model of a grass drying enterprise which determines the harvesting schedule that maximizes net financial return for the season. Though the model is concerned with high-temperature drying, the approach is applicable to any crop which is harvested several times over a season after periods of regrowth. Only a simplified account is presented here; the reader is referred to Audsley (1974) for full details. The harvesting schedule is defined as a list of areas to be cut each week, classified by age since last cutting. The problem is to determine that schedule which maximizes net return such that the constraints on crop area available for cutting, and harvester and drier capacities are satisfied.

The crop constraints are formulated as follows. Let us assume that Week 6 is the last week on which a first-cut of the crop can be taken, and that a recut of the crop can only be taken after 4, 5 or 6 weeks of regrowth. Let $i$ $(= 1, 2, \ldots, m)$ denote the $i$th week, $x_{i,0}$ [ha] denote the area cut in Week $i$ for the first time, and $x_{i,i-j}$ [ha] the area of crop regrowth cut in Week $i$ and previously cut in Week $i$–$j$ (where $i > 4$, and $j = 4$, 5 or 6). Now the total area of first-cut crop cannot exceed the total crop area A [ha], therefore

$$\sum_{i=1}^{6} x_{i,0} \leqslant A \tag{12.26}$$

The area of crop cut, for example, in Weeks 14, 15 and 16, which is 4, 5 and 6 weeks old respectively, is limited by the area that was cut in Week 10. Mathematically this can be stated as

$$x_{10,4} + x_{10,5} + x_{10,6} \geqslant x_{14,10} + x_{15,10} + x_{16,10}$$

Generalizing, we have the crop regrowth constraints:

$$x_{i+4,i}\, x_{i+5,i}\, x_{i+6,i} \leqslant x_{i,0} \quad (i = 1, 2, 3, 4) \tag{12.27a}$$

$$x_{9,5} + x_{10,5} + x_{11,5} \leqslant x_{5,0} + x_{5,1} \tag{12.27b}$$

$$x_{10,6} + x_{11,6} + x_{12,6} \leqslant x_{6,0} + x_{6,1} + x_{6,2} \qquad (12.27c)$$

$$x_{i+4,i} + x_{i+5,i} + x_{i+6,i} \leqslant x_{i,i-6} + x_{i,i-5} + x_{i,i-4} \quad (i = 7, 8, \ldots, m) \qquad (12.27d)$$

The harvester's capacity is assumed to be limited by both its forward speed and throughput. There is thus a limit due to the harvester's forward speed on the area of crop which can be cut in a week, $H_{max}$ [ha], giving

$$x_{i,0} \leqslant H_{max} \quad (i = 1, 2, 3, 4) \qquad (12.28a)$$

$$x_{5,0} + x_{5,1} \leqslant H_{max} \qquad (12.28b)$$

$$x_{6,0} + x_{6,1} + x_{6,2} \leqslant H_{max} \qquad (12.28c)$$

$$x_{i,i-6} + x_{i,i-5} + x_{i,i-4} \leqslant H_{max} \quad (i = 7, 8, \ldots, m) \qquad (12.28d)$$

Furthermore, there is a limit due to harvester throughput on the weight of crop that can be cut in a week, $W_{max}$ [kg]. If $w_{i,0}$ [kg ha$^{-1}$] denotes first-cut yield and $w_{i,i-j}$ [kg ha$^{-1}$] denotes regrowth yield, then

$$w_{i,0} x_{i,0} \leqslant W_{max} \quad (i = 1, 2, 3, 4) \qquad (12.29a)$$

$$w_{5,0} x_{5,0} + w_{5,1} x_{5,1} \leqslant W_{max} \qquad (12.29b)$$

$$w_{6,0} x_{6,0} + w_{6,1} x_{6,1} + w_{6,2} x_{6,2} \leqslant W_{max} \qquad (12.29c)$$

$$w_{i,i-6} x_{i,i-6} + w_{i,i-5} x_{i,i-5} + w_{i,i-4} x_{i,i-4} \leqslant W_{max} \quad (i = 7, 8, \ldots, m) \qquad (12.29d)$$

The area of crop which can be dried is limited by the time the drier can operate each week, $D_{max}$ [hours]. The time needed to dry a unit area of crop $h_{i,i-j}$ [hour ha$^{-1}$], where the subscripts $i$ and $i$–$j$ are as previously defined, is calculated from the moisture content and yield of the crop as

$$h_{i,i-j} = Y_{i,i-j} \left( \frac{m_1}{1-m_1} - \frac{m_2}{1-m_2} \right) \Big/ M$$

where $Y_{i,i-j}$ [kg dry matter ha$^{-1}$] is the dry matter yield of the crop, $M$ [kg hour$^{-1}$] is the amount of water the drier can remove per unit time, and $m_1$ and $m_2$ are the initial and final moisture contents of the crop expressed as a proportion of wet weight. The constraints imposed by the capacity of the drier are therefore

$$h_{i,0} x_{i,0} \leqslant D_{max} \quad (i = 1, 2, 3, 4) \qquad (12.30a)$$

$$h_{5,0} x_{5,0} + h_{5,1} x_{5,1} \leqslant D_{max} \qquad (12.30b)$$

$$h_{6,0} x_{6,0} + h_{6,1} x_{6,1} + h_{6,2} x_{6,2} \leqslant D_{max} \qquad (12.30c)$$

$$h_{i,i-6} x_{i,i-6} + h_{i,i-5} x_{i,i-5} + h_{i,i-4} x_{i,i-4} \leqslant D_{max} \quad (i = 7, 8, \ldots, m) \qquad (12.30d)$$

The objective is to maximize the net return, that is the return from the sale of the dried herbage less drier operating costs. If the operating costs are taken as fixed costs and $v_{i,i-j}$ [£ kg$^{-1}$] denotes the value of the crop, then the objective function to be maximized is

$$Z = \sum_{i=1}^{6} v_{i,0} x_{i,0} + v_{5,1} x_{5,1} + v_{6,1} x_{6,1} + v_{6,2} x_{6,2} + \sum_{i=7}^{m} \sum_{j=4}^{6} v_{i,i-j} x_{i,i-j} \qquad (12.31)$$

Fixed costs are not included in the objective function because they are common to all solutions.

To summarize, the linear programme has $(4m + 1)$ constraints and $(3m - 9)$ decision variables, and its complete statement (using equation (12.26) to (12.31)) is

$$\text{maximize} \quad Z = \sum_{i=1}^{6} v_{i,0}x_{i,0} + v_{5,1}x_{5,1} + v_{6,1}x_{6,1} + v_{6,2}x_{6,2}$$

$$+ \sum_{i=7}^{m} \sum_{j=4}^{6} v_{i,i-j}x_{i,i-j} \qquad \text{[objective function]}$$

$$\text{subject to} \quad \sum_{i=1}^{6} x_{i,0} \leqslant A \qquad \text{[first-cut constraint]}$$

$$x_{i,0} - x_{i+4,i} - x_{i+5,i} - x_{i+6,i} \geqslant 0 \quad (i = 1, 2, 3, 4)$$

$$x_{5,0} + x_{5,1} - x_{9,5} - x_{10,5} - x_{11,5} \geqslant 0$$

$$x_{6,0} + x_{6,1} + x_{6,2} - x_{10,6} - x_{11,6} - x_{12,6} \geqslant 0$$

$$x_{i,i-6} + x_{i,i-5} + x_{i,i-4} - x_{i+4,i} - x_{i+5,i} - x_{i+6,i} \geqslant 0 \quad (i = 7, 8, .., m)$$

[crop regrowth constraints]

$$x_{i,0} \leqslant H_{max} \quad (i = 1, 2, 3, 4)$$

$$x_{5,0} + x_{5,1} \leqslant H_{max}$$

$$x_{6,0} + x_{6,1} + x_{6,2} \leqslant H_{max}$$

$$x_{i,i-6} + x_{i,i-5} + x_{i,i-4} \leqslant H_{max} \quad (i = 7, 8, \ldots, m)$$

[constraints due to harvester's forward speed]

$$w_{i,0}x_{i,0} \leqslant W_{max} \quad (i = 1, 2, 3, 4)$$

$$w_{5,0}x_{5,0} + w_{5,1}x_{5,1} \leqslant W_{max}$$

$$w_{6,0}x_{6,0} + w_{6,1}x_{6,1} + w_{6,2}x_{6,2} \leqslant W_{max}$$

$$w_{i,i-6}x_{i,i-6} + w_{i,i-5}x_{i,i-5} + w_{i,i-4}x_{i,i-4} \leqslant W_{max} \quad (i = 7, 8, \ldots, m)$$

[constraints due to harvester's throughput]

$$h_{i,0}x_{i,0} \leqslant D_{max} \quad (i = 1, 2, 3, 4)$$

$$h_{5,0}x_{5,0} + h_{5,1}x_{5,1} \leqslant D_{max}$$

$$h_{6,0}x_{6,0} + h_{6,1}x_{6,1} + h_{6,2}x_{6,2} \leqslant D_{max}$$

$$h_{i,i-6}x_{i,i-6} + h_{i,i-5}x_{i,i-5} + h_{i,i-4}x_{i,i-4} \leqslant D_{max} \quad (i = 7, 8, \ldots, m)$$

[drier constraints]

and all $x_{i,0}$, $x_{i,i-j} \geqslant 0$ [non-negativity conditions]

In addition to determining harvest schedules, the model can be used as a management aid in planning the number of hectares a drier can cope with and the harvester or drier capacity needed for a given hectarage. Full details of the model's application are given in Audsley (1974).

### GRASSLAND AND FERTILIZER USE WITH SEPARATE GRAZING AND CONSERVATION AREAS

Richards and Hobson (1979) describe a model that optimizes land and nitrogen fertilizer usage in grassland systems where grazing and conservation areas are separate. They address themselves to two problems:

1. Given a fixed total grassland area and fixed fertilizer N usage defined as

$$\text{N usage [kg ha}^{-1}\text{ year}^{-1}] = \frac{\text{total fertilizer N used [kg year}^{-1}]}{\text{total grassland area [ha]}}$$

   how should the area be split between grazing and conservation and what fertilizer N application rates should be used on each of the two subdivisions so as to maximize stocking rate?
2. Given a fixed total grassland area and a fixed stocking rate, how should the area be divided and what fertilizer N rates should be used so as to minimize fertilizer N usage?

In order to answer these questions, the following relationships are postulated which equate the quantity of silage consumed by the cows in winter [kg dry matter] to the quantity produced from the cutting area, and the amount of grass [kg dry matter] consumed during the grazing season to the amount produced from the grazing area:

$$\frac{(365 - L)I_c M}{A_c} = \xi_c G_c \tag{12.32}$$

and

$$\frac{L I_g M}{A_g} = \xi_g G_g \tag{12.33}$$

where

$L$ = the length [days] of the grazing season
$I_c$, $I_g$ = the consumption [kg dry matter cow$^{-1}$ day$^{-1}$] of silage in winter and herbage during the grazing season respectively
$M$ = the number of cows
$A_c$, $A_g$ = the areas [ha] allocated for cutting and grazing respectively
$\xi_c$ = the efficiency of silage utilization (i.e. the proportion of grass grown in the cutting area that is consumed)
$\xi_g$ = the efficiency of grass utilization in the grazing area
$G_c$, $G_g$ = the total annual grass yields [kg dry matter ha$^{-1}$] under cutting and grazing respectively

The total annual grass yields are given by empirically-determined quartic functions of nitrogen supply (= fertilizer N + soil N):

$$G_c = a_0 + a_1 N_c + a_2 N_c^2 + a_3 N_c^3 + a_4 N_c^4 \tag{12.34}$$

$$G_g = b_0 + b_1 N_g + b_2 N_g^2 + b_3 N_g^3 + b_4 N_g^4 \tag{12.35}$$

where $N_c$ and $N_g$ are the nitrogen supply [kg N ha$^{-1}$ year$^{-1}$] on the cutting and grazing areas respectively, and the $a$'s and $b$'s are the coefficients of the quartics.

Substituting equation (12.34) into (12.32) and (12.35) into (12.33) gives

$$\frac{(365 - L)I_c M}{A_c} = \xi_c (a_0 + a_1 N_c + a_2 N_c^2 + a_3 N_c^3 + a_4 N_c^4) \tag{12.36}$$

and

$$\frac{L I_g M}{A_g} = \xi_g (b_0 + b_1 N_g + b_2 N_g^2 + b_3 N_g^3 + b_4 N_g^4) \tag{12.37}$$

$N_c$ and $N_g$ each comprise two components:

$$N_c = F_c + S_c \tag{12.38}$$

and

$$N_g = F_g + S_g \tag{12.39}$$

where $F_c$ and $F_g$ [kg ha$^{-1}$ year$^{-1}$] are the fertilizer N rates applied to the cutting and grazing areas respectively, and $S_c$ and $S_g$ [kg ha$^{-1}$ year$^{-1}$] are the respective soil N levels. Fertilizer N usage is given by the relationship:

$$\text{N usage [kg ha}^{-1}\text{ year}^{-1}] = \frac{F_c A_c + F_g A_g}{A_c + A_g} = \frac{(N_c - S_c)A_c + (N_g - S_g)A_g}{A_c + A_g} \tag{12.40}$$

A complete statement of the model is given by equations (12.36)–(12.40).

Problem (1), in which fertilizer N usage and the total grassland area are fixed, may be solved by rewriting equations (12.36) and (12.37) with $M$ only on the left-hand side. The right-hand sides of the equations can then be equated and, on substituting for $N_c$ using equation (12.40), the resultant quartic can be solved for $N_g$ if values of $A_g$, $S_c$ and $S_g$ are assumed. Having solved for $N_g$, then $N_c$ can be determined from equation (12.40), $F_c$ and $F_g$ from (12.38) and (12.39) respectively and $M$ from (12.36) or (12.37). By stepping through values of $A_c$ and $A_g$ in say 1 ha steps, the corresponding values of $N_g$, $N_c$ and $M$ can be found and the conditions for maximum number of cows determined by inspection. The quartic in $N_g$ has four roots. Any root is rejected if it is: (a) complex, (b) negative or (c) positive but forcing $N_c$ negative. Using these rules, Richards and Hobson found that there was only one acceptable root for values of $A_g$ and $A_c$ in the region of the optimum.

Problem (2), in which the number of cows and the total grassland area are fixed, may be solved in a similar manner. Equations (12.36) and (12.37) can be solved for $N_c$ and $N_g$ respectively for pairs of values of $A_c$ and $A_g$ stepping through 1 ha steps as before. Only solutions with feasible roots are of interest and the area split using the minimum amount of fertilizer N can be found.

Richards and Hobson's results show that the optimum split of area is around 70% grazed and 30% cut, with higher fertilizer N rates on the cut area than on the grazed area.

## Exercises

### Exercise 12.1

A company produces four types of compound fertilizer by mixing nitrates, phosphates, potash and inert ingredients. Nitrates cost the company 225 £ t$^{-1}$, phosphates 220, potash 175 and inert ingredients 25. The amounts of nitrates and phosphates that can be obtained are limited to 3000 and 3250 t month$^{-1}$ respectively; the other ingredients can be bought in unlimited amounts. The composition and price of the compounds produced are shown below.

| *Compound* | *Composition* [g ingredient (100 g compound)$^{-1}$] | | | *Price* [£ t$^{-1}$] |
|---|---|---|---|---|
| | *Nitrates* | *Phosphates* | *Potash* | |
| 1:1:1 | 15 | 15 | 15 | 150 |
| High N | 20 | 10 | 10 | 142 |
| Low N | 10 | 20 | 20 | 157 |
| No N | — | 20 | 20 | 132 |

It costs 35 £ t$^{-1}$ to mix, bag and distribute each compound and the company can sell whatever it produces, but total production cannot exceed 20 000 t month$^{-1}$. Furthermore, the firm has contracts to supply 5000 tonnes of 'High N', 3000 tonnes of 'Low N' and 2500 tonnes of 'No N' next month. What amount of each compound should the company produce in the coming month so as to maximize its profits?

**Exercise 12.2**

An arable farmer has 200 ha of high-grade land and 100 ha of low-grade land available for the coming season, and £150 000 to spend on crops. He also has 5000 man-hours of free labour available (his own and his family's); additional labour can be bought at a cost of 2.50 £ man-hour$^{-1}$. The crops under consideration are spring wheat, spring barley, maincrop potatoes, sugar beet and oilseed rape. The available information on each crop is as follows:

| *Crop* | *Expected yield* [t ha$^{-1}$] | | *Labour requirements* [man-hours ha$^{-1}$] | *Other costs (e.g. seed, fertilizer, sprays)* [£ ha$^{-1}$] | *Expected price* [£ t$^{-1}$] |
|---|---|---|---|---|---|
| | *High-grade land* | *Low-grade land* | | | |
| Spring wheat | 4.6 | 3.4 | 12.4 | 135 | 112 |
| Spring barley | 4.8 | 3.5 | 12.4 | 117 | 107.5 |
| Maincrop potatoes | 40 | 20 | 72.5 | 725 | 60 |
| Sugar beet | 45 | 28 | 67.3 | 365 | 25 |
| Oilseed rape | 2.5 | 1.5 | 14.0 | 185 | 225 |

Determine the crop plan that maximizes his profits.

**Exercise 12.3**

In a hypothetical grazing system, pasture growth [kg herbage $ha^{-1}$ $day^{-1}$] is described by a Gompertz function, with initial specific growth rate parameter $g_0$ and specific growth rate decay parameter $g_D$, and animal consumption [kg herbage $animal^{-1}$ $day^{-1}$] by a rectangular hyperbola, with maximum consumption rate parameter $C_M$ and Michaelis–Menten constant $C_K$. Determine the first-order differential equation describing the dynamics of this system and show that the stocking density $S$ [animals $ha^{-1}$] necessary to maintain herbage biomass at a constant level $H^*$ [kg herbage $ha^{-1}$] is given by

$$S(t) = g_0 (H^* + C_K) \exp(-g_D t)/C_M$$

Chapter 13

# Farm planning and control: II

## Introduction

In this chapter, we continue the account of models designed for application in farm planning and control started in Chapter 12, by considering livestock production. Numerous models of livestock production systems and subsystems have been reported in the literature, most of which are large computer simulations. This is not a well-developed area, despite its vast body of literature. Most of the reported models are highly empirical and of limited predictive ability. They are not easily summarized, pictorially or mathematically. The material presented here therefore takes the form of a brief survey of the more relevant systems simulations, with most consideration being given to those models which are mathematically concise and illustrate a particular mathematical approach.

## The dairy enterprise

The dairy enterprise, despite its agricultural significance, has received relatively little attention from modellers. This is hardly surprising in view of the fact that milk production is the most complex of livestock production enterprises, involving the manipulation and control of complicated physiological processes such as pregnancy, lactation and deposition of body tissue. Crabtree (1970) produced one of the first published accounts of a dairy enterprise model, in which a computer simulation (programmed in the high-level language DYNAMO) of an intensive dairy system is developed. The simulation takes into account grass production, cow nutrition and energy requirements, herd structure and cash flow. These components are generally described by empirical relationships either taken from the literature or determined directly from experimental data.

Bywater and Dent (1976) describe proposals for constructing a detailed model to simulate nutrient intake and partition by the dairy cow, intended primarily for use as an aid to the management of a dairy herd. The paper is an introduction to a proposed series of papers describing in detail the different components of the model. Bywater (1976), in the second paper in the series, gives a conceptual scheme for simulating milk yield and composition based on a review of the literature. No validation or results from this submodel are presented. Further papers in the series, as far as we are aware, have yet to be published. Chudleigh (1977) describes a simple economic model of a smallholder dairying system in Kenya. He suggests that the model may be used as part of a larger cost-benefit analysis designed to assist the allocation of resources for research and development aimed at improving the performance of the smallholder dairy farmers.

Models for aiding dairy enterprise management have been designed for programmable calculators for use on the farm. Published examples include Wood (1979) and France *et al.* (1982). Wood (1979) gives a model for calculating feed supply from specified patterns of milk production and liveweight change. France *et al.* (1982) describe a model that estimates milk production, revenue from milk sales, and concentrate requirements, and allows a comparison of the financial effects of different calving patterns. The gamma curve is used to represent the cow's lactation curve, and milk production over a period is calculated using the incomplete gamma function (p. 222). Linn and Spike (1980) review programmable calculators and their application to feeding and management of dairy cattle and Kirk (1981) gives an account of the use of these machines in mastitis control programmes in the USA.

Seasonality is an important factor in milk production. Milk supply depends largely on the distribution of cow calving dates, which is in turn influenced by seasonal factors, particularly the availability of good grass. Wood (1969, 1970, 1976, 1980, 1981) has analysed seasonal variation in milk yield by fitting the gamma curve to numerous lactation records using regression analysis (*see* Glossary). Wood's results show, for example, that winter calvers tend to produce more in total lactation than spring calvers, daily yield is depressed during the winter months and stimulated during the spring, and the fat and protein concentrations also exhibit seasonal variations.

Milk marketing boards and manufacturers are faced with a steady consumer demand for milk for liquid consumption and perishable milk products, and in order to encourage a reasonably constant inflow of milk to their factories, pay farmers a seasonal milk price. Killen and Keane (1978) constructed a linear programming model for determining the optimum distribution of calving dates for satisfying product demand and establishing a set of seasonal milk prices which will equitably compensate farmers. The linear programme is stated thus:

$$\text{minimize} \quad \sum_{j=1}^{12} c_j x_j + d \sum_{i=1}^{12} \left( \sum_{j=1}^{12} a_{ij} x_j - m_i \right)$$

$$\text{subject to} \quad \sum_{j=1}^{12} a_{ij} x_j \geqslant m_i \quad (i = 1, 2, \ldots, 12)$$

where

$x_j$ = the total output [litres year$^{-1}$] from the $j$th system, defined as those cows calving in month $j$

$c_j$ = the variable cost of production [£ litre$^{-1}$] associated with a cow calving in month $j$

$d$ = the disposal cost for surplus production [£ litre$^{-1}$]

$a_{ij}$ = the proportion of milk produced in month $i$ by a cow that calved in month $j$

$m_i$ = the consumer demand [litres] in month $i$

Killen and Keane determined the $a_{ij}$'s from Wood (1969) and, of course, $\Sigma_{j=1}^{12} a_{ij} = 1$, $j = 1, 2, ..., 12$. The solution to this linear programme gives those values of $x_j$ that represent the optimum volume of intake from each of the 12 systems. Killen and Keane go on to obtain a set of seasonal milk prices by solving the dual problem (*see* Glossary):

$$\text{maximize} \quad \sum_{i=1}^{12} m_i y_i - d \sum_{i=1}^{12} m_i$$

$$\text{subject to} \quad \sum_{i=1}^{12} a_{ij} y_i \leqslant c_j + d \quad (j = 1, 2, ..., 12)$$

where the $y_i$ [£ litre$^{-1}$] denote the dual variables. They interpret $y_i^*$, the optimal values of $y_i$, as those values of monthly intake which maximize the value of total intake to the marketing board (for a general interpretation of dual variables see Taha, 1971, pp. 84–85). The $y_i^*$ are then used as a basis for calculating a seasonal pricing system, the price paid in each month to the farmer, $p_i$ [£ litre$^{-1}$], being calculated as

$$p_i = y_i^* + k \quad (i = 1, 2, ..., 12)$$

where $k$ represents the fixed element paid on every litre and $y_i^*$ represents the bonus paid in month $i$.

Another example of the use of linear programming in dairy production is given by Reyes *et al.* (1981), who describe the construction of a multi-time period linear programming model to evaluate general feeding and management alternatives across a production cycle (calving interval) for dairymen in northeast Texas.

## Beef systems

Published models of beef production systems fall into two categories—those employing computer simulation techniques and those utilizing optimization methods such as linear programming. The level of generality of the models ranges from enterprise component level to national industry level. In this section, some of the more recent contributions to beef systems modelling are outlined. A review of earlier work is given by Joandet and Cartwright (1975).

Most of the models published have been systems simulations. Ryan (1974) describes a simulation model of a group of beef animals passing through a feedlot. The model is programmed in FORTRAN and structured on the daily progress of a group of animals. It is used to evaluate alternative management practices, which include alternative selling criteria, culling prices and the effects of managerial expertise as reflected in growth rates and mortality levels. In addition, the effects of changing cattle prices and ration costs are considered. A computer simulation is also developed by Halter *et al.* (1976) to evaluate different policies for encouraging the transformation of traditional ranches into modern, more profitable ones in Venezuela.

Morley (1977) describes a model of a steer-beef production system which was developed for a Mediterranean environment in Argentina. Curll (1978) went on to evaluate the model for a similar environment in Australia, and subsequently used it to examine the effects of superphosphate application on pasture growth in an average and poor season for different combinations of stocking rates, beef prices and dates of sale. Kelley White *et al.* (1978) describe a simulation model, which incorporates a linear programming routine, used to analyse the Guyanese livestock industry.

A model for comparing beef production systems and sales strategies in an extensive ranching region in South Africa is constructed by Louw, Grosskopf and Groenewald (1979). It simulates revenues from four beef production systems, ranging from inflexible weaner marketing to marketing of three-year-old oxen, and uses regression equations to relate stocking rate and rainfall to calving percentage and weaning mass. Sanders and Cartwright (1979a,b) give a model for simulating beef production under a wide range of environmental and management conditions using cattle of widely differing genotype. The model has been used to simulate beef production enterprises in parts of the USA, South America and Africa. Congleton and Goodwill (1980a, b, c) describe a beef production simulation programmed in DYNAMO. The main components of the model simulate the herd structure of a cow herd and a production herd, composed of calves from the cow herd. The model is used to evaluate mating plans, culling policies and options such as producing replacement heifers or purchasing them.

A stochastic simulation for aiding decision-making in an Australian campaign to eradicate bovine brucellosis is constructed by Beck and Dillon (1980). It uses data on district cattle numbers, disease prevalence and campaign intensity to project workload requirements, culling rates and campaign duration under alternative campaign strategies. Sullivan, Cartwright and Farris (1981) describe the integration of a forage production model and a cattle production model to simulate meat and milk output under tropical conditions in East Africa. The model was used to analyse the effects of improved management practices on a traditional village livestock grazing system. Beck, Harrison and Johnston (1982) give a model for simulating the risks and returns from establishing improved pasture in an extensive beef breeding enterprise. Particular attention is given to modelling the effect of stocking rate policy. Use of the model is illustrated for an enterprise in northeastern New South Wales, Australia.

The majority of optimization models of beef systems given in the literature are formulated as large linear programmes. Wilton *et al.* (1974) describe such a linear programming model of an on-farm integrated beef production enterprise. The model includes cropping, feeding and breeding activities with their requirements for land, labour, animal housing and crop storage facilities. Three enterprises with pure-bred cows of different mature size, each under typical Ontario farm conditions, are analysed using the model. Morris, Parkins and Wilton (1976) went on to use the model to determine the effects of different mature cow weights, levels of potential milk yield and crop feeding on economic efficiency.

Miller, Brinks and Sutherland (1978) construct a linear programming model for determining management policy for the coming year on a typical mountain ranch in southern Colorado. The model calculates the level of each activity (e.g. cow herd size, yearling herd size, area of meadow to be harvested for hay) that maximizes the net return for the ranch subject to resource limitations on land, labour and capital. Options for expanding the enterprise are evaluated by exploring the optimal solution using parametric programming. The linear programme consists of 127 constraints and 274 variables. Miller *et al.* (1980), at a higher level of generality and over a longer planning horizon, describe a linear programming model of the US beef production system for determining the supply of beef cattle and selecting feeding options that maximize prime beef production over a five-year period. The model represents the USA as five beef producing regions with inter-regional transportation of calves for feeding purposes. Yorks *et al.* (1980) use the model to compare minimization of feeding costs and minimization of fossil energy as objectives in US beef production.

There are a few instances in the literature of the use of optimization techniques other than linear programming in beef production systems modelling. Meyer and Newett (1970) and Kennedy (1972) determine optimal weights for buying and selling feedlot cattle and optimal feeding programmes using dynamic programming. Conway (1974a, b) develops an economic decision model for maximizing revenue in excess of variable costs from grazing steers using calculus.

## Lamb and other livestock systems

Computer simulation has been used in the study of lamb, pig, poultry and rabbit production systems. Several examples are reported in the literature, and a representative cross-section is briefly described here.

Edelsten and Newton (1977) describe a stochastic simulation of a lowland sheep flock for determining a farmer's best system of lamb production. Inputs to the model include topography, soil type, climatic data, economic parameters and the farmer's attitude to risk. Geisler, Paine and Geytenbeek (1977) give an account of a computer simulation of an intensified lambing system incorporating two flocks and the rapid remating of ewes. Each flock lambs three times in two years and ewes diagnosed as non-pregnant in one flock are transferred to the other in time for the next mating. The total number of lambs from each lambing is predicted and the value of the lamb crop calculated. Geisler *et al.* (1979) describe a simulation of lamb production from an autumn catch crop. The model used data on the crop of interest and utilization factors for that crop to calculate land-use parameters and gross margins for management policies chosen by the user.

A model of a lamb production system based on empirical relationships derived from field experiments is constructed by Curll and Davidson (1977). It was used to examine the effects on lamb production of reducing the area of pasture grazed by the ewes until the end of mating, with the aims of restricting ewe liveweights at mating and increasing pasture availability during pregnancy so as to ensure liveweight gains. The effects on gross margins were calculated for different combinations of environment, stocking rate and meat price. White and Morley (1977) describe a model for determining optimal stocking rates for Merino sheep. The model simulates the effect of certain compensations on the marginal returns from increased stocking rates. For example, the higher prices paid for wool of reduced fibre diameter compensate for decreases in the ewe's fleece weight at the higher stocking rates.

A simulation of a pig herd is described by Blackie and Dent (1974). It is designed for integration with a management information system, thereby providing a useful management aid for the controller of a pig production enterprise. The authors, in a later paper, go on to use simulation to analyse hog production strategies (Blackie and Dent, 1976). Devindar, Williams and Tung (1980) give a model for assessing the impact of future research and extension activities on the profit of an average swine producer in Hawaii. The model has two main components. The first component simulates production in a typical Hawaiin pig enterprise, and the second uses discounted cash flow analysis to calculate annual revenues, costs and rate of return on capital investment for the period under consideration.

A dynamic model for determining an optimal system of broiler production is described by Greig *et al.* (1977), which takes into account both technical and economic factors. The model accommodates changes in feed prices, dietary composition and strain of bird, and permits the investigation of management factors such as length of feeding period and the desirability of segregating the birds according to sex. White *et al.* (1978) construct a simulation of poultry egg production. The model was used to compare the profitability of feeding high- and low-protein diets when rearing pullets for egg production, and to determine the age at which adult birds should be slaughtered in order to maximize profits.

Walsingham, Edelsten and Brockington (1977) use a dynamic model of a rabbit population to simulate a doe enterprise. The model is constructed to consider the effect of adopting various strategies for the replcement of breeding stock.

## Culling

Modelling has been used to determine replacement policies for continually-operating livestock enterprises—such as when to replace a cow in a dairy herd. The problem is analogous to that in industry of calculating the optimal policy for replacement of machines in a factory. Gartner and Herbert (1979) and Walsingham, Edelsten and Brockington (1977) use simulation to investigate culling and replacement in a dairy herd and a rabbit population respectively. The replacement problem is well-suited to solution by dynamic programming (p. 63), and this approach has been used by White (1959) to determine hen replacement in an egg production enterprise and by McArthur (1973), Smith (1973) and Stewart *et al.* (1977) to investigate replacement policy in dairy herds. The approach is illustrated here using a highly simplified, but instructive, example adapted from Throsby (1964).

Let $g_i$ and $h_i$ denote the annual return and cost respectively of keeping a dairy cow of age $i$ for another year ($i = 0, 1, 2, \ldots$), where age is measured from the commencement of productive life at $i = 0$. Let $s_i$ be the return from culling a cow at age $i$ and assume that the animal is only replaced by one about to begin its productive life (i.e. of age $i = 0$). Let a policy of *keep* be represented by $K$ and one of *cull and replace* by $R$. Define $f_i$ as the financial return from following an optimal policy with a cow of age $i$. Assume $g_i \geqslant h_i$ and $f_i = 0$ for $i > 5$, then $f_i \geqslant 0$ for all $i$. The recurrence relation for this problem is given by

$$f_i = \max_{K,\, R} \begin{bmatrix} K: g_i - h_i + f_{i+1} \\ R: s_i + g_0 - h_0 + f_1 \end{bmatrix}$$

Suppose $g_i$, $h_i$ and $s_i$ take the following values:

| $i$ | $g_i$ | $h_i$ | $s_i$ |
|---|---|---|---|
| 0 | 4 | 2 | 2 |
| 1 | 7 | 3 | 3 |
| 2 | 10 | 3 | 5 |
| 3 | 9 | 4 | 7 |
| 4 | 6 | 4 | 9 |
| 5 | 5 | 4 | 6 |

Applying the recurrence equation gives

$$f_5 = \max \begin{bmatrix} K: 5 - 4 + 0 \\ R: 6 + 4 - 2 + f_1 \end{bmatrix} = \max \begin{bmatrix} 1 \\ 8 + f_1 \end{bmatrix} = 8 + f_1 \rightarrow \text{cull and replace}$$

$$f_4 = \max \begin{bmatrix} K: 6 - 4 + f_5 \\ R: 9 + 4 - 2 + f_1 \end{bmatrix} = \max \begin{bmatrix} 10 + f_1 \\ 11 + f_1 \end{bmatrix} = 11 + f_1 \rightarrow \text{cull and replace}$$

$$f_3 = \max \begin{bmatrix} K: 9 - 4 + f_4 \\ R: 7 + 4 - 2 + f_1 \end{bmatrix} = \max \begin{bmatrix} 16 + f_1 \\ 9 + f_1 \end{bmatrix} = 16 + f_1 \rightarrow \text{keep}$$

The solution therefore is to cull and replace a cow after she has completed four years of productive life (i.e. when $i = 4$). Notice that the recurrence relation computes $f_i$ from $f_{i+1}$ and hence the order of computation is given by $f_5 \rightarrow f_4 \rightarrow f_3$, which is dynamic programming of the *backward* type.

## Rationing

Ration formulation is an important aspect of the management of many livestock enterprises, particularly intensive systems in which the animals are fed large quantities of purchased feedstuffs. The problem is very similar to that of planning least-cost diets for hospitals and other institutions (e.g. Smith, 1963; Gue and Liggett, 1966) and formulating feed mixes for farm livestock (e.g. Chapter 3, p. 49; Dent and Casey, 1967; Beneke and Winterboer, 1973), where the aim is to find the least expensive combination of feedstuffs that will supply prescribed amounts of particular nutrients.

In formulating a ration, the farmer usually seeks to determine the amount of each feedstuff to give an animal each day so that its energy and nutrient requirements for maintenance and production are satisfied as cheaply as possible. For ruminants, the problem may be stated mathematically thus:

$$\text{minimize} \quad Z = \sum_{j=1}^{n} c_j x_j \qquad \text{[objective function]} \qquad (13.1a)$$

$$\text{subject to} \quad \sum_{j=1}^{n} e_j x_j \geqslant E_{\text{req}} \qquad \text{[energy constraint]} \qquad (13.1b)$$

$$\sum_{j=1}^{n} p_j x_j \geqslant P_{\text{req}} \qquad \text{[protein constraint]} \qquad (13.1c)$$

$$\sum_{j=1}^{n} x_j \leqslant F_{\max} \qquad \text{[intake constraint]} \qquad (13.1d)$$

$$x_j \geqslant 0 \quad \text{for all } j \qquad \text{[non-negativity conditions]} \qquad (13.1e)$$

where

$n$ = number of feedstuffs available to the farmer
$x_j$ = quantity of feedstuff $j$ to feed the animal [kg dry matter day$^{-1}$]
$c_j$ = cost of feedstuff $j$ [pence (kg dry matter)$^{-1}$]

$e_j$ = energy concentration of the $j$th feedstuff [MJ (kg dry matter)$^{-1}$]
$p_j$ = protein concentration of the $j$th feedstuff [g protein (kg dry matter)$^{-1}$]
$E_{req}$ = animal's energy requirement [MJ day$^{-1}$]
$P_{req}$ = animal's protein requirement [g day$^{-1}$]
$F_{max}$ = animal's maximum feed intake [kg dry matter day$^{-1}$]

Other constraints may be imposed on the problem, such as the requirements of the ruminant for particular minerals and a constraint specifying a minimum crude fibre content in the ration.

In formulating rations at present in the United Kingdom, energy concentrations and requirements are generally expressed in terms of metabolizable energy (ME), the portion of its food energy that the ruminant can utilize after faecal, urinary and methane losses have been discounted, using the system of calculation set out by the Ministry of Agriculture, Fisheries and Food (1975), and protein concentrations and requirements in terms of crude protein (crude protein = nitrogen × 6.25). We therefore focus our attention here on rationing using these established systems, but conclude the section with an account of some of the ramifications for ration formulation in the United Kingdom of the new protein and revised energy systems proposed by the Agricultural Research Council (1980).

The rationing problem as defined by equations (13.1a–e) is in a form suitable for solution by linear programming, *provided* the coefficients of the $x_j$'s and the RHS values are constants. Unfortunately, this is rarely the case; the RHS values are almost invariably a function of the ruminant's liveweight and its level of production. This difficulty is usually overcome by computing a series of rations corresponding to incremental changes in liveweight and level of production and regarding the RHS values as constants over an increment of change. A further difficulty arises because the ME requirements and maximum feed intake level can vary with the ME concentration of the ration $r$ [MJ (kg dry matter)$^{-1}$]. The efficiency of utilization of dietary ME for growth varies appreciably with $r$ in the case of growing and fattening ruminant animals. This problem is overcome by expressing the energy constraint in terms of net energy instead of ME, using the variable net energy system due to Harkins, Edwards and McDonald (1974). Net energy is defined as ME less the heat increment and represents that part of food energy available to the animal for maintenance and production. The variable net energy system is widely applied in ration formulation, and the reader is referred to the Ministry of Agriculture, Fisheries and Food (1975) for details of its application.

There appears to be no completely satisfactory way of circumventing the maximum intake problem. A ruminant's maximum feed intake level $I$ [kg dry matter (kg liveweight)$^{-1}$ day$^{-1}$] as a function of the energy concentration of the ration $r$ is reasonably well described by a rectangular hyperbola:

$$I(r) = k_1 \frac{r}{r + k_2} - k_3$$

where $k_1$, $k_2$ and $k_3$ are positive constants. As $r = \Sigma_{j=1}^{n} e_j x_j / \Sigma_{j=1}^{n} x_j$, the rationing problem (equations (13.1)) is amenable to solution by separable programming (p. 59) in theory. However, this approach is seldom practicable. A trial-and-error procedure is often adopted in practice. The animal's maximum feed intake $F_{max}$ is given by $wI(r)$, where $w$ is the liveweight [kg] of the animal. An initial value of $F_{max}$ corresponding to say $r$ = 10 MJ (kg dry matter)$^{-1}$ is taken, i.e. $F_{max} = wI(r = 10)$,

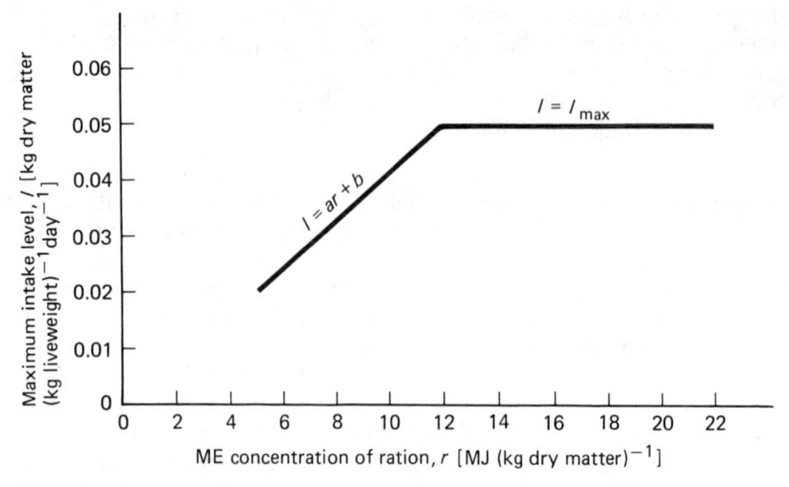


*Figure 13.1* Using a ramp function to describe a cow's maximum feed intake level as a function of the energy concentration of the ration. (The scales along each axis are hypothetical.)

and the linear programme solved. Let the ME concentration of the resulting ration be $r_1$. If intake of this ration $\Sigma x_j$ is greater than or equal to $wI(r = r_1)$, then $F_{max}$ is set equal to $wI(r = r_1)$ and the linear programme resolved. The procedure is repeated until an optimal ration is obtained which satisfies the criterion

$$\sum_{j=1}^{n} x_j \leqslant wI\left(r = \sum_{j=1}^{n} e_j x_j / \Sigma x_j\right)$$

Parametric programming (pp. 55–57) can be us d to improve the computational efficiency of the approach.

A possible way round the problem that might have limited practical application is now put forward. Instead of a rectangular hyperbola, let $I(r)$ be described by a ramp function as illustrated in *Figure 13.1*. The intake restriction can then be expressed as two constraints:

$$\sum_{j=1}^{n} x_j \leqslant (ar + b)w \tag{13.2a}$$

and

$$\sum_{j=1}^{n} x_j \leqslant wI_{max} \tag{13.2b}$$

Let $r^* = \Sigma_{j=1}^{n} e_j x_j/(wI_{max})$, then $r^* \leqslant r$. If the inequality $\Sigma_{j=1}^{n} x_j \leqslant (ar^* + b)w$ is satisfied, then equations (13.2) automatically holds. Therefore, instead of using equations (13.2), let the restriction on intake be

$$\sum_{j=1}^{n} x_j \leqslant (ar^* + b)w \tag{13.3a}$$

and

$$\sum_{j=1}^{n} x_j \leqslant wI_{\max} \tag{13.3b}$$

Putting

$$r^* = \sum_{j=1}^{n} e_j x_j/(wI_{\max})$$

in equation (13.3b) and simplifying gives the intake constraints

$$\sum_{j=1}^{n} (I_{\max} - ae_j)x_j \leqslant bwI_{\max} \tag{13.4a}$$

and

$$\sum_{j=1}^{n} x_j \leqslant wI_{\max} \tag{13.4b}$$

Using an intake restriction of the form of equations (13.4a,b) allows ordinary linear programming to be used and obviates the need to re-solve the linear programme in formulating a particular ration.

In recent years, considerable research has been directed towards a more precise definition of ruminants' protein requirements and the fate of dietary protein in their rations. The outcome of this research has led the Agricultural Research Council (1980) to propose an entirely new system of evaluating dietary protein requirements. The new system expresses requirements in terms of rumen-degradable protein (RDP) and rumen-undegradable protein (UDP). In addition, the Council proposes substantial revisions to the system of evaluating ME requirements. These sets of proposals further complicate the calculation of least-cost rations for ruminants by linear programming. The general mathematical statement of the rationing problem now becomes:

$$\text{minimize} \quad Z = \sum_{j=1}^{n} c_j x_j \qquad \text{[objective function]} \tag{13.5a}$$

$$\text{subject to} \quad \sum_{j=1}^{n} m_j x_j \geqslant M(q) \qquad \text{[ME constraint]} \tag{13.5b}$$

$$\sum_{j=1}^{n} d_j x_j \geqslant D(q) \qquad \text{[RDP constraint]} \tag{13.5c}$$

$$\sum_{j=1}^{n} u_j x_j \geqslant U(q) \qquad \text{[UDP constraint]} \tag{13.5d}$$

$$\sum_{j=1}^{n} x_j \leqslant F(q) \qquad \text{[intake constraint]} \qquad (13.5e)$$

$$x_j \geqslant 0 \quad \text{for all } j \qquad \text{[non-negativity conditions]} \qquad (13.5f)$$

where $c_j$, $x_j$ and $n$ are defined as in equations (13.1), and

$m_j$ = ME content of the $j$th feedstuff [MJ (kg dry matter)$^{-1}$]
$d_j$, $u_j$ = RDP and UDP content respectively of the $j$th feedstuff [g RDP(UDP) (kg dry matter)$^{-1}$]
$q$ = metabolizability of the ration [MJ metabolizable energy (MJ gross energy)$^{-1}$], and is directly proportional to $r$, the ME concentration of the ration
$M(q)$ = animal's ME requirement [MJ day$^{-1}$], which is a function of $q$
$D(q)$, $U(q)$ = animal's RDP and UDP requirement respectively [g day$^{-1}$], both functions of $q$
$F(q)$ = animal's maximum feed intake [kg dry matter day$^{-1}$], also a function of $q$

There is no completely satisfactory method for resolving the problem of the functional dependence of the RHS values on $q$, though Crabtree (1982) offers a practical way round the difficulty in describing an interactive system of least-cost ration formulation for growing and lactating cattle. He constrains the ration to a metabolizability $q$ by introducing a dummy decision variable $X$ with zero cost, and the following three constraints into the linear programme:

$$\sum_{j=1}^{n} x_j - X = 0 \qquad (13.5g)$$

$$\sum_{j=1}^{n} q_j^* x_j - qX \geqslant 0 \qquad (13.5h)$$

$$\sum_{j=1}^{n} q_j^* x_j - (q + \delta q)X \leqslant 0 \qquad (13.5i)$$

where $q_j^*$ is the metabolizability [MJ metabolizable energy (MJ gross energy)$^{-1}$] of the $j$th feedstuff, and $\delta q$ is a small perturbation in $q$ (taken as 0.005). The two inequalities (equations (13.5h,i)) are preferred to the single equality $\Sigma\, q_j^* x_j = qX$ because the equality condition can make the location of feasible solutions extremely difficult when requirements are tightly constrained. The search for the ration metabolizability with the least cost is made by re-solving the linear programme (equations (13.5a–i)) for a range of $q$ in steps of $\Delta q$. For further details, the reader is referred to Crabtree (1982).

There is a number of other accounts of ration formulation in the literature, most of which employ linear programming. Bath and Bennett (1980) and Black and Hlubik (1980), for example, detail the construction of computer programs based on the routine application of linear programming for calculating dairy cattle rations. Similar programs for beef cattle are to be found in Brokken (1971a, b) and McDonough (1971). Glen (1980) describes a more sophisticated approach to beef cattle rationing; this involves using parametric programming to obtain a piecewise linear representation of the cost of the ration as a function of the ration's energy concentration, and then using differential calculus to determine the least-cost ration itself. In contrast, France (1982) gives a non-exact iterative method for rationing beef cattle for use on a programmable calculator. France, Neal and Pollott (1982) also describe a similar approach for rationing pregnant ewes.

## Exercises

### Exercise 13.1

A sheep farmer starts off with a flock of $k$ sheep. At the end of each year, over the next $N$ years, he has to decide how many sheep to sell and how many to keep. If he sells, his profit on a sheep is $r_n$ [£] in year $n$. If he keeps, the number of sheep kept in year $n$ will be doubled in year $n + 1$. The farmer intends to sell completely at the end of $N$ years. Formulate the problem as a dynamic programming problem using a system of backward recurrence.

### Exercise 13.2

Reformulate the problem given in Exercise 13.1 as a dynamic programming problem using a system of forward recurrence. Compare the computational efficiency of the forward and backward approach.

### Exercise 13.3

A 250 kg steer, required to gain at a rate of 0.8 kg day$^{-1}$, has a net energy (NE) requirement of 34.1 MJ day$^{-1}$, a digestible crude protein (DCP) requirement of 447 g day$^{-1}$, and a forage appetite level of 4.7 kg dry matter (DM) day$^{-1}$. Calculate a least-cost ration using a medium quality hay, barley and soyabean meal. The information on each feedstuff is given below. By how much does the cost of barley have to increase for its inclusion in the optimal ration to become uneconomic, assuming that the costs of the other feedstuffs hold constant?

| | *Feedstuff* | | |
|---|---|---|---|
| | *Hay* | *Barley* | *Soya* |
| NE content at the required animal production level [MJ (kg DM)$^{-1}$] | 4.5 | 9.2 | 7.9 |
| DCP content [g (kg DM)$^{-1}$] | 39 | 82 | 453 |
| Substitution rate [kg DM (kg DM hay)$^{-1}$] | 1.0 | 0.7 | 0.7 |
| Dry matter content [g DM (kg fresh weight)$^{-1}$] | 850 | 860 | 900 |
| Cost [£ (tonne fresh weight)$^{-1}$] | 20 | 100 | 130 |

**Exercise 13.4**

A Friesian cow weighing 600 kg yields 25 kg of milk a day in the second week of lactation. Her metabolizable energy (ME) requirement is 185 MJ $day^{-1}$ and her appetite level is 15 kg dry matter of forage $day^{-1}$. The available feeds are silage (ME content = 9 MJ (kg dry matter)$^{-1}$) and a dairy compound (ME content = 12.5 MJ (kg dry matter)$^{-1}$). Calculate the ration which minimizes concentrate usage. (Assume each kg dry matter of compound consumed reduces silage intake by 0.72 kg dry matter.)

# Glossary

### Binomial distribution

The word binomial means consisting of two terms (L. *bis*, twice + *nomen*, name). The binomial distribution is used to describe the outcome from a number of trials, there being only two possible results from each trial. We employ an application of this distribution used in the text (p. 108), as an illustration.

Consider the rainfall during a period of $n$ days. Suppose that the rainfall during each day is independent of the rainfall on every other day, and, if any rain falls at all, suppose that $h$ mm of rain fall. Let

$$p = \text{probability of no rain during a day}$$

and

$$q = \text{probability of h mm during that day}$$

Then, since $p$ and $q$ include all the possibilities there are, therefore

$$p + q = 1$$

For $n$ days, all possible outcomes may be described by expanding $(p + q)^n$, to give

$$1 = (p + q)^n = p^n + np^{n-1}\, q + \frac{n(n-1)}{1 \times 2}\, p^{n-2}\, q^2 + \dots + \frac{n!}{(n-r)!\, r!}\, p^{n-r}\, q^r + \dots + npq^{n-1} + q^n$$

The first term, $p^n$, is the probability of no rain on any of the $n$ days; the second term, $np^{n-1}q$, is the probability of rain falling on just one day; and

$$\frac{n!}{(n-r)!\, r!}\, p^{n-r}\, q^r = P_r$$

is the probability of rain falling on $r$ days, so that $rh$ mm of rain fall in all, during the period of $n$ days.

If $E(R_n)$ is the expected value of the total rainfall during the $n$-day period, $R_n$, then

$$E(R_n) = h \sum_{r=0}^{n} rP_r$$

By writing out in full the series, it is easily seen that the first term vanishes, because $r = 0$, and that $n$ and $q$ are common factors in the remaining $n - 1$ terms, so the series becomes $(p + q)^{n-1}$ after removing $nq$. Thus

$$E(R_n) = nqh$$

To find the variance $V(R_n)$, we need to evaluate

$$V(R_n) = h^2 \sum_{r=0}^{n} (r - nq)^2 P_r$$

$$= h^2 \sum_{r=0}^{n} (r^2 - 2nrq + n^2 q^2) P_r$$

The last term is

$$h^2 n^2 q^2$$

since

$$\sum_{r=0}^{n} P_r = 1$$

We have shown above that

$$\sum_{r=0}^{n} r P_r = nq$$

so the second term becomes

$$h^2(-2nq)nq$$

The first term is evaluated by writing

$$r^2 = r(r - 1) + r$$

and by writing out the series in detail as before, to give

$$\sum_{r=0}^{n} r^2 P_r = n(n - 1)q^2 + nq$$

Gathering the terms, therefore

$$V(R_n) = h^2 nq(1 - q)$$

The coefficient of variation ($CV$) is thus given by

$$CV = \frac{[V(R_n)]^{1/2}}{E(R_n)} = \left(\frac{1 - q}{nq}\right)^{1/2}$$

Further reading: Weatherburn (1968), Feller (1968).

**Diffusion**

Many physical phenomena are described by the diffusion (or heat) equation; for example, the diffusion of a solute in a solvent, and vorticity transport in fluid dynamics. The diffusion equation also occurs in probability theory, especially the description of Brownian motion (*see* Feller, 1968). It was originally developed by Fourier to describe heat conduction, and here we give a simple derivation by considering the description of temperature in a rod of conducting material whose cylindrical surface is insulated.

Let the rod have a uniform cross-section and assume that the temperature does not vary from point to point on a section, i.e. temperature depends only on position $x$ and time $t$ (*Figure G.1*). Applying the law of energy conservation to a slice of rod lying between $x$ and $x + \Delta x$ gives

rate of heat storage = inflow − outflow + heat generation

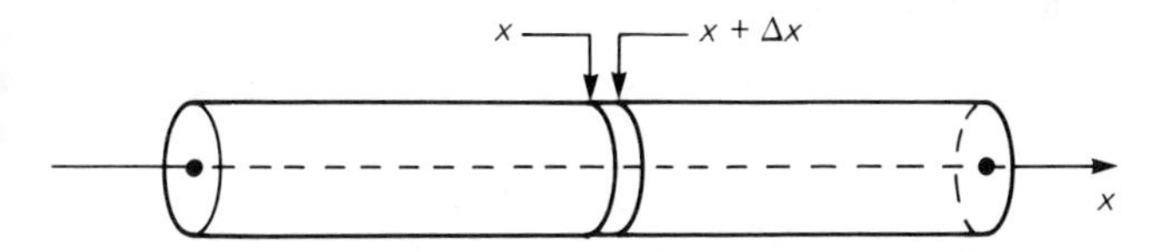


*Figure G.1* Heat conduction in an insulated rod of uniform cross section

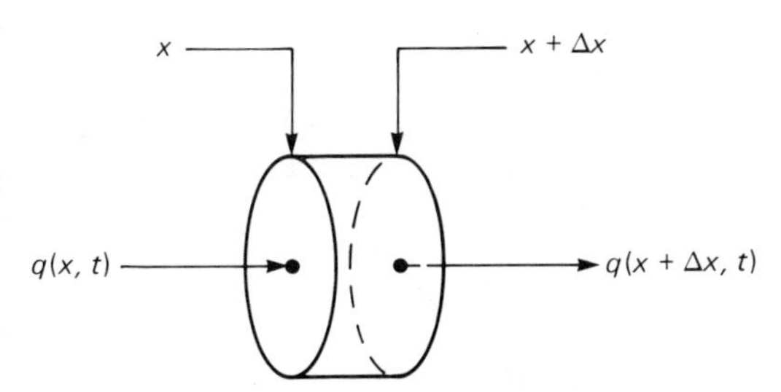


*Figure G.2* Heat flow $q(x,t)$ at point $x$ at time $t$

Let $q(x,t)$ [$\text{J m}^{-2}\,\text{s}^{-1}$] be the flow of heat at point $x$ at time $t$ (*Figure G.2*). The rate at which heat enters the slice through the surface at $x$ is $A\,q(x,t)$ [$\text{J s}^{-1}$], where $A$ [$\text{m}^2$] is the area of cross section. Similarly, the rate at which heat leaves the slice at $x + \Delta x$ is $A\,q(x + \Delta x,t)$. Let $r$ [$\text{J m}^{-3}\,\text{s}^{-1}$] be the rate of heat generation per unit volume, then the rate of heat generation in the slice is $A\Delta xr$ [$\text{J s}^{-1}$]. The rate of heat storage in the slice [$\text{J s}^{-1}$] is given by

$$\rho c A \Delta x \frac{\partial u}{\partial t}$$

where $\rho$ [$\text{kg m}^{-3}$] is the density, $c$ [$\text{J K}^{-1}\,\text{kg}^{-1}$] the heat capacity per unit mass, and $u$ [K] the temperature. Applying the laws of energy conservation:

$$A\,q(x,t) - A\,q(x + \Delta x,t) + A\Delta xr = A\Delta x\rho c\frac{\partial u}{\partial t}$$

Simplifying and rearranging:

$$-\left[\frac{q(x + \Delta x,t) - q(x,t)}{\Delta x}\right] + r = \rho c\frac{\partial u}{\partial t}$$

In the limit as $\Delta x \to 0$, we have (partial differentiation (*q.v.*))

$$-\frac{\partial q}{\partial x} + r = \rho c \frac{\partial u}{\partial t}$$

Now Fourier's law of heat conduction in one dimension gives

$$q = -\kappa \frac{\partial u}{\partial x}$$

where $\kappa$ is the thermal conductivity [$\mathrm{J\,m^{-1}\,s^{-1}\,K^{-1}}$], which on substituting into the heat balance equation yields

$$\frac{\partial}{\partial x}\left(\kappa \frac{\partial u}{\partial x}\right) + r = \rho c \frac{\partial u}{\partial t}$$

If $\kappa$ is a constant and there is no heat generation (i.e. $r = 0$), the balance equation reduces to

$$\frac{\partial^2 u}{\partial x^2} = \frac{\rho c}{\kappa} \frac{\partial u}{\partial t}$$

The quantity $\kappa/(\rho c)$ is usually written as $k$, the diffusivity:

$$\frac{\partial^2 u}{\partial x^2} = \frac{1}{k} \frac{\partial u}{\partial t}$$

Further reading: Carslaw and Jaeger (1960), Crank (1975).

**Duality**

For any given linear programme:

minimize $\boldsymbol{c}^{\mathrm{T}}\boldsymbol{x}$ subject to $\boldsymbol{Dx} \geqslant \boldsymbol{d}$, $\boldsymbol{x} \geqslant 0$ (Problem I)

there exists a linear programme:

maximize $\boldsymbol{d}^{\mathrm{T}}\boldsymbol{w}$ subject to $\boldsymbol{D}^{\mathrm{T}}\boldsymbol{w} \leqslant \boldsymbol{c}$, $\boldsymbol{w} \geqslant 0$ (Problem II)

where $\boldsymbol{c}$ and $\boldsymbol{x}$ are $n$-vectors, $\boldsymbol{D}$ is an $m \times n$ matrix, $\boldsymbol{d}$ and $\boldsymbol{w}$ are $m$-vectors, and the superscript T denotes transposition. Problem I is called the dual of Problem II and vice versa. It should be noted that the optimal values of the objective functions for the two problems are equal, that is

minimum $\boldsymbol{c}^{\mathrm{T}}\boldsymbol{x}$ = maximum $\boldsymbol{d}^{\mathrm{T}}\boldsymbol{w}$

Furthermore, the values of the decision variables (as opposed to the slack variables) in the dual problem are the shadow prices on the constraints of the primal problem. Duality is of considerable importance in the theory of linear programming. Any linear programme can be expressed in the form of Problem I by replacing each equality constraint such as $\Sigma_j D_{ij} x_j = d_i$ by the two constraints $\Sigma_j D_{ij} x_j \geqslant d_i$ and $\Sigma_j D_{ij} x_j \leqslant d_i$, and converting each $\leqslant$ constraint to a $\geqslant$ constraint by multplying both sides of the inequality by $-1$.

Further reading: Taha (1971), Wagner (1969).

**Eigenvalues**

If $\boldsymbol{A}$ is a square matrix of order $(n \times n)$ and $\boldsymbol{I}$ is the unit matrix of the same order, then the matrix

$$\boldsymbol{B} = \boldsymbol{A} - \lambda\boldsymbol{I}$$

is called the *characteristic* matrix of $\boldsymbol{A}$, $\lambda$ being a parameter. The equation

$$|\boldsymbol{B}| = |\boldsymbol{A} - \lambda\boldsymbol{I}| = 0$$

where $|\boldsymbol{B}|$ denotes the determinant of matrix $\boldsymbol{B}$, is called the characteristic equation of $\boldsymbol{A}$ and is in general an equation of degree $n$ in $\lambda$. The $n$ roots of this equation are called the *eigenvalues* (or characteristic roots) of $\boldsymbol{A}$. With reference to p. 30, if the dependence of the residual $R$ on the two parameters $P_1$ and $P_2$ is given by

$$R = P_1^2 + 3P_2^2$$

then the Hessian matrix $\boldsymbol{H}$, whose elements are given by

$$H_{ij} = \frac{\partial^2 R}{\partial P_i \partial P_j} \quad \text{with} \quad i, j = 1,2$$

is

$$\boldsymbol{H}_P = \begin{pmatrix} 2 & 0 \\ 0 & 6 \end{pmatrix}$$

This has eigenvalues of 2 and 6. On the other hand, if

$$R = 2p_1^2 + 2p_2^2 + 2p_1 p_2$$

then the Hessian matrix is

$$\boldsymbol{H}_p = \begin{pmatrix} 4 & 2 \\ 2 & 4 \end{pmatrix}$$

Note that $|\boldsymbol{H}_P| = |\boldsymbol{H}_p| = 12$. The eigenvalues of $\boldsymbol{H}_p$ are given by the equation

$$|\boldsymbol{H}_p - \lambda\boldsymbol{I}| = 0$$

that is

$$\begin{vmatrix} (4-\lambda) & 2 \\ 2 & (4-\lambda) \end{vmatrix} = 0$$

giving

$$(4 - \lambda)^2 - 4 = 0$$

which has the solution $\lambda = 2$, $\lambda = 6$; these are the eigenvalues of $\boldsymbol{H}_p$. Finally, the parameter transformation $P_1 = (p_1 - p_2)/\sqrt{2}$ and $P_2 = (p_1 + p_2)/\sqrt{2}$ will convert the two forms into each other, but without essentially changing the problem.

Further reading: Stephenson (1961), Hildebrand (1965).

**$F$-distribution**

If $x_1$ and $x_2$ possess independent $\chi^2$-distributions with $\nu_1$ and $\nu_2$ degrees of freedom respectively, then

$$F = \frac{x_1/\nu_1}{x_2/\nu_2}$$

has the $F$-distribution, with $(\nu_1, \nu_2)$ degrees of freedom, whose probability density function is given by

$$f(F) = cF^{(\nu_1 - 2)/2}\,(\nu_2 + \nu_1 F)^{-(\nu_1 + \nu_2)/2}$$

where

$$c = \nu_1^{\nu_1/2}\,\nu_2^{\nu_2/2}\,\Gamma\left(\frac{\nu_1 + \nu_2}{2}\right)\Big/\left[\Gamma\left(\frac{\nu_1}{2}\nu\right)\Gamma\left(\frac{\nu_2}{2}\nu\right)\right]$$

and $\Gamma$ denotes the gamma function.

A property of the $F$-distribution is that the inverse ratio

$$F^{-1} = \frac{x_2/\nu_2}{x_1/\nu_1}$$

has an $F$-distribution with $(\nu_2, \nu_1)$ degrees of freedom. The lower $100\alpha\%$ point on $(\nu_1, \nu_2)$ degrees of freedom is therefore the reciprocal of the upper $100(1 - \alpha)\%$ point on $(\nu_2, \nu_1)$ degrees of freedom, that is

$$F_\alpha(\nu_1, \nu_2) = 1/F_{1-\alpha}(\nu_2, \nu_2)$$

This means that there is no need to tabulate both ends of the $F$-distribution, and only the upper percentage points are generally tabulated; a short table of 95% points is given in *Table G.1*. As an example of using tables of $F$-values, consider the $F$-distribution with $\nu_1 = 6$ and $\nu_2 = 10$. Using *Table G.1*,

$$F_{0.95}(6, 10) = 3.22$$

Now

$$F_{0.05}(6, 10) = 1/F_{0.95}(10, 6) = 1/4.06 = 0.25$$

**TABLE G.1. Variance ratio: 95% points of the $F$-distribution**

| | $\nu_1$ = *degrees of freedom for the greater estimate of variance* | | | | | | | | | | |
|---|---|---|---|---|---|---|---|---|---|---|---|
| $\nu_2$ | 2 | 4 | 6 | 8 | 10 | 15 | 20 | 30 | 40 | 50 | 100 |
| 2 | 19.0 | 19.2 | 19.3 | 19.4 | 19.4 | 19.4 | 19.4 | 19.5 | 19.5 | 19.5 | 19.5 |
| 4 | 6.94 | 6.39 | 6.16 | 6.04 | 5.96 | 5.86 | 5.80 | 5.75 | 5.72 | 5.70 | 5.66 |
| 6 | 5.14 | 4.53 | 4.28 | 4.15 | 4.06 | 3.94 | 3.87 | 3.81 | 3.77 | 3.75 | 3.71 |
| 8 | 4.46 | 3.84 | 3.58 | 3.44 | 3.35 | 3.22 | 3.15 | 3.08 | 3.04 | 3.02 | 2.97 |
| 10 | 4.10 | 3.48 | 3.22 | 3.07 | 2.98 | 2.85 | 2.77 | 2.70 | 2.66 | 2.64 | 2.59 |
| 15 | 3.68 | 3.06 | 2.79 | 2.64 | 2.54 | 2.40 | 2.33 | 2.25 | 2.20 | 2.18 | 2.12 |
| 20 | 3.49 | 2.87 | 2.60 | 2.45 | 2.35 | 2.20 | 2.12 | 2.04 | 1.99 | 1.97 | 1.91 |
| 30 | 3.32 | 2.69 | 2.42 | 2.27 | 2.16 | 2.01 | 1.93 | 1.84 | 1.79 | 1.76 | 1.70 |
| 40 | 3.23 | 2.61 | 2.34 | 2.18 | 2.08 | 1.92 | 1.84 | 1.74 | 1.69 | 1.66 | 1.59 |
| 50 | 3.18 | 2.56 | 2.29 | 2.13 | 2.03 | 1.87 | 1.78 | 1.69 | 1.63 | 1.60 | 1.52 |
| 100 | 3.09 | 2.46 | 2.19 | 2.10 | 1.93 | 1.77 | 1.68 | 1.57 | 1.52 | 1.48 | 1.39 |

Therefore, there is a 90% probability that the calculated $F(6,10)$ value lies in the range

$$0.25 \leqslant \text{calculated } F(6,10) \leqslant 3.22$$

The $F$-test is often used to test the hypothesis that two normal populations have the same parent variance. Here, we give a description of the test which is relevant to the application in Chapter 2 (p. 30). Let $R_e$ be the error residual sum of squares with $\nu_e$ degrees of freedom, and after fitting the model to the data, $R_\ell$ is the lack of fit residual sum of squares with $\nu_\ell$ degrees of freedom. We wish to know if the lack of fit residual indicates whether or not the model is an acceptable fit to the data. Assume that $R_\ell/\nu_\ell > R_e/\nu_e$. The $F$-ratio is defined by

$$F = \frac{R_\ell/\nu_\ell}{R_e/\nu_e}$$

*Table G.1* gives 95% points for the $F$-distribution. If the calculated $F$ is less than the value in the table, the model is, in terms of its ability to fit the data, an acceptable fit. Exercise 2.4 gives an example of the technique.

Further reading: Weatherburn (1968), Snedecor and Cochran (1980).

**Fourier analysis**

Consider the two variables $x$ and $y$, where $x$ is an independent variable lying in the range

$$0 \leqslant x \leqslant x_m$$

where $x_m$ is the maximum value of $x$, and $y$ is a dependent variable. Suppose that for $n$ equally spaced values of $x$, $y$ is measured (*Figure G.3*), giving the data set $(x_1, y_1), (x_2 y_2), \ldots, (x_n, y_n)$. The 'distance' between all adjacent pairs of measurements is $x_m/n$, and also the distance from the last measurement at $x_n$ to the first measurement at $x_1$ is equal to $x_m/n$ (it is assumed that $x = 0$ and $x = x_m$ are equivalent, so that the distance from $x_n$ to $x_1$ is $x_m - x_n + x_1$ which is equal to $x_m/n$).

As the first measurement of $y$ is at $x = x_1$, the second measurement is at

$$x_2 = x_1 + \frac{x_m}{n}$$

and the $i$th measurement at

$$x_i = x_1 + (i-1)\frac{x_m}{n} \quad (i = 1, 2, 3, \ldots, n)$$

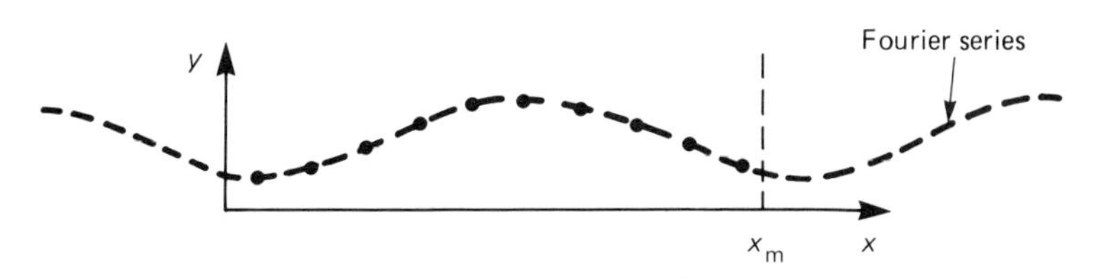


*Figure G.3* Representing data by a Fourier series

Note that the data set $(x_1, y_1), (x_2, y_2), \ldots, (x_n, y_n)$ contains only $n + 1$ independent data values, namely, $x_1$, and $y_1, y_2, \ldots, y_n$.

Next, we represent the data by a Fourier series

$$\begin{aligned} y = a_0 &+ a_1 \cos\left[\left(\frac{x}{x_m}\right)360°\right] + b_1 \sin\left[\left(\frac{x}{x_m}\right)360°\right] \\ &+ a_2 \cos\left[2\left(\frac{x}{x_m}\right)360°\right] + b_2 \sin\left[2\left(\frac{x}{x_m}\right)360°\right] \\ &+ \ldots \\ &+ a_j \cos\left[\left(\frac{x}{x_m}\right)360°\right] + b_j \sin\left[j\left(\frac{x}{x_m}\right)360°\right] \\ &+ \ldots \end{aligned}$$

$a_0, a_1, \ldots$ and $b_1, b_2, \ldots$ are called the Fourier coefficients. The coefficients may be estimated by

$$a_0 = \frac{1}{n} \sum_{i=0}^{n} y_i$$

and for $j \geqslant 1$

$$a_j = \frac{2}{n} \sum_{i=1}^{n} y_i \cos\left[j\left(\frac{x_i}{x_m}\right)360°\right]$$

$$b_j = \frac{2}{n} \sum_{i=1}^{n} y_i \sin\left[j\left(\frac{x_i}{x_m}\right)360°\right]$$

The last two expressions are only approximate, although the approximation is reasonable if $n$ is not too small.

If $n$ is even, and if terms in the Fourier series up to $j = n/2$ are included, then the $n + 1$ independent measurements are exactly matched by $n + 1$ Fourier coefficients, and the Fourier representation will go exactly through the measured values. It is usually only appropriate to use a Fourier series if the first few terms of the series dominates. For instance, when describing light receipt or temperature throughout a day or a year, one might find that the terms $a_0$, $a_1$, and $b_1$ give a sufficiently accurate description of the environmental variable for most purposes (p. 100, p. 106).

Sometimes, it is convenient to use an alternative form for the Fourier series, namely

$$\begin{aligned} y = a_0 &+ c_1 \cos\left[\left(\frac{x - \Delta x_1}{x_m}\right)360°\right] + c_2 \cos\left[2\left(\frac{x - \Delta x_2}{x_m}\right)360°\right] + \ldots \\ &+ c_j \cos\left[j\left(\frac{x - \Delta x_j}{x_m}\right)360°\right] + \ldots \end{aligned}$$

The coefficients $c_j$ and the phase shifts $\Delta x_j$ are obtained from $a_j$ and $b_j$ by means of

$$c_j^2 = a_j^2 + b_j^2 \quad \text{and} \quad \tan\left[j\left(\frac{\Delta x_j}{x_m}\right)360°\right] = \frac{b_j}{a_j}$$

If $a_j < 0$, then it is necessary to take the negative square root for $c_j$. Also, $\Delta x_j$ lies in the range $-\frac{1}{4}x_m \leqslant \Delta x_j \leqslant \frac{1}{4}x_m$ for all $a_j$.

Further reading: Churchill (1963), Stephenson (1961).

**Least squares**

*See* Regression.

**Likelihood**

Let $f(x;\ \lambda)$ be the frequency (or probability density) function of the random variable $x$, where $\lambda$ is the parameter to be estimated. Suppose $n$ observations of $x$ are made. Let $x_1, x_2, \ldots, x_n$ denote these $n$ observations. The function

$$L(x_1, x_2, \ldots, x_n;\ \lambda) = \prod_{i=1}^{n} f(x_i, \lambda)$$

defines a function of the random variables $x_1, x_2, \ldots, x_n$ and the parameter $\lambda$ and is called the *likelihood function*. If the observed values are obtained from $n$ independent trials of an experiment for which $f(x;\ \lambda)$ is the frequency function of a discrete random variable $x$ then, for any particular set of observed values, $L$ gives the probability of obtaining that set of values, including their order of occurrence. If $x$ is a continuous random variable, then $L$ gives the probability density at the sample point $(x_1, x_2, \ldots, x_n)$, where the sample space is thought of as $n$-dimensional.

A maximum likelihood estimator $\lambda$ of the parameter $\lambda$ in the frequency function $f(x;\ \lambda)$ is an estimator that maximizes the likelihood function $L(x_1, x_2, \ldots, x_n;\ \lambda)$ as a function of $\lambda$. If $x_1, x_2, \ldots, x_n$ are taken as fixed, the likelihood function $L$ becomes a function of $\lambda$ only. The problem of finding a maximum likelihood estimator is therefore that of finding $\lambda$ which maximizes $L(\lambda)$. Maximum likelihood estimators can usually be determined using the calculus.

As an example, consider the problem of estimating $\lambda$ in the exponential distribution density function:

$$\begin{aligned} f(x;\lambda) &= \lambda e^{-\lambda x} \quad x > 0 \\ &= 0 \quad x \leqslant 0 \end{aligned}$$

Six observations of $x$ give $x_1 = 0.1$, $x_2 = 1.5$, $x_3 = 0.5$, $x_4 = 1.2$, $x_5 = 0.7$, $x_6 = 1.0$. The likelihood function is given by

$$\begin{aligned} L(\lambda) &= \lambda e^{-0.1\lambda}\, \lambda e^{-1.5\lambda}\, \lambda e^{-0.5\lambda}\, \lambda e^{-1.2\lambda}\, \lambda e^{-0.7\lambda}\, \lambda e^{-1.0\lambda} \\ &= \lambda^6 e^{-5.0\lambda} \end{aligned}$$

The value of $\lambda$ that maximizes $L(\lambda)$ is given by

$$\frac{\partial L}{\partial \lambda} = \lambda^5 e^{-5.0\lambda}(6 - 5\lambda) = 0$$

The non-trivial solution to this equation gives the estimate $\lambda = 1.2$.

Further reading: Hoel (1962), Edwards (1972).

**Markov process**

Let $\bar{y}_N$ be the expected value of the independent variable $y$ during the $N$th time interval (for example, $N$ could be the day number). It is assumed that $y_N$ is distributed in some manner about $\bar{y}_N$ so that

$$y_N = \bar{y}_N + \epsilon_N$$

where $\epsilon_N$ is a random variable with zero mean and standard deviation $\sigma$. Equally well we could take

$$\ln(y_N/\bar{y}_N) = \epsilon_N$$

if it were felt to be more appropriate. For quantities such as temperature, which do not have a lower bound as far as we are concerned, the first expression may be preferred, whereas for other quantities such as light receipt or rainfall, which cannot fall below zero, the second recommends itself.

In a simple Markov process, it is assumed that the value of the random variable for the $N$th time interval is influenced by the outcome during the $(N - 1)$th time interval, and is unaffected by any additional knowledge of the previous history of the system. The two expressions above become

$$y_N = \bar{y}_N + \epsilon_N + \rho\epsilon_{N-1} \quad \text{and} \quad \ln(y_N/\bar{y}_N) = \epsilon_N + \rho\epsilon_{N-1}$$

where $\rho$ is an autocorrelation coefficient. In words, if yesterday was more sunny, rainy, or colder than average, this increases the probability that today will also be more sunny, rainy, or colder than average. We write both of the above expressions as

$$z_N - \bar{z}_N = \epsilon_N + \rho\epsilon_{N-1}$$

where $z_N = y_N$ or $z_N = \ln(y_N)$. For the previous day

$$z_{N-1} - \bar{z}_{N-1} = \epsilon_{N-1} + \rho\epsilon_{N-2}$$

Taking expected values, therefore

$$E(z_N) = \bar{z}_N \qquad E[(z_N - \bar{z}_N)^2] = \sigma^2$$

$$E[(z_N - \bar{z}_N)(z_{N-1} - \bar{z}_{N-1})] = \rho\sigma^2$$

These are approximations valid for a reasonably large number of measurements.

Further reading: Feller (1968).

**Monte Carlo methods**

Monte Carlo methods are simulation techniques for problems having a probabilistic basis; they comprise that branch of mathematics which is concerned with experiments on random numbers. The methods are applied to two types of problem. First, in certain problems involving some kind of stochastic process, Monte Carlo methods have been used to simulate most of the well-known theoretical probability distributions as well as many empirical ones. Secondly, in certain deterministic mathematical problems which cannot readily be solved analytically, approximate solutions may sometimes be obtained by simulating a stochastic process whose moments, probability density function, or cumulative distribution function satisfy the functional relationships or solution requirements of the deterministic problem. Solutions to higher-order difference equations and multiple integrals are often obtained in this way.

We illustrate the Monte Carlo approach by computing the area of the first quadrant of a circle with radius unity (*Figure G.4*). Any pair of uniformly distributed random numbers $(r_x, r_y)$ defined over the interval (0,1) corresponds to a point within the unit square of *Figure G.4*. Let $f(x) = \sqrt{(1 - x^2)}$. If $f(r_x) \geqslant r_y$ for the generated random number pair, then the point $(r_x, r_y)$ lies under or on the

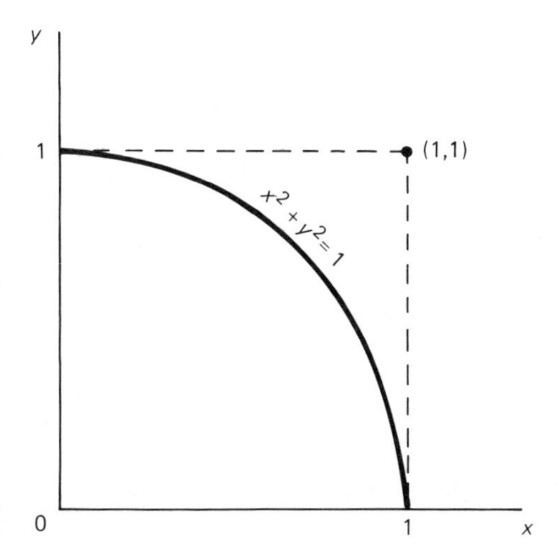


*Figure G.4* Using Monte Carlo methods for numerical integration

circle. If $f(r_x) < r_y$, then $(r_x, r_y)$ lies above the curve. Accepting and counting the first type of random occurrence and dividing this count by the total number of pairs generated, we obtain a ratio which corresponds to the proportion of the area of the unit square lying under the circle. This ratio tends to $\pi/4$ as the number of pairs generated increases.

Further reading: Naylor *et al.* (1966), Hammersley and Handscomb (1964).

**Partial differentiation**

Suppose $f(x,y)$ is a real single-valued function of two independent variables $x$ and $y$. Then the partial derivative of $f(x,y)$ with respect to $x$ is defined as

$$\frac{\partial f}{\partial x} = \lim_{\Delta x \to 0} \left[ \frac{f(x + \Delta x, y) - f(x,y)}{\Delta x} \right]$$

Similarly, the partial derivative of $f(x,y)$ with respect to $y$ is defined as

$$\frac{\partial f}{\partial y} = \lim_{\Delta y \to 0} \left[ \frac{f(x, y + \Delta y) - f(x,y)}{\Delta y} \right]$$

In other words, the partial derivative of $f(x,y)$ with respect to $x$ may be thought of as the ordinary derivative of $f(x,y)$ with respect to $x$ obtained by treating $y$ as a constant. Similarly, the partial derivative of $f(x,y)$ with respect to $y$ may be found by treating $x$ as constant and evaluating the ordinary derivative of $f(x,y)$ with respect to $y$. For example, consider the equation

$$z = xy + x^3$$

Partial differentiation with respect to $x$ gives

$$\frac{\partial z}{\partial x} = y + 3x^2$$

and partial differentiation with respect to $y$ gives

$$\frac{\partial z}{\partial y} = x$$

Further reading: Stephenson (1961), Sneddon (1957).

**Partial fractions**

If one wishes to evaluate an integral such as

$$\int \frac{\mathrm{d}x}{x(A + x)}$$

where $A$ is a constant, it is usual to use the method of partial fractions to split the integrand into components which can then be easily integrated. We write

$$\frac{1}{x(A + x)} = \frac{a}{x} + \frac{b}{A + x}$$

where $a$ and $b$ are constants to be determined. Using the identity

$$1 \equiv a(A + x) + bx$$

therefore

$$a = \frac{1}{A} \quad \text{and} \quad b = -a = -\frac{1}{A}$$

Thus

$$\frac{1}{x(A + x)} = \frac{1}{A}\left(\frac{1}{x} - \frac{1}{A + x}\right)$$

Both of the terms on the right of this equality can be directly integrated, giving

$$\int \frac{\mathrm{d}x}{x(A + x)} = \frac{1}{A}\,[\ln x - \ln(A + x)]$$

A similar approach may be used to put into an integrable form expressions such as

$$\frac{1}{(A + x)(B + x)} = \frac{a}{A + x} + \frac{b}{B + x}$$

$$\frac{1}{(A + x)(B + x)(C + x)} = \frac{a}{A + x} + \frac{b}{B + x} + \frac{c}{C + x}$$

and also quadratic forms of the type

$$\frac{1}{(A + x)(B + Cx + Dx^2)} = \frac{a}{A + x} + \frac{bx + c}{B + Cx + Dx^2}$$

Tables of standard integrals often list many of these forms.

Further reading: Stephenson (1961), Siddons, Snell and Morgan (1952).

**Regression**

The *linear regression model* assumes that there is one stochastic variable $y$ and one deterministic variable $x$, and that $E(y \mid x)$, the expected value of $y$ given $x$, is linearly dependent on $x$:

$$E(y \mid x) = \beta_0 + \beta_1 x$$

and the variance $V(y \mid x)$ is constant:

$$V(y \mid x) = \sigma^2$$

The variable $y$ is known as the *dependent* variable and $x$ as the *independent* (or *regressor*) variable. The straight line $y = \beta_0 + \beta_1 x$ is called the *regression line of $y$ on $x$* and the parameter $\beta_1$ is called the *regression coefficient* (or *slope*). Suppose we have the observations $(X_1, Y_1)$, $(X_2, Y_2)$, ..., $(X_n, Y_n)$, it is more convenient to write the linear regression model in the form

$$E(y \mid x) = \beta_0 + \beta_1 (x - \bar{X})$$

$$V(y \mid x) = \sigma^2$$

where $\bar{X} = \Sigma_{i=1}^{n} X_i/n$. Estimates of $\beta_0$ and $\beta_1$ are determined by the method of least squares, whereby $S$, the sum of squared differences between the observed values of $y$ and the expected values given by the regression line, is formed, that is

$$S(\beta_0, \beta_1) = \sum_{i=1}^{n} [Y_i - \beta_0 - \beta_1 (X_i - \bar{X})]^2$$

and the values of $\beta_0$ and $\beta_1$ which minimize $S$ are found. These values satisfy the normal equations:

$$\frac{\partial S}{\partial \beta_0} = \frac{\partial S}{\partial \beta_1} = 0$$

The solution to these equations is given by

$$\beta_0 = \bar{Y} = \sum_{i=1}^{n} Y_i/n \quad \text{and} \quad \beta_1 = \sum_{i=1}^{n} (X_i - \bar{X})Y_i \Big/ \sum_{i=1}^{n} (X_i - \bar{X})^2$$

Sometimes the assumption of constant variance in the linear regression model is invalid (i.e. $\sigma^2 \neq$ a constant). In such situations, least squares estimates of $\beta_0$ and $\beta_1$ can be found by minimizing

$$S(\beta_0, \beta_1) = \sum_{i=1}^{n} w_i[Y_i - \beta_0 - \beta_1 (X_i - \bar{X})]^2$$

where the $w_i$ are appropriate weights. This is known as *weighted linear regression.* If several observations of $y$ have been made at each selected $x$, the weights are often chosen as the reciprocal of the variance of the observed $y$'s at the value of $x$ in question.

The *multiple linear regression model* is an extension of the simple linear case to include several independent variables $x_1, x_2, ..., x_k$ and may be stated thus:

$$E(y \mid x_1, x_2, ..., x_k) = \beta_0 + \beta_1(x_1 - \overline{X}_1) + \beta_2(x_2 - \overline{X}_2) + ... + \beta_k(x_k - \overline{X}_k)$$
$$V(y \mid x_1, x_2, ..., x_k) = \sigma^2$$

The parameters $\beta_1, \beta_2, ..., \beta_k$ are called the partial regression coefficients and the $\overline{X}$'s are computed from the $n$ observations $(Y_1, X_{11}, X_{21}, ..., X_{k1})$, ..., $(Y_n, X_{1n}, X_{2n}, ..., X_{kn})$. The sum of squares of the $Y_i$'s from their expectations is

$$S(\beta_0, \beta_1, ..., \beta_k) = \sum_{i=1}^{n} [Y_i - \beta_0 - \beta_1(X_{1i} - \overline{X}_1) - ... - \beta_k(X_{ki} - \overline{X}_k)]^2$$

and the least squares estimates of the parameters $\beta_0, \beta_1, ..., \beta_k$ are the solutions of the normal equations

$$\frac{\partial S}{\partial \beta_0} = \frac{\partial S}{\partial \beta_1} = \frac{\partial S}{\partial \beta_2} = ... = \frac{\partial S}{\partial \beta_k} = 0$$

The linear and multiple regression models are linear in the parameters $\beta_0, \beta_1, ..., \beta_k$. Any regression model that is nonlinear in the parameters is called a nonlinear model. A nonlinear model which can be transformed into a form which is linear in the parameters (for example, by taking natural logarithms) is said to be *intrinsically linear*. A nonlinear model which cannot be converted into such a form is said to be *intrinsically nonlinear*. In the case of intrinsically nonlinear models, the solution of the normal equations can be extremely difficult and iterative methods generally have to be employed.

Further reading: Snedecor and Cochran (1980), Draper and Smith (1981).

**$t$-distribution**

If $x_1$ is normally distributed with zero mean and unit variance and $x_2^2$ has a $\chi^2$ distribution with $v$ degrees of freedom, and $x_1$ and $x_2$ are independently distributed, then the variable

$$t = \frac{x_1 \sqrt{v}}{x_2}$$

has a $t$-distribution, with $v$ degrees of freedom, whose probability density function is given by

$$f(t) = c\left(1 + \frac{t^2}{v}\right)^{-(v+1)/2}$$

where

$$\Gamma[(v+1)/2]/[\sqrt{(\pi v)}\,\Gamma(v/2)]$$

and $\Gamma$ denotes the gamma function.

A common application of the $t$-distribution is for finding confidence limits for a mean. Let $y$ be normally distributed with mean $\mu$ and variance $\sigma^2$ and let $\bar{y}$ and $s^2$ be their respective estimates based on a random sample of size $n$. Then

$$x_1 = \frac{\bar{y} - \mu}{\sigma/\sqrt{n}}$$

and

$$x_2^2 = \frac{(n-1)s^2}{\sigma^2}$$

satisfy the requirements of $x_1$ and $x_2$ in the $t$-distribution, therefore

$$t = \frac{(\bar{y} - \mu)\sqrt{n}}{s}$$

has a $t$-distribution with $n - 1$ degrees of freedom. Let $t_{\beta,v}$ denote the $100 \times (1 - \beta)\%$ point of the mod. $t$-distribution for $v$ degrees of freedom, then, where $Pr$ denotes probability,

$$Pr[-t_{\beta,v} \leqslant \frac{(\bar{y} - \mu)\sqrt{n}}{s} \leqslant t_{\beta,v}] = 1 - \beta$$

This can be expressed as an inequality on $\mu$,

$$Pr[\bar{y} - t_{\beta,v}s/\sqrt{n} \leqslant \mu \leqslant \bar{y} + t_{\beta,v}s/\sqrt{n}] = 1 - \beta$$

therefore $100(1 - \beta)\%$ confidence intervals for $\mu$ are

$$\bar{y} \pm t_{\beta,v}s/\sqrt{n}$$

The $t$-statistic is used in this way for calculating the confidence limits of parameters estimated by fitting a model to experimental data (p. 31). The 90% confidence interval is given by $\beta = 0.1$, and values of $t_{0.1,v}$ are

| $v$ | 2 | 3 | 5 | 10 | 20 | 50 | $\infty$ |
|---|---|---|---|---|---|---|---|
| $t_{0.1,v}$ | 2.92 | 2.35 | 2.02 | 1.81 | 1.73 | 1.68 | 1.64 |

Further reading: Weatherburn (1968), Snedecor and Cochran (1980).

### Time series

A time series is a set of ordered observations on a quantity taken at different points of time. Usually, these points are equidistant in time. The essential quality of the series is the order of the observations according to the time variable. Methods of time series analysis embrace a whole gamut of techniques ranging through polynomial fitting, moving averages and exponential smoothing. Here, we confine our attention to a brief description of one of the most commonly used forecasting models, the Box–Jenkins model.

Essentially, the Box–Jenkins model combines autocorrelation with polynomial regression using finite differences. Application of the method is as follows.

Consider the time series $X_0, X_1, X_2, \ldots, X_n$. The method begins by constructing a backward difference scheme:

| $i$ | $X_i$ | $\nabla X_i$ | $\nabla^2 X_i$ | ..... | $\nabla^d X_i$ |
|---|---|---|---|---|---|
| 0 | $X_0$ | | | | |
| 1 | $X_1$ | $X_1 - X_0$ | | | |
| 2 | $X_2$ | $X_2 - X_1$ | $\nabla X_2 - \nabla X_1$ | | |
| . | . | . | . | . | |
| . | . | . | . | . | |
| . | . | . | . | . | |
| $d-1$ | $X_{d-1}$ | . | . | . | |
| $d$ | $X_d$ | . | . | | $\nabla^{d-1}X_d - \nabla^{d-1}X_{d-1}$ |
| $d+1$ | $X_{d+1}$ | . | . | | . |
| . | . | . | . | | . |
| . | . | . | . | | . |
| . | . | . | . | | . |
| $n-2$ | $X_{n-2}$ | $X_{n-2} - X_{n-3}$ | $\nabla X_{n-2} - \nabla X_{n-3}$ | | $\nabla^{d-1}X_{n-2} - \nabla^{d-1}X_{n-3}$ |
| $n-1$ | $X_{n-1}$ | $X_{n-1} - X_{n-2}$ | $\nabla X_{n-1} - \nabla X_{n-2}$ | | $\nabla^{d-1}X_{n-1} - \nabla^{d-1}X_{n-2}$ |
| $n$ | $X_n$ | $X_n - X_{n-1}$ | $\nabla X_n - \nabla X_{n-1}$ | ... | $\nabla^{d-1}X_n - \nabla^{d-1}X_{n-1}$ |

where $\nabla$ is the backward difference operator such that

$$\nabla X_i = X_i - X_{i-1}$$

and

$$\nabla^m X_i = \nabla^{m-1} X_i - \nabla^{m-1} X_{i-1} \quad (m = 2,3,4, \text{etc.})$$

The differencing scheme continues until there is no autocorrelation in the $\nabla^{d+1}X_i$ column (i.e. there is no correlation between $\nabla^{d+1}X_i$ and $\nabla^{d+1}X_{i-1}$), thus determining a value for the scale of difference parameter $d$.

The relationship

$$\nabla^d X_i = \beta_0 + \beta_1 \nabla^d X_{i-1} + \beta_2 \nabla^d X_{i-2} + \ldots + \beta_p \nabla^d X_{i-p}$$

is then fitted to the $d$th differences using multiple linear regression (*q.v.*). $\nabla^d X_{n+1}$, and hence the next value in the time series $X_{n+1}$, can then be estimated.

Further reading: Kendall (1973), Box and Jenkins (1976).

# Solutions to exercises

## Chapter 1

### Solution 1.1

Differentiating $y = ax/(b + x)$ gives

$$\frac{dy}{dx} = \frac{a(b + x) - ax}{(b + x)^2} = \frac{ab}{(b + x)^2} \qquad \text{(S1.1.1)}$$

At $x = 0$, $dy/dx = a/b$, and as $x \to \infty$, $y \to ax/x = a$. A second differentiation of equation (S1.1.1) gives

$$\frac{d^2y}{dx^2} = -\frac{2ab}{(b + x)^3} \qquad \text{(S1.1.2)}$$

The magnitude of this is maximum for $x \geqslant 0$ is when $x = 0$ with the value $-2a/b^2$. Note that the least linear part of the rectangular hyperbola is in the region of the origin.

### Solution 1.2

Differentiating

$$W = \frac{at}{b + t} \qquad \text{(S1.2.1)}$$

gives

$$\frac{dW}{dt} = \frac{ab}{(b + t)^2} \qquad \text{(S1.2.2)}$$

From equation (S1.2.1), therefore

$$t = \frac{bW}{a - W} \qquad \text{(S1.2.3)}$$

Substituting equation (S1.2.3) into equation (S1.2.2) gives

$$\frac{dW}{dt} = \frac{(a - W)^2}{ab} \qquad \text{(S1.2.4)}$$

## Chapter 2

### Solution 2.1

This problem can be worked either way: by differentiating the growth equation twice and eliminating the time variable $t$; or, more easily, by integrating the two differential equations with the given boundary conditions.

The exponential quadratic growth equation can be written (p. 90)

$$\ln x = a_0 + a_1 t + a_2 t^2 \qquad \text{(S2.1.1)}$$

Differentiation gives

$$\frac{1}{x}\frac{dx}{dt} = a_1 + 2a_2 t \quad \text{with} \quad \ln x = a_0 \quad \text{at } t = 0 \qquad \text{(S2.1.2)}$$

We introduce a second state variable $y$ by means of

$$y = a_2 t \qquad \text{(S2.1.3)}$$

and differentiation gives

$$\frac{dy}{dt} = a_2 \quad \text{with} \quad y = 0 \quad \text{at } t = 0 \qquad \text{(S2.1.4)}$$

The two differential equations specifying the problem are thus

$$\frac{1}{x}\frac{dx}{dt} = a_1 + 2y \quad \text{and} \quad \frac{dy}{dt} = a_2$$

$$\text{with at } t = 0 \quad \ln x = a_0 \quad \text{and} \quad y = 0 \qquad \text{(S2.1.5)}$$

Integration of these two equations to recover equation (S2.1.1) is straightforward.

A biological view of equations (S2.1.5) is that at time $t = 0$, the specific growth rate $(1/x)(dx/dt)$ of the organism has the value $a_1$, and as time progresses, some factor (parameterized in $a_2$) causes the specific growth rate to change (usually decreasing) linearly with time. There is an alternative interpretation of equation (S2.1.1) (p. 91).

### Solution 2.2

Euler's formula, given in equation (2.5), may be written

$$y(t + \Delta t) = y(t) + \Delta t \frac{dy}{dt} \qquad \text{(S2.2.1)}$$

Taking, for example, $t = 0.1$, a table can be constructed, line by line, working from left to right:

| $t$ | $y$ | $dy/dt$ | $\Delta t\, dy/dt$ | $\exp(-t)$ |
|---|---|---|---|---|
| 0 | 1.000 | −1.000 | −0.1000 | 1.000 |
| 0.1 | 0.900 | −0.900 | −0.0900 | 0.905 |
| 0.2 | 0.810 | −0.810 | −0.0810 | 0.819 |
| 0.3 | 0.729 | −0.729 | −0.0729 | 0.741 |
| etc. | | | | |

Integration of $dy/dt = -y$ with $y = 1$ at $t = 0$ gives

$$y = e^{-t} \tag{S2.2.2}$$

which is given in the fifth column of the table.

**Solution 2.3**

For $dy/dt = f(y) = -y$, we define

$$f_1 = f(y) = -y \qquad y_1 = y + \Delta t \frac{dy}{dt}(y)$$

$$f_2 = \frac{dy}{dt}(y_1) = -y_1 \qquad g = \frac{1}{2}(f_1 + f_2) \quad \text{and} \quad \Delta y = g\Delta t \tag{S2.3.1}$$

Updating is achieved by

$$t \to t + \Delta t \quad \text{and} \quad y \to y + \Delta y \tag{S2.3.2}$$

As before, we construct a table, working line by line from left to right (with $\Delta t = 0.1$):

| $t$ | $y$ | $f_1$ | $y_1$ | $f_2$ | $g$ | $\Delta y$ |
|---|---|---|---|---|---|---|
| 0 | 1.000 | −1.000 | 0.900 | −0.900 | −0.950 | −0.095 |
| 0.1 | 0.905 | −0.905 | 0.815 | −0.815 | −0.860 | −0.086 |
| 0.2 | 0.819 | etc. | | | | |

Comparing the second column with the exact solutions in the last column of the solution to Exercise 2.2, it is seen that the accuracy is much improved.

**Solution 2.4**

The degrees of freedom (p. 31) = 84 − 4 = 80. The mean residual sum of squares (equation (2.18)) is

$$\sigma_t^2 = R/(n - m) = 0.8/80 = 0.01$$

Thus $\ln(y/Y) = (0.01)^{1/2} = \pm\, 0.1$, and $y/Y \approx 0.9$ or 1.1. On average, there is a 10% difference between predicted and experimental values.

The error variance is

$$\sigma^2 = 0.4/50 = 0.008$$

Hence

$$F = \sigma_t^2/\sigma^2 = 1.25 \quad \text{with} \quad \nu_1 = 80 \quad \text{and} \quad \nu_2 = 50$$

The 10% probability level is given by the 95% point of the *F*-distribution (since we have constrained *F* to be larger than unity, and lie in one half of the distribution), and by interpolation in *Table G.1* (p. 274) this is about 1.5. Therefore, the model gives an acceptable fit to the data at the 10% level.

### Solution 2.5

We give the program without comment cards:

```
TITLE          SIMULATION MODEL FOR EXPONENTIAL GROWTH
INCON          XO = 1
PARAMETER      K = 1
METHOD         RECT
DYNAMIC
    DXDT      =K * X
    X         =INTGRL (XO, DXDT)
PRINT          DXDT, X
TIMER DELT =0.1, PRDEL = 1, FINTIM = 50
END
STOP
ENDJOB
```

## Chapter 3

### Solution 3.1

The graphical solution is shown in *Figure S3.1*, the shaded area being the feasible region. The highest permissible line of equal profit is the line $x_1 + 2x_2 = 32$, which touches the feasible region at the point P, with coordinates (10, 11). The optimal solution is therefore $z = 32$, $x_1 = 10$, $x_2 = 11$.

### Solution 3.2

Let $X$ be the amount of concentrate mix consumed [(kg DM of concentrates) day$^{-1}$] and $Y$ be the amount of hay consumed [(kg DM of hay) day$^{-1}$]. The linear programme is

minimize $z = X$

subject to $12.78X + 7.74Y \geqslant 16$ [ME constraint]

$159X + 116Y \geqslant 160$ [CP constraint]

$0.63X + Y \leqslant 1.02$ (= $12.79 \times 80/1000$)

[intake constraint]

$X, Y \geqslant 0$ [non-negativity conditions]

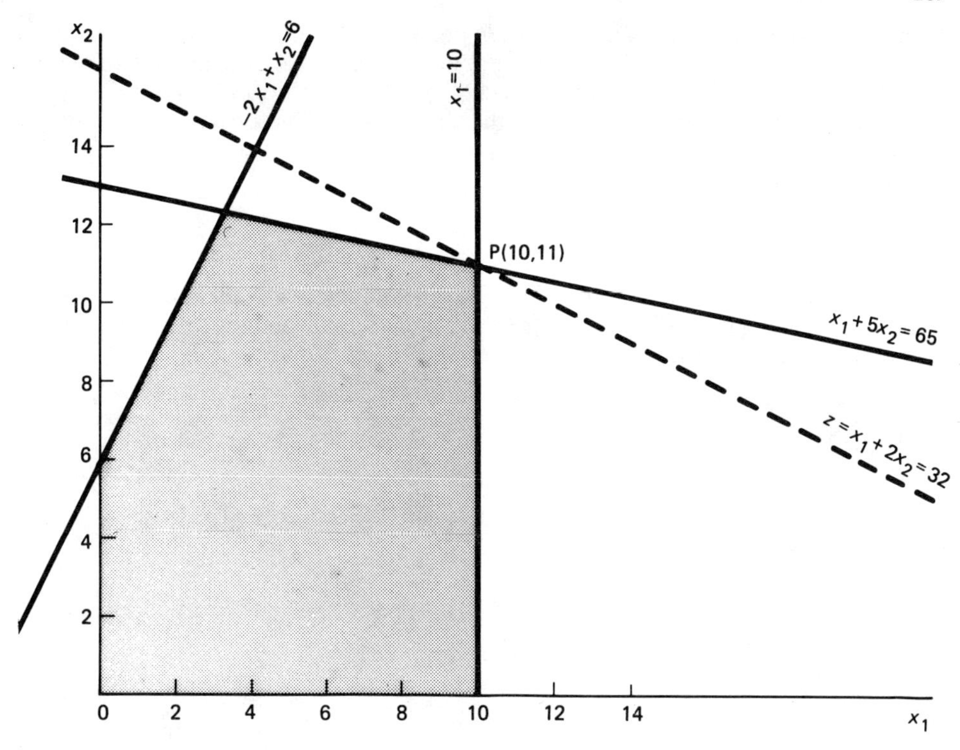


*Figure S3.1* Graphical solution to Exercise 3.1

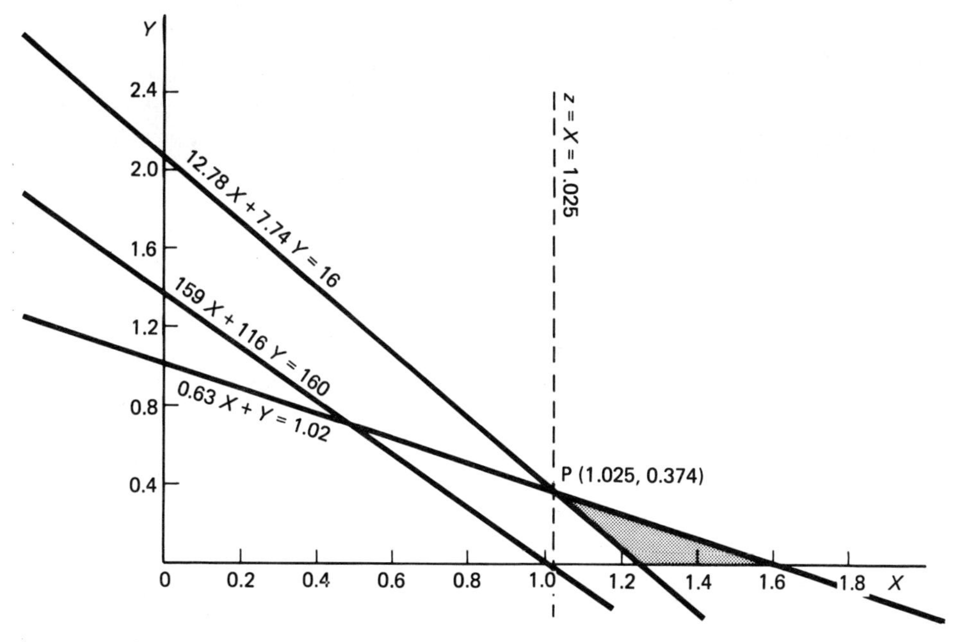


*Figure S3.2* Graphical solution to Exercise 3.2

It can be seen by inspection that the crude protein constraint is redundant, as values of $X$ and $Y$ satisfying the energy constraint automatically satisfy this constraint, and can therefore be eliminated from the problem. The graphical solution is shown in *Figure S3.2*, the feasible region being the shaded area. The protein constraint has been retained for illustration. The optimal solution is $z = 1.025$, $X = 1.025$, $Y = 0.374$, which occurs at the vertex P.

**Solution 3.3**

Let 1 kg of finishing mix contain $x_1$ kg of soyabean meal, $x_2$ kg of weatings, $x_3$ kg of maize meal, $x_4$ kg of barley and $x_5$ kg of mineral-vitamin supplement. Let $z$ denote the cost (in pence) of a kilogram of mix. The linear programme is

$$\text{minimize} \quad z = 14.5x_1 + 11x_2 + 16x_3 + 13x_4 + 40x_5$$

$$\text{subject to} \quad x_1 + x_2 + x_3 + x_4 + x_5 = 1 \quad \text{[total weight constraint]}$$

$$15x_1 + 11.9x_2 + 14.5x_3 + 12.7x_4 \geqslant 13 \quad \text{[energy constraint]}$$

$$41x_1 + 9.9x_2 + 7.3x_3 + 7.7x_4 \geqslant 12 \quad \text{[protein constraint]}$$

$$28x_1 + 6.4x_2 + 2.6x_3 + 3.2x_4 \geqslant 6 \quad \text{[lysine constraint]}$$

$$5.2x_1 + 7.5x_2 + 2x_3 + 4.6x_4 \leqslant 5 \quad \text{[fibre constraint]}$$

$$x_5 \geqslant 0.02 \quad \text{[min.–vit. constraint]}$$

$$x_1, x_2, x_3, x_4, x_5 \geqslant 0 \quad \text{[non-negativity conditions]}$$

The optimal solution is $z = 13.7$, $x_1 = 0.280$, $x_2 = 0.112$, $x_3 = 0$, $x_4 = 0.589$, $x_5 = 0.020$. The reduced cost of the nonbasic variable $x_3$ in the optimal solution is 0.3664, indicating that the cost of maize meal would have to fall by 3.67 £ $t^{-1}$ before it is used in the optimal mix.

**Solution 3.4**

The problem can be formulated as a mixed-integer programme:

$$\text{minimize} \quad z = 300x_1 + 295x_2 + 296x_3 + 9.9\delta_1 + 16.8\delta_2 + 16\delta_3$$

$$\text{subject to} \quad \left.\begin{array}{l} x_1 - 2.2\delta_1 \leqslant 0 \\ x_2 - 2.8\delta_2 \leqslant 0 \\ x_3 - 3.2\delta_3 \leqslant 0 \end{array}\right\} \quad \text{[site capacity constraints]}$$

$$\delta_1 + \delta_2 + \delta_3 = 1 \quad \text{[site selection constraint]}$$

$$x_1 + x_2 + x_3 \geqslant 2 \quad \text{[factory size constraint]}$$

$$x_1, x_2, x_3, \delta_1, \delta_2, \delta_3 \geqslant 0 \quad \text{[non-negativity conditions]}$$

$$\delta_1, \delta_2, \delta_3 = \text{integer } 0\text{–}1 \quad \text{[integer restrictions]}$$

where $z$ is the cost of the development [£ × $10^3$], $x_i$ is the size of factory [ha] to be built at Site $i$ ($i = 1, 2, 3$), and $\delta_i$ is a zero-one variable which must be either 0 or 1 ($i = 1, 2, 3$). The optimal solution is $z = 606.8$, $\delta_2 = 1$, $x_2 = 2.0$, $\delta_1 = \delta_3 = x_1 = x_3 = 0$; thus the optimal decision is to select Site 2 and build a 2 ha factory there.

## Solution 3.5

Choose $(0, 2)$ as the interval of interest for $x$ and the points $P_1(0,0)$, $P_2(1,1)$, $P_3(2,4)$ as the vertices of a piecewise linear approximation to $f(x) = x^2$. Do exactly the same for $y$. The LP400 separable programming formulation is then

$$
\begin{array}{rll}
\text{maximize} & z = x + y & \text{[objective function]} \\
\text{subject to} & x^2 + y^2 \leqslant 4 & \text{[constraint]} \\
\lambda_1{}^{(x)} + \lambda_2{}^{(x)} + \lambda_3{}^{(x)} & = 1 & \text{[convexity row } X\text{]} \\
0\lambda_1{}^{(x)} + 1\lambda_2{}^{(x)} + 2\lambda_3{}^{(x)} - x & = 0 & \text{[reference row } X\text{]} \\
0\lambda_1{}^{(x)} + 1\lambda_2{}^{(x)} + 4\lambda_3{}^{(x)} - x^2 & = 0 & \text{[function row } X\text{]} \\
\lambda_1{}^{(y)} + \lambda_2{}^{(y)} + \lambda_3{}^{(y)} & = 1 & \text{[convexity row } Y\text{]} \\
0\lambda_1{}^{(y)} + 1\lambda_2{}^{(y)} + 2\lambda_3{}^{(y)} - y & = 0 & \text{[reference row } Y\text{]} \\
0\lambda_1{}^{(y)} + 1\lambda_2{}^{(y)} + 4\lambda_3{}^{(y)} - y^2 & = 0 & \text{[function row } Y\text{]} \\
x, y, \lambda_1{}^{(x)}, \lambda_2{}^{(x)}, \lambda_3{}^{(x)}, \lambda_1{}^{(y)}, \lambda_2{}^{(y)}, \lambda_3{}^{(y)} & \geqslant 0 & \text{[non-negativity conditions]}
\end{array}
$$

$$\lambda_1{}^{(x)}, \lambda_2{}^{(x)}, \lambda_3{}^{(x)} = \text{a set of special variables}$$

$$\lambda_1{}^{(y)}, \lambda_2{}^{(y)}, \lambda_3{}^{(y)} = \text{a set of special variables}$$

```
INPUT
TITLE     SEPARABLE PROGRAMMING PROBLEM
ROWS
  F OBJECTIV
  P CONSTRNT
 CZ CONX
  Z REFX
  Z FUNX
 CZ CONY
  Z REFY
  Z FUNY
COLUMNS
    X          OBJECTIV  1          REFX      -1
    Y          OBJECTIV  1          REFY      -1
    XSQ        CONSTRNT  1          FUNX      -1
    YSQ        CONSTRNT  1          FUNY      -1
 H1 X0000HHH   CONX                 REFX
    X0000HHH   FUNX
 HF X000000F   FUNX                 'X2'       1
  W X0000000   REFX      0
  W X0000001   REFX      1
  W X0000002   REFY      2
 H1 Y0000HHH   CONY                 REFY
    Y0000HHH   FUNY
 HF Y000000F   FUNY                 'X2'       1
  W Y0000000   REFY      0
  W Y0000001   REFY      1
  W Y0000002   REFY      2
RHS
    RHSIDE     CONSTRNT  4          CONX       1
    RHSIDE     CONY      1
ENDATA
```

*Figure S3.3* The EXECUTOR file for Exercise 3.5

```
         PROGRAM
         INITIALZ
         INPUT
         MOVE      ZSCALE,-1.0
         CALL      SUBR(ZNPNTS=4)
         CALL      SUBR(ZNPNTS=0)
         SOLUTION
         EXIT
STEP     STEP
SUBR     NOP
         SETUP
         INVERT
         NORMAL
         STARTSEP
         INTERP
         ENDSEP
         RETURN
         END
```

*Figure S3.4* A COMPILER program for Exercise 3.5

The EXECUTOR file is shown in *Figure S3.3* and a COMPILER program, which automatically refines the piecewise linear approximations in the region of the optimal, is given in *Figure S3.4*. The optimal solution to the problem is $z = 2\sqrt{2}$, $x = y = \sqrt{2}$.

## Solution 3.6

Let $n$ be the number of stages *remaining* and $i$ the individual state at each stage. The problem with towns denoted by these stage and state variables is shown in

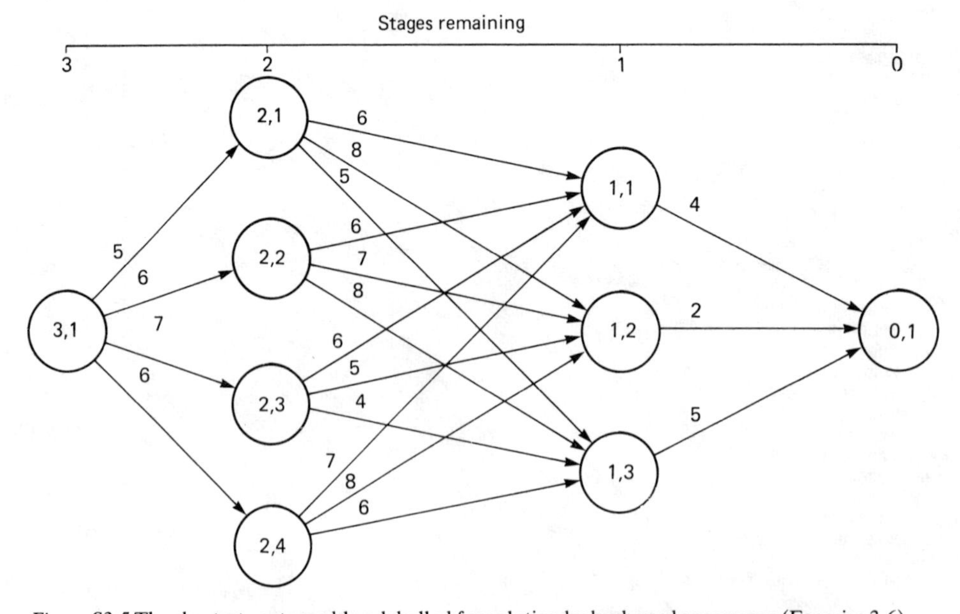


*Figure S3.5* The shortest route problem labelled for solution by backward recurrence (Exercise 3.6)

*Figure S3.5*. Define $f(n,i)$ and $r(n-1, j:n,i)$ as before. Again, the recurrence relation for the problem is

$$f(n,i) = \underset{j}{\text{minimum}}\, [f(n-1,j) + r(n-1,j:n,i)]$$

where the terminal state $f(0,1) = 0$.

Repeated application of the recurrence relation yields

$$\begin{aligned}
\text{Stage 1: } & f(1,1) = f(0,1) + r(0,1:1,1) = 4 \\
& f(1,2) = f(0,1) + r(0,1:1,2) = 2 \\
& f(1,3) = f(0,1) + r(0,1:1,3) = 5 \\
\text{Stage 2: } & f(2,1) = 10 \\
& f(2,2) = f(1,2) + r(1,2:2,2) = 9 \\
& f(2,3) = f(1,2) + r(1,2:2,3) = 7 \\
& f(2,4) = f(1,2) + r(1,2:2,4) = 10 \\
\text{Stage 3: } & f(3,1) = f(2,3) + r(2,3:3,1) = 14
\end{aligned}$$

Backward substitution gives

$$\begin{aligned}
f(3,1) &= f(2,3) + r(2,3:3,1) \\
&= f(1,2) + r(1,2:2,3) + r(2,3:3,1) \\
&= f(0,1) + r(0,1:1,2) + r(1,2:2,3) + r(2,3:3,1)
\end{aligned}$$

The optimal policy is therefore $(3,1) \rightarrow (2,3) \rightarrow (1,2) \rightarrow (0,1)$, i.e. $A \rightarrow B_3 \rightarrow C_2 \rightarrow D$.

## Chapter 4

### Solution 4.1

Each term must have the same units as $dw/dt$, i.e. kg day$^{-1}$. $k$ therefore has units of kg day$^{-2}$. The argument of an exponential is unit-free, and thus $b$ has units of day$^{-1}$. $n$ must be a unit-free number, and $a$ has units of kg day$^{-1-n}$. Note that the argument of any function such as $e^x$ or $\sin x$ which can be expanded as a series (e.g. $1 + x + \frac{1}{2}x^2 + \ldots$) must be unit-free.

### Solution 4.2

The units of $f_N$ are kg nitrogen (kg total matter)$^{-1}$. These units cannot be simplified; it is not, for instance, permissible to cancel out kg.

The units of LAI are m$^2$ leaf (m$^2$ ground)$^{-1}$ and, again, m$^2$ cannot be cancelled out.

### Solution 4.3

If the price of a good is $p$, and $x$ goods are sold (per unit time), then demand elasticity $E$ may be defined as

$$E = \frac{\delta x/x}{\delta p/p} = \frac{\delta(\ln x)}{\delta(\ln p)} \qquad \text{(S4.3.1)}$$

where the prefix $\delta$ indicates a small increment in the variable. This is equivalent to our model sensitivity index $S(Y, P_i)$ of equation (4.6) (p. 73), which measures the sensitivity of the predicted quantity $Y$ to the parameter $P_i$.

## Chapter 5

### Solution 5.1

Using $S = W_f - W$ from equation (5.6) (p. 77) with $S_f = 0$, therefore

$$W_f \int_{W_0}^{W} \frac{dW}{(W_f - W)^2} = \int_0^t k\,dt \tag{S5.1.1}$$

which gives

$$\frac{W}{W_f} = \frac{W_0 + kt(W_f - W_0)}{W_f + kt(W_f - W_0)} \tag{S5.1.2}$$

This equation describes a rectangular hyperbola, similar to equation (1.1) (*Figure 1.1*). It does not pass through the origin, but through the point $t = 0$, $W = W_0$. It approaches an asymptote at $W = W_f$ as $t \to \infty$. There is no point of inflexion. It resembles the monomolecular equation (*Figure 5.3a*) in shape, but with a much slower approach to the asymptote.

### Solution 5.2

Using equation (5.6) with $S_f = 0$, the growth rate equation becomes

$$\frac{dW}{dt} = (W_f - W)\,[k_1 + k_2\,(W_f - W)/W_f] \tag{S5.2.1}$$

which can be divided into partial fractions (*see* Glossary) to give

$$\frac{k_1}{k_2} \int_0^t dt = \int_{W_0}^{W} \left[\frac{1}{k_2\,(W_f - W)} - \frac{1}{(k_1 + k_2)\,W_f - k_2\,W}\right] dW \tag{S5.2.2}$$

After integration and rearrangement, therefore

$$\frac{W}{W_f} = \frac{[(k_1 + k_2)W_f - k_2\,W_0]e^{k_1 t} - (k_1 + k_2)\,(W_f - W_0)}{[(k_1 + k_2)W_f - k_2\,W_0]e^{k_1 t} - k_2\,(W_f - W_0)} \tag{S5.2.3}$$

This equation shares the qualitative properties of the rectangular hyperbola of equation (S5.1.2) and the monomolecular equation (5.16), but lies between the two curves, and thus can have either a fast or a slow approach to the asymptote.

With $k_2 = 0$, equation (S5.2.3) becomes identical to equation (5.16) with $k = k_1$. Taking the limit $k_1 \to 0$ (write $\exp(k_1 t) \approx 1 + k_1 t$), equation (S5.2.3) becomes identical to equation (S5.1.2) with $k = k_2$. The initial slope of the growth equation is (from equation (S5.2.1) with $W = W_0$)

$$\frac{dW}{dt}\,(t = 0) = (W_f - W_0)\,[k_1 + k_2\,(W_f - W_0)/W_f] \tag{S5.2.4}$$

**Solution 5.3**

Differentiating equation (5.43) with respect to time gives

$$\frac{d^2W}{dt^2} = \mu e^{-Dt}\left[\frac{dW}{dt}\left(1-\frac{2W}{B}\right) - DW\left(1-\frac{W}{B}\right)\right] \tag{S5.3.1}$$

which, with substitution for $dW/dt$ from equation (5.43) becomes

$$\frac{d^2W}{dt^2} = \mu We^{-Dt}\left[\mu\left(1-\frac{W}{B}\right)\left(1-\frac{2W}{B}\right)e^{-Dt} - D\left(1-\frac{W}{B}\right)\right] \tag{S5.3.2}$$

Equating this to zero for a point of inflexion, therefore

$$0 = \mu\left(1-\frac{2W}{B}\right)e^{-Dt} - D \tag{S5.3.3}$$

From equation (5.44), it may be shown that

$$e^{-Dt} = 1 - \frac{D}{\mu}\ln\left[\left(\frac{B-W_0}{B-W}\right)\left(\frac{W}{W_0}\right)\right] \tag{S5.3.4}$$

which, with equation (S5.3.3), gives

$$0 = \mu\left(1-\frac{2W}{B}\right)\left\{1-\frac{D}{\mu}\ln\left[\left(\frac{B-W_0}{B-W}\right)\left(\frac{W}{W_0}\right)\right]\right\} - D \tag{S5.3.5}$$

Although an analytic solution to this equation cannot be obtained, the limiting logistic and Gompertz cases are easily verified. For $D = 0$, $W = B/2 = W_f/2$ (equation (5.24) with $W_f = B$). For $B \rightarrow \infty$, $W = W_0 \exp[(\mu - D)/D]$, in agreement with equations (5.35) and (5.34).

**Solution 5.4**

In place of equation (5.17) from which the logistic is derived,

$$\frac{dW}{dt} = \mu_0 W\left(\frac{S}{K+S}\right) \tag{S5.4.1}$$

where $\mu_0$ and $K$ are constants, is taken. Since growth will continue until the substrate is exhausted and $S_f = 0$ in equation (5.6), therefore

$$W_f = S + W \tag{S5.4.2}$$

where $W_f$ is the final weight. Putting this into equation (S5.4.1) gives

$$\frac{dW}{dt} = \mu_0 W\left(\frac{W_f - W}{K + W_f - W}\right) \tag{S5.4.3}$$

which can be written

$$\frac{1}{W_f}\int_{W_0}^{W}\left(\frac{K+W_f}{W} + \frac{K}{W_f - W}\right)dW = \int_0^t \mu_0\, dt \tag{S5.4.4}$$

After integration, the equation connecting $W$ and $t$ is

$$\frac{1}{W_f}\left[(K + W_f)\ln\left(\frac{W}{W_0}\right) + K\ln\left(\frac{W_f - W_0}{W_f - W}\right)\right] = \mu_0 t \tag{S5.4.5}$$

To obtain the logistic, one takes the limit

$$\mu_0 \rightarrow \infty, K \rightarrow \infty \quad \text{and} \quad \mu_0/K \rightarrow \mu/W_f$$

and equations (S5.4.3) and (S5.4.5) become identical to equations (5.20) and (5.21).

The point of inflexion of the curve is obtained by differentiating equation (S5.4.3), equating to zero, and taking the negative root of the resulting quadratic (the other root being $> W_f$ and therefore non-physiological), to give

$$W(\text{inflexion}) = (K + W_f) - [K(K + W_f)]^{1/2}$$

The time of inflexion is obtained by substituting this value into equation (S5.4.5).

### Solution 5.5

Instead of equation (S5.4.1), assume

$$\frac{dW}{dt} = \mu_0 W\left(\frac{S}{K + S}\right)e^{-Dt} \tag{S5.5.1}$$

However, in contrast with the last exercise, growth need not now continue until the substrate is exhausted, so in place of equation (S5.4.2) therefore

$$S + W = \text{constant} = A$$

so that equation (S5.5.1) becomes

$$\frac{dW}{dt} = \mu_0 W\left(\frac{A - W}{K + A - W}\right)e^{-Dt} \tag{S5.5.2}$$

Proceeding as in the previous example, integration gives an equation very similar to equation (S5.4.5), namely

$$\frac{1}{A}\left[(K + A)\ln\left(\frac{W}{W_0}\right) + K\ln\left(\frac{A - W_0}{A - W}\right)\right] = \frac{\mu_0}{D}(1 - e^{-Dt}) \tag{S5.5.3}$$

The final weight $W_f$ may be found by substituting $W = W_f$ in equation (S5.5.3) and solving for $W_f$.

## Chapter 6

### Solution 6.1

$$\text{hours} = \text{integer part } (24t_d) \tag{S6.1.1}$$

$$\text{minutes} = \text{integer part } (60 \times 24t_d - 60 \times \text{hours}) \tag{S6.1.2}$$

and

$$\text{seconds} = \text{integer part } (60 \times 60 \times 24t_d - 60 \times 60 \times \text{hours} - 60 \times \text{minutes}) \tag{S6.1.3}$$

For conversion in the opposite direction

$$t_d = \frac{1}{24}\left(\text{hours} + \frac{\text{minutes}}{60} + \frac{\text{seconds}}{3600}\right) \tag{S6.1.4}$$

**Solution 6.2**

| *First day of* | *Jan* | *Feb* | *Mar* | *Apr* | *May* | *Jun* | *Jul* | *Aug* | *Sep* | *Oct* | *Nov* | *Dec* |
|---|---|---|---|---|---|---|---|---|---|---|---|---|
| *N* | 307 | 338 | 1 | 32 | 62 | 93 | 123 | 154 | 185 | 215 | 246 | 276 |

**Solution 6.3**

Calculations were performed for a latitude of 50.75° (Glasshouse Crops Research Institute, Littlehampton, West Sussex, UK) with the following results:

| *Climatological day, N* | *Day of month, D* | *Month of year, M* | *Declination* [degrees] | *Zenith angle at solar noon, z* [degrees] | *Daylength (fractional) for different definitions of zenith angle* | | | | |
|---|---|---|---|---|---|---|---|---|---|
| | | | | | *90.00°* | *90.83°* | *96.00°* | *102.00°* | *108.00°* |
| 1 | 1 | 3 | −7.8 | 58.6 | 0.45 | 0.45 | 0.50 | 0.55 | 0.61 |
| 11 | 11 | 3 | −4.0 | 54.7 | 0.47 | 0.48 | 0.53 | 0.58 | 0.63 |
| 21 | 21 | 3 | −0.0 | 50.8 | 0.50 | 0.51 | 0.55 | 0.61 | 0.66 |
| 31 | 31 | 3 | 3.9 | 46.8 | 0.53 | 0.53 | 0.58 | 0.64 | 0.69 |
| 41 | 10 | 4 | 7.7 | 43.0 | 0.55 | 0.56 | 0.61 | 0.67 | 0.73 |
| 51 | 20 | 4 | 11.3 | 39.4 | 0.58 | 0.59 | 0.64 | 0.70 | 0.77 |
| 61 | 30 | 4 | 14.6 | 36.2 | 0.60 | 0.61 | 0.66 | 0.73 | 0.81 |
| 71 | 10 | 5 | 17.5 | 33.3 | 0.63 | 0.63 | 0.69 | 0.76 | 0.85 |
| 81 | 20 | 5 | 19.9 | 30.9 | 0.65 | 0.65 | 0.71 | 0.79 | 0.91 |
| 91 | 30 | 5 | 21.7 | 29.1 | 0.66 | 0.67 | 0.73 | 0.82 | 1.00 |
| 101 | 9 | 6 | 22.9 | 27.9 | 0.67 | 0.68 | 0.75 | 0.84 | 1.00 |
| 111 | 19 | 6 | 23.4 | 27.3 | 0.68 | 0.69 | 0.75 | 0.85 | 1.00 |
| 121 | 29 | 6 | 23.3 | 27.5 | 0.68 | 0.69 | 0.75 | 0.85 | 1.00 |
| 131 | 9 | 7 | 22.4 | 28.3 | 0.67 | 0.68 | 0.74 | 0.83 | 1.00 |
| 141 | 19 | 7 | 21.0 | 29.8 | 0.66 | 0.66 | 0.72 | 0.81 | 0.96 |
| 151 | 29 | 7 | 18.9 | 31.9 | 0.64 | 0.65 | 0.70 | 0.78 | 0.88 |
| 161 | 8 | 8 | 16.3 | 34.4 | 0.62 | 0.62 | 0.68 | 0.75 | 0.83 |
| 171 | 18 | 8 | 13.3 | 37.5 | 0.59 | 0.60 | 0.65 | 0.72 | 0.79 |
| 181 | 28 | 8 | 9.9 | 40.8 | 0.57 | 0.58 | 0.62 | 0.68 | 0.75 |
| 191 | 7 | 9 | 6.3 | 44.5 | 0.54 | 0.55 | 0.60 | 0.65 | 0.72 |
| 201 | 17 | 9 | 2.5 | 48.3 | 0.52 | 0.52 | 0.57 | 0.62 | 0.68 |
| 211 | 27 | 9 | −1.4 | 52.2 | 0.49 | 0.50 | 0.54 | 0.60 | 0.65 |
| 221 | 7 | 10 | −5.3 | 56.1 | 0.46 | 0.47 | 0.52 | 0.57 | 0.62 |
| 231 | 17 | 10 | −9.1 | 59.8 | 0.44 | 0.44 | 0.49 | 0.54 | 0.60 |
| 241 | 27 | 10 | −12.6 | 63.4 | 0.41 | 0.42 | 0.47 | 0.52 | 0.57 |
| 251 | 6 | 11 | −15.9 | 66.6 | 0.39 | 0.40 | 0.44 | 0.50 | 0.55 |
| 261 | 16 | 11 | −18.6 | 69.4 | 0.36 | 0.37 | 0.42 | 0.48 | 0.53 |

| *Climatological day, N* | *Day of month, D* | *Month of year, M* | *Declination* [degrees] | *Zenith angle at solar noon, z* [degrees] | *Daylength (fractional) for different definitions of zenith angle* | | | | |
|---|---|---|---|---|---|---|---|---|---|
| | | | | | *90.00°* | *90.83°* | *96.00°* | *102.00°* | *108.00°* |
| 271 | 26 | 11 | −20.9 | 71.6 | 0.35 | 0.35 | 0.41 | 0.46 | 0.52 |
| 281 | 6 | 12 | −22.5 | 73.2 | 0.33 | 0.34 | 0.39 | 0.45 | 0.51 |
| 291 | 16 | 12 | −23.3 | 74.1 | 0.32 | 0.33 | 0.39 | 0.45 | 0.50 |
| 301 | 26 | 12 | −23.4 | 74.1 | 0.32 | 0.33 | 0.39 | 0.45 | 0.50 |
| 311 | 5 | 1 | −22.7 | 73.4 | 0.33 | 0.34 | 0.39 | 0.45 | 0.51 |
| 321 | 15 | 1 | −21.2 | 72.0 | 0.34 | 0.35 | 0.40 | 0.46 | 0.52 |
| 331 | 25 | 1 | −19.1 | 69.9 | 0.36 | 0.37 | 0.42 | 0.48 | 0.53 |
| 341 | 4 | 2 | −16.4 | 67.2 | 0.38 | 0.39 | 0.44 | 0.49 | 0.55 |
| 351 | 14 | 2 | −13.3 | 64.0 | 0.41 | 0.41 | 0.46 | 0.52 | 0.57 |
| 361 | 24 | 2 | −9.7 | 60.5 | 0.43 | 0.44 | 0.49 | 0.54 | 0.59 |

**Solution 6.4**

From equation (6.15b)

$$\cos^2 a_0 = 1 - \cos^2\delta \sin^2 h_0 \tag{S6.4.1}$$

which, using

$$\sin^2 h_0 = 1 - \cos^2 h_0 = 1 - \tan^2\phi \tan^2\delta \tag{S6.4.2}$$

from equation (6.12), becomes

$$\begin{aligned}\cos^2 a_0 &= 1 - \cos^2\delta(1 - \tan^2\phi \tan^2\delta) \\ &= 1 - \cos^2\delta + \tan^2\phi \sin^2\delta \\ &= \sin^2\delta(1 + \tan^2\phi) \\ &= \sin^2\delta/\cos^2\phi\end{aligned}$$

Taking the negative square root, therefore

$$\cos a_0 = -\sin\delta/\cos\phi \tag{S6.4.3}$$

**Solution 6.5**

Application of equations (6.8), (6.9) and (6.14) for a latitude $\phi = 50.75°$, gives

$$g_{113} = 0.7517 \text{ day} \quad \text{and} \quad g_{296} = 0.3859 \text{ day} \tag{S6.5.1}$$

for the longest and shortest days. Assuming a sine wave, then the expression for $g_N$ becomes

$$g_N = 0.5688 \times 0.1829 \sin\left[\left(\frac{N - 21}{365}\right) 360\right] \tag{S6.5.2}$$

This is shown in *Figure 6.1* (p. 99), together with the more accurate results from the eighth column of the table in Solution 6.3.

**Solution 6.6**

The 10-year average of the mean daily light receipts for the months of December and June are 0.87 and 9.09 MJ $m^{-2}$ respectively. In December and June, the radiation is at the bottom and top of the sine curve, and is changing least quickly. One day is approximately equivalent to one degree (in fact, 360/365°), and as cos 15° = 0.97, to assume that the monthly means apply to the values on 21 December and 21 June may produce an error of about 3%. Since the coefficient of variation is 15%, this is acceptable. An estimate of the coefficients of the equation

$$\bar{J}_N = a + b \sin\left[\left(\frac{N-21}{365}\right)360\right] \tag{S6.6.1}$$

is obtained by taking

$$a = \frac{1}{2}(0.87 + 9.09) \quad \text{and} \quad b = \frac{1}{2}(9.09 - 0.87) \tag{S6.6.2}$$

The resulting curve is drawn in *Figure 6.2* (p. 100), for

$$\bar{J}_N = 4.98 + 4.11 \sin\left[\left(\frac{N-21}{365}\right)360\right] \tag{S6.6.3}$$

and is compared directly with the average monthly means, which are placed at the middle of each calendar month.

Better agreement could be obtained by adding extra terms to the Fourier series, replacing equation (S6.6.1) by

$$\begin{aligned}\bar{J}_N = a_0 &+ a_1 \cos\left[\left(\frac{N-21}{365}\right)360\right] + a_2 \cos\left[2\left(\frac{N-21}{365}\right)360\right] \\ &+ a_3 \cos\left[3\left(\frac{N-21}{365}\right)360\right] + \ldots + b_1 \sin\left[\left(\frac{N-21}{365}\right)360\right] \\ &+ b_2 \sin\left[2\left(\frac{N-21}{365}\right)360\right] + b_3 \sin\left[3\left(\frac{N-21}{365}\right)360\right] + \ldots\end{aligned} \tag{S6.6.4}$$

With only 12 monthly means to fit, one should not include too many terms in the series. The calculation of the coefficients of equation (S6.6.4) is explained in the Glossary. Using these methods for the first three terms ($a_0$, $a_1$, $b_1$) in equation (S6.6.4) and simplifying, one obtains the equation

$$\bar{J}_N = 4.59 + 4.04 \sin\left[\left(\frac{N-25}{365}\right)360\right] \tag{S6.6.5}$$

instead of equation (S6.6.3). This is also drawn in *Figure 6.2*, and it can be seen that it gives a more accurate representation of the light values, but uses the same number of parameters.

**Solution 6.7**

First work out logarithms to the base e:

$$\ln J_n \quad \text{for } N = 103, 104, \ldots, 123 \tag{S6.7.1}$$

Then estimate the mean $\bar{J}$, and the standard deviation $\sigma$, by

$$J = \frac{1}{21} \sum_{N=103}^{123} \ln J_N \qquad \text{(S6.7.2)}$$

and

$$\sigma^2 = \frac{1}{20} \sum_{N=103}^{123} \ln(J_N/\bar{J})^2 \qquad \text{(S6.7.3)}$$

Finally, evaluate $\rho_1$ by means of

$$\rho_1 \, \sigma_1^2 = \frac{1}{20} \sum_{N=104}^{123} \ln(J_N/\bar{J}) \, (J_{N-1}/\bar{J}) \qquad \text{(S6.7.4)}$$

The answers are $\sigma_1 = 0.47$ and $\rho_1 = 0.21$.

**Solution 6.8**

Application of the method outlined in the Glossary (*see* Fourier analysis) gives

$$T_a = 10.4 + 6.1 \sin\left[\left(\frac{N - 60}{365}\right) 360\right] \qquad \text{(S6.8.1)}$$

and

$$T_s = 11.6 + 5.2 \sin\left[\left(\frac{N - 76}{365}\right) 360\right] \qquad \text{(S6.8.2)}$$

These are best worked out by writing a short computer program.

**Solution 6.9**

The mean monthly rainfall for October is 61 mm with a standard deviation of 42 mm, giving a coefficient of variation of 0.69. The mean number of rainy days in the 31-day month is 10, with a standard deviation of 5.2.

From equation (6.43), $q = 10/32 = 0.323$. From equation (6.46) with $CV = 0.69$, $n_{eff} = 4.41$ day, which, with equation (6.47) gives $n_p = 31/4.41 = 7.03$ day. The probability of the month of October passing without rain is $(q)^{n_{eff}} = 0.007$, or rather less than 1% according to this rainfall model.

## Chapter 7

**Solution 7.1**

Applying equation (7.1) with $I/I_o = 0.1$ and $k = 0.6$, therefore $0.1 = \exp(-0.6L)$, giving $L = 3.84$.

If the intercepted light (0.9 of the total) is spread over a leaf area of 3.84, then the mean light level is $0.9/3.84 = 0.23$.

**Solution 7.2**

Let $x$ be the fraction of light that is intercepted by unit foliage layer, so that

$$s = 1 - x \tag{S7.2.1}$$

Monteith's equation becomes

$$I(L) = [1 - x(1 - m)]^L I_o \tag{S7.2.2}$$

Now assume that all the foliage elements are independent (rather than Monteith's implicit assumption that each group of foliage elements within unit leaf area index layer acts as an independent unit, but within each unit layer there may or may not be independent action by the foliage elements). Let the elemental layer $\Delta L$ intercept $y\Delta L$ of the radiation incident on it, and take $n\Delta L = L$. Equation (S7.2.2) becomes (replacing $x$ by $y\Delta L$, and $L$ by $n$ on the right-hand side)

$$I(L) = [1 - y\Delta L(1 - m)]^n I_o \tag{S7.2.3}$$

In the limit $n \to \infty$, $\Delta L \to 0$, $n\Delta L = L$, therefore

$$I(L) = e^{-y(1-m)L} I_o \tag{S7.2.4}$$

This is identical to equation (7.1) (p. 115), with

$$k = y(1 - m) \tag{S7.2.5}$$

**Solution 7.3**

With $P = fP_{max}$ where $f$ is the fraction of the maximal rate, equation (7.14) (p. 119) becomes

$$fP_{max} = \frac{\alpha I_\ell P_{max}}{\alpha I_\ell + P_{max}} \tag{S7.3.1}$$

which is rewritten to give

$$I_\ell = \frac{f}{1 - f} \frac{P_{max}}{\alpha} \tag{S7.3.2}$$

With $f = 0.9$ and the given values of $P_{max}$ and $\alpha$, therefore

$$I_\ell(f = 0.9) = 831 \text{ J m}^{-2} \text{ s}^{-1} \text{ PAR} \tag{S7.3.3}$$

This unrealistically high value is a measure of the inadequacies of equation (7.14) as a leaf response equation.

Defining the photosynthetic rate estimated using the approximation as

$$P(\text{app}) = \alpha I_\ell \tag{S7.3.4}$$

the error $\epsilon$ is given by

$$\epsilon = \frac{P(\text{app}) - P_\ell}{P_\ell} \tag{S7.3.5}$$

which, with equations (7.14) and (S7.3.4) becomes

$$\epsilon = \frac{\alpha I_\ell}{P_{max}} \tag{S7.3.6}$$

For a 10% error, therefore

$$I_\ell = 0.1P_{max}/\alpha = 9\ \mathrm{J\ m^{-2}\ s^{-1}\ PAR} \tag{S7.3.7}$$

Again, this low value results from the strong curvature of equation (7.14) near the origin, which is not realistic.

### Solution 7.4

A detailed solution is not given here for this problem, which is discussed at length by Acock, Thornley and Warren Wilson (1971, pp. 66–71). The main points are that at low light and low LAI, a high value of $k$ intercepts the available light efficiently, whereas at high light and high LAI, a low value of $k$ distributes the available light more evenly over the canopy so that it is used more efficiently.

### Solution 7.5

Comparison of equations (7.22) and (7.35) gives

$$k = 1 - Y_g \quad \text{and} \quad c = Y_g m \tag{S7.5.1}$$

With equations (7.23), therefore

$$Y_g = 0.75 \quad \text{and} \quad m = 0.017\ \mathrm{day^{-1}} \tag{S7.5.2}$$

## Chapter 8

### Solution 8.1

Dividing the radiation receipt by the latent heat of water, $\lambda = 2.5\ \mathrm{MJ\ kg^{-1}}$, the quantity of water evaporated is 18/2.5 = 7.2 $\mathrm{kg\ m^{-2}\ day^{-1}}$. This is equivalent to a depth of 7.2 $\mathrm{mm\ day^{-1}}$.

7.2 kg water corresponds to dry matter production of 7.2 × 0.0015 = 0.011 kg dry matter $\mathrm{m^{-2}\ day^{-1}}$.

### Solution 8.2

The psychrometric constant at 20 °C and atmospheric pressure is $\gamma = 66.8\ \mathrm{Pa\ °C^{-1}}$. From equation (8.9a) (p. 140), the correction factor is simply $s/(s + \gamma) = 0.68$.

### Solution 8.3

Differentiating the yield equation with respect to $N$ and $P$ gives

$$\frac{\partial Y}{\partial N} = a_1 + 2a_2 N + hP \tag{S8.3.1}$$

and

$$\frac{\partial Y}{\partial P} = b_1 + 2b_2 P + hN$$

Equating these to zero, as in equations (8.32) (p. 145), gives

$$2a_2 N + hP = -a_1 \tag{S8.3.2}$$

and

$$hN + 2b_2 P = -b_1$$

Solving for $N$ and $P$ to give maximum crop yield, therefore

$$N_{max} = \frac{b_1 h - 2b_2 a_1}{4a_2 b_2 - h^2} \quad \text{and} \quad P_{max} = \frac{a_1 h - 2a_2 b_1}{4a_2 b_2 - h^2} \tag{S8.3.3}$$

For the most economic levels of fertilizer, combining equations (8.35) with equations (S8.3.1), therefore

$$2a_2 N + hP = -(a_1 - c_N/p) \tag{S8.3.4}$$

and

$$hN + 2b_2 P = -(b_1 - c_P/p)$$

where $c_N$ and $c_P$ are the unit costs of nitrogen and phosphate fertilizer, and $p$ the price obtained for the harvested crop. Solving to obtain the economic fertilizer dressing gives

$$N_{ec} = \frac{(b_1 - c_P/p)h - 2b_2(a_1 - c_N/p)}{4a_2 b_2 - h^2}$$

and (S8.3.5)

$$P_{ec} = \frac{(a_1 - c_N/p)h - 2a_2(b_1 - c_P/p)}{4a_2 b_2 - h^2}$$

**Solution 8.4**

The total development time $t_{dev}$ is

$$t_{dev} = t_c + t_a \tag{S8.4.1}$$

Substituting for $t_a$ from equation (8.76) (p. 154), therefore

$$t_{dev} = t_c + p_a + \frac{p^2}{t_c - p_c} \tag{S8.4.2}$$

Differentiating gives

$$\frac{dt_{dev}}{dt_c} = 1 - \frac{p^2}{(t_c - p_c)^2} \tag{S8.4.3}$$

and equating this to zero, hence

$$p^2 = (t_c - p_c)^2$$

and taking the positive square root (which is on the branch of the hyperbola that is of interest), thus

$$t_c = p_c + p \tag{S8.4.4}$$

Substitution of equation (S8.4.4) into equation (8.76) now gives

$$t_a = p_a + p \tag{S8.4.5}$$

and finally, substituting equations (S8.4.4) and (S8.4.5) into equation (S8.4.1) leads to

$$t_{dev} = p_a + p_c + 2p \tag{S8.4.6}$$

**Solution 8.5**

Since

$$k = \exp(-E_a/RT) \tag{S8.5.1}$$

therefore

$$\frac{dk}{dT} = A\exp(-E_a/RT)\left(\frac{E_a}{RT^2}\right) \tag{S8.5.2}$$

and

$$\frac{d^2k}{dT^2} = A\exp(-E_a/RT)\left[\left(\frac{E_a}{RT^2}\right)^2 - \frac{2E_a}{RT^3}\right] \tag{S8.5.3}$$

Equating equation (S8.5.3) to zero, hence at the inflexion point $(T^*, k^*)$

$$T^* = \frac{E_a}{2R} \quad \text{or} \quad \frac{E_a}{RT^*} = 2 \tag{S8.5.4}$$

At $T = T^*$, from equation (S8.5.1) therefore

$$k^* = Ae^{-2} \tag{S8.5.5}$$

Substituting with equation (S8.5.4) in equation (S8.5.2) gives

$$\frac{dk}{dT}(T = T^*) = Ae^{-2}\left(\frac{4R}{E_a}\right) \tag{S8.5.6}$$

and the straight line with gradient of equation (S8.5.6) through the point $(T^*, k^*)$ is

$$k - Ae^{-2} = Ae^{-2}\frac{4R}{E_a}\left(T - \frac{E_a}{2R}\right)$$

which simplifies to equation (8.81), namely

$$k = \frac{4AR}{E_a e^2}\left(T - \frac{E_a}{4R}\right) \tag{S8.5.7}$$

# Chapter 9

## Solution 9.1

Define variables

$\overline{T}_5$ = five-day average temperature [°C]
$S_{10}$ = rainfall totalled over the past 10 days [mm]
$T_{min}$ = minimum daily temperature [°C]
$n$ = number of blight-favourable days
$T_{1 \text{ to } 5}$ = array for five daily temperature means [°C]
$h_{1 \text{ to } 10}$ = array for ten daily rainfall values [mm]

Initialize all variables to zero.
Input the last day's values of $\overline{T}$, $T_{min}$ and $h$ into

$$T_5, \quad T_{min}, \quad h_{10}$$

Compute $\overline{T}_5$ and $S_{10}$ by means of (':=' means 'is assigned the value of')

$$\overline{T}_5 := \left( \sum_{i=1}^{5} T_i \right) \Big/ 5$$

$$S_{10} := \sum_{i=1}^{10} h_i$$

Test and accumulate blight-favourable days

if $\overline{T}_5 < 25.5$ and $S_{10} \geqslant 30$ and $T_{min} \geqslant 7.2$ then $n := n + 1$

if $n = 10$ then print 'Blight warning'

Move the arrays along one position

for $i = 1$ to 4, $T_i := T_{i+1}$

for $i = 1$ to 9, $h_i := h_{i+1}$

Go to 'Input the last day's values ...'.

## Solution 9.6

Consider the fate of the cohort of organisms $n(t, a)\delta a$ (*Figure S9.1*). After a period of time $\delta t$ this cohort is now of age $a + \delta t$, and the surviving organisms belong to $n(t + \delta t, a + \delta t)\delta a$. The number that has died is, by the definition of the death rate $D(t, a)$, equal to $D(t, a)n(t, a)\delta a \delta t$. Thus, we can write

$$n(t + \delta t, a + \delta t) = n(t, a) - D(t, a)\, n(t, a)\delta t \qquad \text{(S9.6.1)}$$

where the common factor of $\delta a$ has been cancelled throughout. To first order, using a two-dimensional Taylor series

$$n(t + \delta t, a + \delta t) = n(t, a) + \frac{\partial n}{\partial t}\delta t + \frac{\partial n}{\partial a}\delta t \qquad \text{(S9.6.2)}$$

Substitution into equation (S9.6.1) gives the required results.

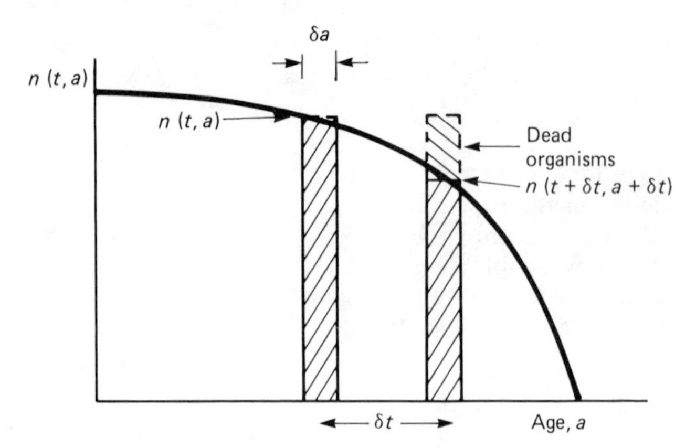


*Figure S9.1* The fate of a cohort of organisms, $n(t, a)\delta a$ (Exercise 9.6)

## Chapter 10

### Solution 10.1

The potentially degradable part is the only component of dynamic interest in solving this problem. Let $\pi_2(t)$ be the amount of potentially degradable feed protein in the rumen at time $t$ [days]. The instantaneous rate of degradation of the potentially degradable component is $c\pi_2$ [kg day$^{-1}$], and the instantaneous rate of outflow from the rumen is $k\pi_2$ [kg day$^{-1}$]. The differential equation describing the dynamics of the situation is therefore

$$\frac{d\pi_2}{dt} = -(c + k)\pi_2 \quad 0 \leqslant t \leqslant \infty$$

This differential equation is easily integrated analytically to give

$$\pi_2(t) = \pi_2(0)e^{-(c+k)t} \quad 0 \leqslant t \leqslant \infty$$

As the instantaneous rate of feed protein degradation in the rumen is $c\pi_2$, the total amount of the potentially degradable component degraded in the rumen is

$$\begin{aligned}\int_0^\infty c\pi_2\,dt &= c\pi_2(0)\int_0^\infty e^{-(c+k)t}\,dt \\ &= c\pi_2(0)\left[-\frac{1}{(c+k)}\,e^{-(c+k)t}\right]_{t=0}^{t=\infty} \\ &= \frac{c}{c+k}\,\pi_2(0)\end{aligned}$$

Therefore, the fraction of the ingested feed protein which is degraded in the rumen is given by

$$f_{deg} = \left[\pi_1(0) + \frac{c}{c+k}\,\pi_2(0)\right]\Big/\pi(0)$$

### Solution 10.2

The differential equation describing the dynamics of the situation this time is

$$\frac{d\pi_2}{dt} = -k\pi_2 \qquad 0 \leqslant t \leqslant t_0 \tag{S10.2.1}$$

$$= -(c+k)\pi_2 \qquad t_0 < t \leqslant \infty \tag{S10.2.2}$$

The analytical solution to equation (S10.2.1) is

$$\pi_2(t) = \pi_2(0)e^{-kt} \quad 0 \leqslant t \leqslant t_0 \tag{S10.2.3}$$

and to equation (S10.2.2) is

$$\pi_2(t) = \pi_2(t_0)\exp[-(c+k)(t-t_0)] \qquad t_0 < t \leqslant \infty \tag{S10.2.4}$$

From (S10.2.3), $\pi_2(t_0) = \pi_2(0)e^{-kt_0}$. Equation (S10.2.4) therefore becomes

$$\pi_2(t) = \pi_2(0)\exp[-(c+k)t + ct_0] \qquad t_0 < t \leqslant \infty$$

The total amount of the potentially degradable component degraded in the rumen is

$$\int_{t_0}^{\infty} c\pi_2\,dt = c\pi_2(0)e^{ct_0}\int_{t_0}^{\infty} e^{-(c+k)t}\,dt$$

$$= c\pi_2(0)e^{ct_0}\left[-\frac{1}{(c+k)}\,e^{-(c+k)t}\right]_{t=t_0}^{t=\infty}$$

$$= \frac{c}{c+k}\,e^{-kt_0}\,\pi_2(0)$$

Therefore,

$$f_{deg} = \left[\pi_1(0) + \frac{c}{c+k}\,e^{-kt_0}\,\pi_2(0)\right]\Big/\pi(0)$$

## Chapter 11

### Solution 11.1

We are given estimates of the length of lactation $t_f = 308$ days, total milk yield $Y =$ 6500 kg, time to maximum yield $t_m = 56$ days, and the average relative rate of decline $\bar{r} = -0.03/7$ per day $= -0.0043$ per day. The parameters $a$, $b$ and $c$ are determined using equations (11.15), (11.16) and (11.18), namely

$$c = \bar{r}(t_m + t_f)/(t_m - t_f)$$

$$b = ct_m$$

$$a = Yc^{b+1}/\gamma(b+1, ct_f)$$

These equations give $c = 0.0062$, $b = 0.35$, and

$$a = 6500 \times 0.001\,047/\gamma(1.35, 1.9)$$

From *Table 11.1*, $\gamma(1.35, 1.9) = 0.676429$; therefore $a = 10.06$. A cow's milk production over the time interval $t_{i-1}$ to $t_i$ is given by equation (11.7), namely

$$Y_i = \frac{a}{c^{b+1}} [\gamma(b + 1, ct_i) - \gamma)(b + 1, ct_{i-1})]$$

Applying this equation iteratively and using *Table 11.1* (intermediate values of the incomplete gamma function being determined by linear interpolation) gives

| *Month* | $t_i$ | $\gamma(1.35, 0.0062t_i)$ | $Y_i$ [kg] |
|---|---|---|---|
| 1 | 28 | 0.063 272 | 608 |
| 2 | 56 | 0.146 135 | 797 |
| 3 | 84 | 0.229 874 | 805 |
| 4 | 112 | 0.309 238 | 763 |
| 5 | 140 | 0.382 075 | 700 |
| 6 | 168 | 0.447 845 | 632 |
| 7 | 196 | 0.506 436 | 563 |
| 8 | 224 | 0.558 260 | 498 |
| 9 | 252 | 0.603 768 | 437 |
| 10 | 280 | 0.643 545 | 382 |
| 11 | 308 | 0.678 190 | 333 |
| | | | Total = 6518 |

The error in total yield computed from the sum of the four-weekly yields is due to a combination of rounding, interpolating and the use of approximations for determining the parameters of the gamma curve.

**Solution 11.2**

Let $z = ae^{-bx}$, then

$$dz = -abe^{-bx}dx \quad \text{that is} \quad dx = -\frac{1}{bz}dz$$

On substituting for $x$,

$$\int_0^t e^{-a\exp(-bx)} e^{-cx} dx = \int_a^{a\exp(-bt)} e^{-z}\left(\frac{z}{a}\right)^{c/b}\left(-\frac{1}{bz}\right)dz$$

$$= \frac{1}{ba^{c/b}} \int_{a\exp(-bt)}^{a} e^{-z} z^{(c/b)-1} dz$$

$$= \frac{1}{ba^{c/b}} \left[\int_0^a e^{-z} z^{(c/b)-1} dz - \int_0^{a\exp(-bt)} e^{-z} z^{(c/b)-1} dz\right]$$

$$= \frac{1}{ba^{c/b}} [\gamma(c/b, a) - \gamma(c/b, ae^{-bt})]$$

Putting $t = \infty$, $a = 2$, $b = 1$, and $c = 1.2$, the integral becomes

$$[\gamma(1.2, 2) - \gamma(1.2, 0)]/2^{1.2}$$

*Table 11.1* gives $\gamma(1.2, 2) = 0.750784$ and $\gamma(1.2, 0) = 0$, thus the value of the integral is 0.3268.

### Solution 11.3

The left-hand side of the differential equation has units of (kg egg) day$^{-1}$, therefore each term on the right-hand side must have these units. The units of $k_1$ are (kg egg) (g amino acid)$^{-1}$ and the units of $k_2$ are (kg egg) (kg liveweight)$^{-0.67}$ day$^{-1}$.

## Chapter 12

### Solution 12.1

The objective is to maximize profits, so the profit on a tonne of each fertilizer must be calculated before the linear programme can be formulated. One tonne of '1 : 1 : 1' comprises 15% nitrates, 15% phosphates, 15% potash and 55% inert ingredients. The cost of '1 : 1 : 1' is therefore:

$$0.15 \times 225 + 0.15 \times 220 + 0.15 \times 175 + 0.55 \times 25 + 35 = 141.75 \text{ £ t}^{-1}$$

Profit on '1 : 1 : 1' = price − cost = 150 − 141.75 = 8.25 £ t$^{-1}$. Similarly, profit on 'High N' = 7.5 £ t$^{-1}$, profit on 'Low N' = 8 £ t$^{-1}$, and profit on 'No N' = 3 £ t$^{-1}$.

Let $x_1, x_2, x_3, x_4$ [t] be the decision variables denoting next month's production of '1 : 1 : 1', 'High N', 'Low N', and 'No N' respectively, and let $z$ [£] denote next month's profit. The linear programme is therefore

$$\begin{aligned}
\text{maximize} \quad & z = 8.25x_1 + 7.5x_2 + 8x_3 + 3x_4 \\
\text{subject to} \quad & 0.15x_1 + 0.2x_2 + 0.1x_3 + 0x_4 \leqslant 3000 \quad \text{[nitrate availability]} \\
& 0.15x_1 + 0.1x_2 + 0.2x_3 + 0.2x_4 \leqslant 3250 \quad \text{[phosphate availability]} \\
& x_1 + x_2 + x_3 + x_4 \leqslant 20000 \quad \text{[production capacity]} \\
& x_2 \geqslant 5000 \quad \text{['High N' lower bound]} \\
& x_3 \geqslant 3000 \quad \text{['Low N' lower bound]} \\
& x_4 \geqslant 2500 \quad \text{['No N' lower bound]} \\
& x_1, x_2, x_3, x_4 \geqslant 0 \quad \text{[non-negativity conditions]}
\end{aligned}$$

The optimal solution is $z = 147\,375$, $x_1 = 9500$, $x_2 = 5000$, $x_3 = 3000$, $x_4 = 2500$.

**Solution 12.2**

The expected profit on a crop [£ $ha^{-1}$] = expected price − labour cost − other costs, e.g. profit on spring wheat grown on high-grade land = 112 × 4.6 − 12.4 × 2.50 − 135 = 349.2 £ $ha^{-1}$.

Let $X_{HW}$ be the area [ha] of high-grade land allocated to spring wheat, $X_{LW}$ low-grade to wheat, $X_{HB}$ high-grade to barley, $X_{LB}$ low-grade to barley, $X_{HP}$ high-grade to potatoes, $X_{LP}$ low-grade to potatoes, $X_{HS}$ high-grade to sugar beet, $X_{LS}$, low-grade to sugar beet, $X_{HO}$ high-grade to oilseed rape, $X_{LO}$ low-grade to oilseed rape, and $Z$ be the profit [£]. The linear programme is

$$\text{maximize} \quad Z = 349.2X_{HW} + 214.8X_{LW} + 368X_{HB} + 228.25X_{LB} + 1493.75X_{HP} + 293.75X_{LP} + 591.75X_{HS} + 166.75X_{LS} + 342.5X_{HO} + 117.5X_{LO} \quad \text{[objective function]}$$

$$\text{subject to} \quad X_{HW} + X_{HB} + X_{HP} + X_{HS} + X_{HO} \leqslant 200 \quad \text{[high-grade land constraint]}$$

$$X_{LW} + X_{LB} + X_{LP} + X_{LS} + X_{LO} \leqslant 100 \quad \text{[low-grade land constraint]}$$

$$135(X_{HW} + X_{LW}) + 117(X_{HB} + X_{LB}) + 725(X_{HP} + X_{LP}) + 365(X_{HS} + X_{LS}) + 185(X_{HO} + X_{LO}) \leqslant 150000 \quad \text{[constraint on capital available for costs other than labour]}$$

$$166(X_{HW} + X_{LW}) + 148(X_{HB} + X_{LB}) + 906.25(X_{HP} + X_{LP}) + 533.25(X_{HS} + X_{LS}) + 220(X_{HO} + X_{LO}) \leqslant 162500 \quad \text{[capital constraint]}$$

The constraints on capital are arrived at as follows. The farmer has available 5000 man-hours of free labour, which is equivalent to £12500 of extra capital that can only be spent on labour. The farmer therefore has £162500 of capital but at most £150000 of which can be spent on costs other than labour.

The optimal crop plan is to allocate 155.75 ha of high-grade land to potatoes, 44.25 ha of high-grade land to barley and 100 ha of low-grade land to barley. This yields a profit of £271764.36.

**Solution 12.3**

Let $H$ [kg herbage $ha^{-1}$] denote herbage biomass at time $t$ [days], then

$$\frac{dH}{dt} = \text{growth} - \text{consumption}$$

$$= g_0 H e^{-g_D t} - C_M \frac{HS}{H + C_K}$$

For equilibrium at $H^*$, $dH/dt = 0$ and $H = H^*$, therefore

$$g_0 H^* e^{-g_D t} - C_M \frac{H^* S}{H^* + C_K} = 0$$

that is

$$S(t) = g_0 (H^* + C_K) e^{-g_D t} / C_M$$

# Chapter 13

## Solution 13.1

Each year corresponds to a stage, which is completed with the sale of a portion of the farmer's flock. Let $n$ be the stage variable denoting the number of stages completed ($n = 1, 2, \ldots, N$). Let the number of sheep kept at the end of year $n$ be $x_n$, the number sold be $y_n$, and $z_n = x_n + y_n$. We have

$$z_1 = 2k \quad \text{and} \quad z_n = 2x_{n-1} \quad (n = 2, 3, \ldots, N)$$

The states of the system in stage $n$ can be described by the state variable $z_n$, the number of sheep available at the end of stage $n$ for allocation to stage $n + 1$. Note that $x_n$, $y_n$ and $z_n$ are non-negative integers, $y_n$ cannot exceed $z_n$, and $z_n$ cannot exceed $2^n k$. Let $f(n, z_n)$ represent the total financial return from following an optimal policy in stages $n, n + 1, \ldots, N$ given $z_n$. The backward recurrence relation is then

$$f(N, z_N) = \underset{y_N = z_N \leqslant 2^N k}{\text{maximum}} [r_N y_N]$$

$$f(n, z_n) = \underset{0 \leqslant y_n \leqslant z_n \leqslant 2^n k}{\text{maximum}} [r_n y_n + f(n + 1, 2(z_n - y_n))] \quad (n = 1, 2, \ldots, N - 1)$$

## Solution 13.2

We define our stage variable $n$ as before but make $x_n$, the number of sheep actually allocated to stage $n + 1$ at the end of stage $n$, our state variable. Let $g(n, x_n)$ represent the total financial return from following an optimal policy in stages $1, 2, \ldots, n$ given $x_n$. The forward recurrence relation is

$$g(1, x_1) = \underset{y_1 = 2k - x_1}{\text{maximum}} [r_1 y_1]$$

$$g(n, x_n) = \underset{\substack{y_n \leqslant 2^n k - x_n \\ (x_n + y_n)/2 = \text{an integer}}}{\text{maximum}} \left[r_n y_n + g\left(n - 1, \frac{x_n + y_n}{2}\right)\right] \quad (n = 2, 3, \ldots, N)$$

where $x_n$ and $y_n$ are non-negative integers.

The backward procedure is easier to apply than the forward one. In the backward procedure, the state transformation in going from stage $n + 1$ to stage $n$ is given by $z_{n+1} = 2(z_n - y_n)$, which allows $f(n, z_n)$ to be calculated quite easily. In the forward procedure, however, the state transformation in going from stage $n - 1$ to stage $n$, given by $x_{n-1} = (x_n + y_n)/2$, has the added complication that $(x_n + y_n)/2$ must be integral. This means that, in addition to selecting integer values for $x_n$ and $y_n$ satisfying $x_n + y_n \leqslant 2^n k$, the two values must also satisfy the condition

$$\frac{x_n + y_n}{2} = \text{a non-negative integer}$$

This increases the computational workload.

**Solution 13.3**

Let $x_1$ be the amount of hay fed [kg dry matter day$^{-1}$], $x_2$ the amount of barley, $x_3$ the amount of soya, and $z$ the cost of the ration [pence day$^{-1}$]. The linear programme is

$$\begin{aligned}
\text{minimize } z &= \tfrac{2}{0.85}x_1 + \tfrac{10}{0.86}x_2 + \tfrac{13}{0.9}x_3 && \text{[objective function]}\\
\text{subject to } \quad 4.5x_1 + 9.2x_2 + 7.9x_3 &\geq 34.1 && \text{[energy constraint]}\\
39x_1 + 82x_2 + 453x_3 &\geq 447 && \text{[protein constraint]}\\
x_1 + 0.7x_2 + 0.7x_3 &\leq 4.7 && \text{[intake constraint]}\\
x_1, x_2, x_3 &\geq 0 && \text{[non-negativity conditions]}
\end{aligned}$$

The optimal solution is $z = 34.323$, $x_1 = 3.144$, $x_2 = 1.840$, and $x_3 = 0.383$. The optimum daily ration is therefore 3.70 kg fresh weight of hay, 2.14 kg fresh weight of barley, and 0.43 kg fresh weight of soyabean meal. Cost ranging shows that the price of barley has to rise to 17.95 pence (kg dry matter)$^{-1}$ (i.e. 152.58 £ (tonne fresh weight)$^{-1}$) before the values of the decision variables in the optimal solution change. On reaching this price, $x_2$ drops out of the optimal basis, being replaced by the PROTEIN slack. Therefore the cost of barley has to rise by 52.58 £ (tonne fresh weight)$^{-1}$ before its inclusion in the optimal ration becomes uneconomic.

**Solution 13.4**

Let $x_1$ and $x_2$ be the amount fed [kg dry matter day$^{-1}$] of silage and dairy compound respectively. The linear programme is

$$\begin{aligned}
\text{minimize } \quad z &= x_2 && \text{[objective function]}\\
\text{subject to } \quad 9x_1 + 12.5x_2 &\geq 185 && \text{[metabolizable energy constraint]}\\
x_1 + 0.72x_2 &\leq 15 && \text{[appetite constraint]}\\
x_1, x_2 &\geq 0 && \text{[non-negativity conditions]}
\end{aligned}$$

The optimal solution is $x_1 = 9.02$, $x_2 = 8.31$.

# Bibliography

## Chapter 1
## Role of mathematical models in agriculture and agricultural research

### Further reading and references

BROCKINGTON, N.R. (1979). *Computer Modelling in Agriculture.* Oxford: Oxford University Press
DOYLE, C.J. (1981). The role of economics in agricultural research. In *Annual Report 1980 of the Grassland Research Institute* (ed. W.A.D. Donaldson and K.M. Down), pp. 127–133. Hurley, Maidenhead: Grassland Research Institute
DOYLE, C.J. and THORNLEY, J.H.M. (1982). An economic model of research and development expenditure. *Oxford Agrarian Studies* **11**, 173–187
FORRESTER, J.W. (1961). *Industrial Dynamics.* Cambridge, Massachusetts: MIT Press
POPPER, K.R. (1982). *The Open Universe: an Argument for Indeterminism.* London: Hutchinson
SPEDDING, C.R.W. (1979). *An Introduction to Agricultural Systems.* London: Applied Science Publishers
THORNLEY, J.H.M. (1976). *Mathematical Models in Plant Physiology.* London: Academic Press
THORNLEY, J.H.M. (1980). Research strategy in the plant sciences. *Plant, Cell and Environment* **3**, 233–236

## Chapter 2
## Techniques: dynamic deterministic modelling

BROCKINGTON, N.R. (1979). *Computer Modelling in Agriculture.* Oxford: Oxford University Press
FOX, L. and MYERS, D.F. (1968). *Computing Methods for Scientists and Engineers.* Oxford: Oxford University Press
GEAR, C.W. (1971). *Numerical Initial Value Problems in Ordinary Differential Equations.* Englewood Cliffs, New Jersey: Prentice-Hall
SPECKHART, F.H. and GREEN, W.L. (1976). *A Guide to using CSMP—the Continuous System Modeling Program.* Englewood Cliffs, New Jersey: Prentice-Hall
WILSON, A.G. and KIRBY, M.J. (1980). *Mathematics for Geographers and Planners*, second edition. Oxford: Oxford University Press

## Chapter 3
## Techniques: mathematical programming

BEALE, E.M.L. (1968). *Mathematical Programming in Practice.* London: Pitman
BELLMAN, R.E. (1957). *Dynamic Programming.* Princeton: Princeton University Press

BELLMAN, R.E. and DREYFUS, S.E. (1962). *Applied Dynamic Programming.* Princeton: Princeton University Press
DANTZIG, G.B. (1951). Maximization of a linear function of variables subject to linear inequalities. In *Activity Analysis of Production and Allocation* (ed. T.C. Koopmans), Chapter XXI. New York: Wiley
DANTZIG, G.B. (1953). *Computational Algorithm of the Revised Simplex Method.* RM–1266, Rand Corporation
DANTZIG, G.B. (1954). *Notes on Linear Programming: Parts VIII, IX, X—Upper Bounds, Secondary Constraints, and Block Triangularity in Linear Programming.* RM–1367, Rand Corporation
DANTZIG, G.B. (1963). *Linear Programming and Extensions.* Princeton: Princeton University Press
GOMORY, R.E. (1958). Outline of an algorithm for integer solutions to linear programs. *Bulletin of the American Mathematical Society* **64**, 275–278
HADLEY, G. (1962). *Linear Programming.* Reading, Massachusetts: Addison-Wesley
HADLEY, G. (1964). *Nonlinear and Dynamic Programming.* Reading, Massachusetts: Addison-Wesley
HASTINGS, N.A.J. (1973). *Dynamic Programming with Management Applications.* London: Butterworths
HEADY, E.O. and CANDLER, W. (1958). *Linear Programming Methods.* Ames: Iowa State University Press
INTERNATIONAL COMPUTERS LIMITED (1970). *Linear Programming 400.* Technical Publication 4523. London: International Computers Limited
LAND, A.H. and DOIG, A.G. (1960). An automatic method of solving discrete programming problems. *Econometrica* **28**, 497–520
MILLER, C.E. (1963). The simplex method for local separable programming. In *Recent Advances in Mathematical Programming* (ed. R.L. Graves and P. Wolfe), pp. 311–317. New York: McGraw-Hill
ORCHARD-HAYS, W. (1954). *Background, Development and Extensions of the Revised Simplex Method.* RM–1433, Rand Corporation

## Chapter 4
## Testing and evaluation of models

PENNING DE VRIES, F.W.T. (1977). Evaluation of simulation models in agriculture and biology: conclusions of a workshop. *Agricultural Systems* **2**, 99–107
POPPER, K.R. (1958). *The Logic of Scientific Discovery.* London: Hutchinson
THORNLEY, J.H.M., HURD, R.G. and POOLEY, A. (1981). A model of growth of the fifth leaf of tomato. *Annals of Botany* **48**, 327–340

## Chapter 5
## Growth functions

CAUSTON, D.R., ELIAS, C.O. and HADLEY, P. (1978). Biometrical studies of plant growth. I. The Richards function, and its application in analysing the effects of temperature on leaf growth. *Plant, Cell and Environment* **1**, 163–184
CAUSTON, D.R. and VENUS, J.C. (1981). *The Biometry of Plant Growth.* London: Arnold
CHANTER, D.O. (1976). *Mathematical Models in Mushroom Research and Production.* D.Phil. Thesis, University of Sussex
HUGHES, A.P. and FREEMAN, P.R. (1967). Growth analysis using frequent small harvests. *Journal of Applied Ecology* **4**, 553–560
HUNT, R. (1982). *Plant Growth Curves.* London: Arnold
HUNT, R. and PARSONS, I.T. (1974). A computer program for deriving growth-functions in plant growth-analysis. *Journal of Applied Ecology* **11**, 297–307
HURD, R.G. (1977). Vegetative plant growth analysis in controlled environments. *Annals of Botany* **41**, 779–787
HUXLEY, J.S. (1924). Constant differential growth ratios. *Nature, London* **114**, 895–896
NICHOLLS, A.O. and CALDER, D.M. (1973). Comments on the use of regression analysis for the study of plant growth. *The New Phytologist* **72**, 571–581

PEARSALL, W.H. (1927). Growth studies. VI. On the relative sizes of growing plant organs. *Annals of Botany* **41**, 549–556
RICHARDS, F.J. (1959). A flexible growth function for empirical use. *Journal of Experimental Botany* **10**, 290–300
RICHARDS, F.J. (1969). The quantitative analysis of growth. In *Plant Physiology VA* (ed. F.C. Steward), pp. 3–76. New York: Academic Press
ROYAL SOCIETY (1975). *Quantities, Units and Symbols.* A report of the Symbols Committee of the Royal Society, London, 54pp
VON BERTALANFFY, L. (1957). Quantitative laws for metabolism and growth. *Quarterly Reviews of Biology* **32**, 217–231

# Chapter 6
# Weather

CHARLES-EDWARDS, D.A. and ACOCK, B. (1977). Growth response of a *Chrysanthemum* crop to the environment. II. A mathematical analysis relating photosynthesis and growth. *Annals of Botany* **41**, 49–58
LARSEN, G.A. and PENGE, R.B. (1981). *Stochastic Simulation of Daily Climatic Data.* USDA/SRS Staff Report No. AGES-810831, 62 pp. Washington DC: Research Division, Statistical Reporting Service, US Department of Agriculture
MONTEITH, J.L. (1973). *Principles of Environmental Physics.* London: Arnold
REES, A.R. and THORNLEY, J.H.M. (1973). A simulation of tulip growth in the field. *Annals of Botany* **37**, 121–131
RICHARDSON, C.W. (1981). Stochastic simulation of daily precipitation, temperature, and solar radiation. *Water Resources Research* **17**, 182–190
SELLERS, W.D. (1965). *Physical Climatology.* Chicago: University of Chicago Press
STERN, R.D. and COE, R. (1982). The use of rainfall models in agricultural planning. *Agricultural Meteorology* **26**, 35–60
STERN, R.D., DENNETT, M.D. and DALE, I.C. (1982a). Analysing daily rainfall measurements to give agronomically useful results. I. Direct methods. *Experimental Agriculture* **18**, 223–236
STERN, R.D., DENNETT, M.D. and DALE, I.C. (1982b). Analysing daily rainfall measurements to give agronomically useful results. II. A modelling approach. *Experimental Agriculture* **18**, 237–253
STUFF, R.G. and DALE, R.F. (1973). A simple method of calendar conversion in computer applications. *Agricultural Meteorology* **12**, 441–442
THORNLEY, J.H.M. (1974). Light fluctuations and photosynthesis. *Annals of Botany* **38**, 363–374
USHER, M.B. (1970). An algorithm for estimating the length and direction of shadows with reference to the shadows of shelter belts. *Journal of Applied Ecology* **7**, 141–145
WEST, E.S. (1952). A study of the annual soil temperature wave. *Australian Journal of Scientific Research* **A5**, 303–314

# Chapter 7
# Plant and crop processes

## Further reading

CANNY, M.J. (1973). *Phloem Translocation.* Cambridge: Cambridge University Press
CHARLES-EDWARDS, D.A. (1981). *The Mathematics of Photosynthesis and Productivity.* London: Academic Press
DE WIT, C.T. and others (1978). *Simulation of Assimilation, Respiration and Transpiration.* Wageningen: Pudoc
JOHNSON, C.B. (ed.) (1981). *Physiological Processes Limiting Plant Productivity.* London: Butterworths
MOORBY, J. (1981). *Transport Systems in Plants.* London: Longman
ROSE, D.A. and CHARLES-EDWARDS, D.A. (ed.) (1981). *Mathematics and Plant Physiology.* London: Academic Press
THORNLEY, J.H.M. (1976). *Mathematical Models in Plant Physiology.* London: Academic Press

## Light interception

ACOCK, B., THORNLEY, J.H.M. and WARREN WILSON, J. (1970). Spatial variation of light in the canopy. In *Prediction and Measurement of Photosynthetic Productivity* (ed. I. Setlik), pp. 91–102. Wageningen: Pudoc

ANDERSON, M.C. (1966). Stand structure and light penetration. II. A theoretical analysis. *Journal of Applied Ecology* **3**, 41–54

BIKHELE, Z.N., MOLDAU, Kh.A. and ROSS, Yu.K. (1980). *Mathematical Modelling of Transpiration and Photosynthesis of Plants with Soil-Water Stress* [in Russian]. Leningrad: Gidrometeoizdat

CHARLES-EDWARDS, D.A. and THORNLEY, J.H.M. (1973). Light interception by an isolated plant: a simple model. *Annals of Botany* **37**, 919–928

CHARLES-EDWARDS, D.A. and THORPE, M.R. (1976). Interception of diffuse and direct-beam radiation by a hedgerow apple orchard. *Annals of Botany* **40**, 603–613

CHARTIER, P. (1966). Etude de microclimat lumineux dans la végétation. *Annales Agronomiques* **17**, 571–602

COWAN, I.R. (1968). The interception and absorption of radiation in plant stands. *Journal of Applied Ecology* **5**, 367–379

DE WIT, C.T. (1965). Photosynthesis of leaf canopies. *Agricultural Research Reports (Wageningen)* **663**, 1–57

IDSO, S.B. and DE WIT, C.T. (1970). Light relations in plant canopies. *Applied Optics* **9**, 177–184

JACKSON, J.E. and PALMER, J.W. (1979). A simple model of light transmission and interception by discontinuous canopies. *Annals of Botany* **44**, 381–383

MONSI, M. and SAEKI, T. (1953). Uber den Lichtfaktor in den Pflanzengesellschaften und seine Bedeutung für die Stoffproduktion. *Japanese Journal of Botany* **14**, 22–52

MONTEITH, J.L. (1965). Light distribution and photosynthesis in field crops. *Annals of Botany* **29**, 17–37

NILSON, T. (1971). A theoretical analysis of the frequency of gaps in plant stands. *Agricultural Meteorology* **8**, 25–38

NORMAN, J.M. and JARVIS, P.G. (1975). Photosynthesis in sitka spruce (*Picea sitchenis* (Bong.) Carr.). V. Radiation penetration theory and a test case. *Journal of Applied Ecology* **12**, 839–878

PALMER, J.W. (1977). Diurnal light interception and a computer model of light interception by hedgerow apple orchards. *Journal of Applied Ecology* **14**, 601–614

SAEKI, T. (1960). Interrelationships between leaf amount, light distribution and total photosynthesis in a plant community. *Botanical Magazine, Tokyo* **73**, 55–63

THORNLEY, J.H.M. (1976) *Mathematical Modelling in Plant Physiology*. London: Academic Press

WHITFIELD, D.M. (1980). Interaction of single tobacco plants with direct beam light. *Australian Journal of Plant Physiology* **7**, 435–447

WHITFIELD, D.M. and CONNOR, D.J. (1980). Architecture of individual plants in a field-grown tobacco crop. *Australian Journal of Plant Physiology* **7**, 415–433

## Photosynthesis

ACOCK, B., HAND, D.W., THORNLEY, J.H.M. and WARREN WILSON, J. (1976). Photosynthesis in stands of green peppers. An application of empirical and mechanistic models to controlled-environment data. *Annals of Botany* **40**, 1293–1307

ACOCK, B., THORNLEY, J.H.M. and WARREN WILSON, J. (1971). Photosynthesis and energy conversion. In *Potential Crop Production* (ed. P.F. Wareing and J.P. Cooper), pp. 43–75. London: Heinemann

HESKETH, J.D. and JONES, J.W. (ed.) (1980). *Predicting Photosynthesis for Ecosystem Models*, Volumes I and II. Florida: CRC Press

THORNLEY, J.H.M. (1976). *Mathematical Models in Plant Physiology*. London: Academic Press

## Respiration

BARNES, A. and HOLE, C.C. (1978). A theoretical basis of growth and maintenance respiration. *Annals of Botany* **42**, 1217–1221

McCREE, K.J. (1970). An equation for the rate of respiration of white clover plants grown under controlled conditions. In *Prediction and Measurement of Photosynthetic Productivity* (ed. I. Setlik), pp. 221–229. Wageningen: Pudoc

PENNING DE VRIES, F.W.T. (1975a). The costs of maintenance processes in plant cells. *Annals of Botany* **39**, 77–92

PENNING DE VRIES, F.W.T. (1975b). Use of assimilates in higher plants. In *Photosynthesis and Productivity in Different Environments* (ed. J.P. Cooper), pp. 459–480. Cambridge: Cambridge University Press

PIRT, S.J. (1965). The maintenance energy of bacteria in growing cultures. *Proceedings of the Royal Society* **B163**, 224–231

THORNLEY, J.H.M. (1970). Respiration, growth and maintenance in plants. *Nature, London* **227**, 304–305

THORNLEY, J.H.M. (1971). Energy, respiration and growth in plants. *Annals of Botany* **35**, 721–728

THORNLEY, J.H.M. (1976). *Mathematical Models in Plant Physiology.* London: Academic Press

THORNLEY, J.H.M. (1977). Growth, maintenance and respiration: a re-interpretation. *Annals of Botany* **41**, 1191–1203

THORNLEY, J.H.M. (1982). Interpretation of respiration coefficients. *Annals of Botany* **49**, 257–259

THORNLEY, J.H.M. and HESKETH, J.D. (1972). Growth and respiration in cotton bolls. *Journal of Applied Ecology* **9**, 315–317

## Partitioning of dry matter and its components

BROUWER, R. (1962). Distribution of dry matter in the plant. *Netherlands Journal of Agricultural Science* **10**, 361–376

CHARLES-EDWARDS, D.A. (1976). Shoot and root activities during steady-state plant growth. *Annals of Botany* **40**, 767–772

COOPER, A.J. and THORNLEY, J.H.M. (1976). Response of dry matter partitioning, and carbon and nitrogen levels in the tomato plant to changes in root temperature: experiment and theory. *Annals of Botany* **40**, 1139–1142

CURRY, R.B., BAKER, C.H. and STREETER, J.G. (1975). SOYMOD I: a dynamic simulator of soybean growth and development. *Transactions of the American Society of Agricultural Engineers* **18**, 673–674

DAVIDSON, R.L. (1969). Effect of root/leaf temperature differentials on root/shoot ratios in some pasture grasses and clover. *Annals of Botany* **33**, 561–569

HOLT, D.A., BULA, R.J., MILES, G.E., SCHREIBER, D.P. and PEART, R.M. (1975). Environmental physiology, modeling and simulation of alfalfa growth. I. Conceptual development of SIMED. *Research Bulletin of Purdue University Agricultural Experiment Station*, No. 907, 26 pp

PATEFIELD, W.M. and AUSTIN, R.B. (1971). A model for the simulation of growth of *Beta vulgaris* L. *Annals of Botany* **35**, 1227–1250

REYNOLDS, J.F. and THORNLEY, J.H.M. (1982). A shoot:root partitioning model. *Annals of Botany* **49**, 585–597

SCAIFE, M.A. and SMITH, R. (1973). The phosphorus requirement of lettuce. II. A dynamic model of phosphorus uptake and growth. *Journal of Agricultural Science, Cambridge* **80**, 353–361

THORNLEY, J.H.M. (1972a). A model to describe the partitioning of photosynthate during vegetative plant growth. *Annals of Botany* **36**, 419–430

THORNLEY, J.H.M. (1972b). A balanced quantitative model for root:shoot ratios in vegetative plants. *Annals of Botany* **36**, 431–441

THORNLEY, J.H.M. (1976). *Mathematical Models in Plant Physiology.* London: Academic Press

THORNLEY, J.H.M. (1977). Root:shoot interactions. *Symposium of the Society for Experimental Biology* **31**, 367–389

WHITE, H.L. (1937). The interaction of factors in the growth of *Lemna.* XII. The interaction of nitrogen and light intensity in relation to root length. *Annals of Botany* **1**, 649–654

## Nutrient uptake

CLARKSON, D.T. (1981). Nutrient interception and transport by root systems. In *Physiological Processes Limiting Plant Productivity* (ed. C.B. Johnson), pp. 307–330. London: Butterworths

NYE, P.H. and TINKER, P.B. (1977). *Solute Movement in the Soil–Root System.* Oxford: Blackwell

PASSIOURA, J.B. (1963). A mathematical model for the uptake of ions from the soil solution. *Plant and Soil* **18**, 225–238

RUSSELL, R.S. (1977). *Plant Root Systems.* London: McGraw-Hill

# Chapter 8
# Crop responses and models

## Water use and uptake

JARVIS, P.G. (1981). Stomatal conductance, gaseous exchange and transpiration. In *Plants and their Atmospheric Environment* (ed. J. Grace, E.D. Ford and P.G. Jarvis), pp. 175–204. Oxford: Blackwell

JARVIS, P.G., EDWARDS, W.R.N. and TALBOT, H. (1981). Models of plant and crop water use. In *Mathematics and Plant Physiology* (ed. D.A. Rose and D.A. Charles-Edwards), pp. 151–194. London: Academic Press

LARCHER, W. (1980). *Physiological Plant Ecology*. Berlin: Springer

MONTEITH, J.L. (1965). Evaporation and environment. *Symposium of the Society for Experimental Biology* **29**, 205–234

MONTEITH, J.L. (1973). *Principles of Environmental Physics*. London: Arnold

PENMAN, H.L. (1948). Natural evaporation from open water, bare soil and grass. *Proceedings of the Royal Society* **A193**, 120–145

THOM, A.S. (1968). The exchange of momentum, mass and heat between an artificial leaf and the airflow in a wind-tunnel. *Quarterly Journal of the Royal Meteorological Society* **94**, 44–55

THOM, A.S. (1975). Momentum, mass and heat exchange of plant communities. In *Vegetation and the Atmosphere*, Volume 1 (ed. J.L. Monteith), pp. 57–109. London: Academic Press

## Fertilizer responses

COLWELL, J.D. (1974). *The Computation of Optimal Rates of Application of Fertilizers from Quadratic Response Functions.* Technical Paper, No. 21, 17 pp. Canberra: Division of Soils, CSIRO

COLWELL, J.D. (1977). *National Soil Fertility Project. I. Objectives and Procedures.* Canberra: CSIRO

GREENWOOD, D.J., CLEAVER, T.J. and TURNER, M.K. (1974). Fertiliser requirements of vegetable crops. *Proceedings of the Fertiliser Society*, No. 145

THORNLEY, J.H.M. (1977). Root:shoot interactions. *Symposium of the Society for Experimental Biology* **31**, 367–389

THORNLEY, J.H.M. (1978). Crop response to fertilizers. *Annals of Botany* **42**, 817–826

## Development

CHARLES-EDWARDS, D.A. and REES, A.R. (1974). A simple model for the cold requirement of the tulip. *Annals of Botany* **38**, 401–408

LANDSBERG, J.J. (1977). Effect of weather on plant development. In *Environmental Effects on Crop Physiology* (ed. J.J. Landsberg and C.V. Cutting), pp. 289–307. London: Academic Press

REES, A.R., TURQUAND, E.D. and BRIGGS, J.B. (1972). Interrelations of bulb storage treatment and housing date on flowering date, stem length and flower differentiation in tulip. *Experimental Horticulture* **23**, 52–63

ROBERTSON, G.W. (1968). A biometeorological time scale for a cereal crop involving day and night temperature and photoperiod. *International Journal of Biometeorology* **12**, 191–223

## Crop yield: planting density responses

BERRY, G. (1967). A mathematical model relating plant yield with arrangement for regularly spaced crops. *Biometrics* **23**, 505–515

BLEASDALE, J.K.A. and NELDER, J.A. (1960). Plant population and crop yield. *Nature, London* **188**, 342

COOPER, A.J. and THORNLEY, J.H.M. (1976). Response of dry matter partitioning, growth, and carbon and nitrogen levels in the tomato plant to changes in root temperature: experiment and theory. *Annals of Botany* **40**, 1139–1152

DIXON, M. and WEBB, E.C. (1964). *Enzymes*, second edition. London: Longman

DUNCAN, W.G. (1958). The relationship between crop population and yield. *Agronomy Journal* **50**, 82–84
FARAZDAGHI, H. and HARRIS, P.M. (1968). Plant competition and crop yield. *Nature, London* **217**, 289–290
GREENWOOD, D.J., CLEAVER, T.J. and TURNER, M.K. (1974). Fertiliser requirements of vegetable crops. *Proceedings of the Fertilizer Society*, No. 145
HOLLIDAY, R. (1960a). Plant population and crop yield. *Nature, London* **186**, 22–24
HOLLIDAY, R. (1960b). Plant population and crop yield: part I. *Field Crop Abstracts* **13**, 159–167
HOLLIDAY, R. (1960b). Plant population and crop yield: part II. *Field Crop Abstracts* **13**, 247–254
HUDSON, H.G. (1941). Population studies with wheat. III. Seed rates in nursery trials and field plots. *Journal of Agricultural Science* **31**, 138–144
KIRA, T., OGAWA, H. and SAKAZAKI, N. (1953). Intraspecific competition among higher plants. 1. Competition–yield–density interrelationship in regularly dispersed populations. *Journal of the Institute of Polytechnics, Osaka City University* **4**(1) Series D, 1–16
PANT, M.M. (1979). Dependence of plant yield on density and planting pattern. *Annals of Botany* **44**, 513–516
SHARPE, P.R. and DENT, J.B. (1968). The determination and economic analysis of relationships between plant population and yield of main crop potatoes. *Journal of Agricultural Science* **70**, 123–129
SHINOZAKI, K. and KIRA, T. (1956). Intraspecific competition among higher plants. VII. Logistic theory of the C–D effect. *Journal of the Institute of Polytechnics, Osaka City University* **7**, Series D, 35–72
THORNLEY, J.H.M. (1983). Crop yield and planting density. *Annals of Botany* **52**, 257–259
WARNE, L.G.G. (1951). Spacing experiments on vegetables. II. The effect of the thinning distance on the yields of globe beet, long beet, carrots and parsnips grown at a standard inter-row distance in Cheshire, 1948. *Journal of Horticultural Science* **26**, 84–97
WILLEY, R.W. and HEATH, S.B. (1969). Plant population and crop yield. *Advances in Agronomy* **21**, 281–321

## Models of crop growth

ANGUS, J.F., KORNHER, A. and TORSSELL, B.W.R. (1980) *A Systems Approach to Estimation of Swedish Ley Production. Progress Report 1979/80.* Report 85. Uppsala: Swedish University of Agricultural Sciences
ARKIN, G.F., RICHARDSON, C.W. and MAAS, S.J. (1978). Forecasting grain sorghum yields using probability functions. *Transactions of the American Society of Agricultural Engineers* **21**, 874–880
ARKIN, G.F., VANDERLIP, R.L. and RITCHIE, J.T. (1976). The dynamic grain sorghum growth model. *Transactions of the American Society of Agricultural Engineers* **19**, 622–630
BAKER, D.N. (1980). Simulation for research and crop management. In *Proceedings of World Soybean Research Conference II* (ed. F.T. Corbin), pp. 533–546. Boulder, Colorado: Westview Press
BRIDGE, D.W. (1976). A simulation model approach for relating effective climate to winter wheat yields on the great plains. *Agricultural Meteorology* **17**, 185–194
CHARLES-EDWARDS, D.A. and ACOCK, B. (1977). Growth response of a *Chrysanthemum* crop to the environment. II. A mathematical analysis relating photosynthesis and growth. *Annals of Botany* **41**, 49–58
DE WIT, C.T. and others (1978). *Simulation of Assimilation, Respiration and Transpiration of Crops.* Wageningen: Pudoc
FICK, G.W., WILLIAMS, W.A. and LOOMIS, R.S. (1973). Computer simulation of dry matter distribution during sugar beet growth. *Crop Science* **13**, 413–417
FICK, G.W., LOOMIS, R.S. and WILLIAMS, W.A. (1975). Sugar beet. In *Crop Physiology* (ed. L.T. Evans), pp. 259–295. Cambridge: Cambridge University Press
FITZPATRICK, E.A. and NIX, H.A. (1970). The climatic factor in Australian grassland ecology. In *Australian Grasslands* (ed. R.M. Moore), pp. 3–26. Canberra: Australian National University Press
FUKAI, S. and SILSBURY, J.H. (1978). A growth model for *Trifolium subterraneum* L. swards. *Australian Journal of Agricultural Research* **29**, 51–65
HAUN, J.R. (1974). Prediction of spring wheat yields from temperature and precipitation data. *Agronomy Journal* **66**, 405–409
HAUN, J.R. (1979). Wheat yield models based on daily plant–environment relationships. In *Proceedings of the Crop Modeling Workshop*, Columbia, Missouri, 1977, pp. 1–43. Washington DC: US Department of Commerce

HOLT, D.A., BULA, R.J., MILES, G.E., SCHREIBER, M.M. and PEART, R.M. (1975). Environmental physiology, modeling and simulation. I. Conceptual development of SIMED. *Research Bulletin of Purdue University Agricultural Experiment Station*, No. 907, 26 pp

IDSO, S.B., PINTER, P.J., HATFIELD, J.L., JACKSON, R.D. and REGINATO, R.J. (1979). A remote sensing model for the prediction of wheat yields prior to harvest. *Journal of Theoretical Biology* **77**, 217–228

JOHNSON, I.R., AMEZIANE, T.E. and THORNLEY, J.H.M. (1983). A model of grass growth. *Annals of Botany* **51**, 599–609

LEAFE, E.L., STILES, W. and DICKINSON, S.E. (1974). The physiological processes influencing the pattern of productivity of the intensively managed grass sward. *Proceedings of the 12th International Grassland Congress* (ed. V.G. Iglovikov and A.P. Movsissyants), Vol. 1, pp. 442–457

LEGG, B.J. (1981). Aerial environment and crop growth. In *Mathematics and Plant Physiology* (ed. D.A. Rose and D.A. Charles-Edwards), pp. 129–149. London: Academic Press

LEGG, B.J., DAY, W., LAWLOR, D.W. and PARKINSON, K.J. (1979). The effects of drought on barley growth: models and measurements showing the relative importance of leaf area and photosynthetic rate. *Journal of Agricultural Science* **92**, 703–716

McKINION, J.M., JONES, J.W. and HESKETH, J.D. (1975). A system of growth equations for the continuous simulation of plant growth. *Transactions of the American Society of Agricultural Engineers* **18**, 975–984

MEYER, G.E., CURRY, R.B., STREETER, J.G. and MEDERSKI, H.J. (1979). A dynamic simulator of soybean growth, development, and seed yield: I. Theory, structure and validation. *Research Bulletin of the Ohio Agricultural Research and Development Center*, No. 1113

MURATA, Y. (1975). Estimation and simulation of rice yield from climatic factors. *Agricultural Meteorology* **15**, 117–131

PITTER, R.L. (1977). The effect of weather and technology on wheat yields in Oregon. *Agricultural Meteorology* **18**, 115–131

SAKAMOTO, C.M. (1978). The *Z*-index as a variable for crop yield estimation. *Agricultural Meteorology* **19**, 305–313

SEIF, E. and PEDERSON, D.G. (1978). Effect of rainfall on the grain yield of spring wheat, with an application. *Australian Journal of Agricultural Research* **29**, 1107–1115

SHEEHY, J.E., COBBY, J.M. and RYLE, G.J.A. (1980). The use of a model to investigate the influence of some environmental factors on the growth of perennial ryegrass. *Annals of Botany* **46**, 343–365

SWEENEY, D.G., HAND, D.W., SLACK, G. and THORNLEY, J.H.M. (1981). Modelling the growth of winter lettuce. In *Mathematics and Plant Physiology* (ed. D.A. Rose and D.A. Charles-Edwards), pp. 217–229. London: Academic Press

THORNLEY, J.H.M. and HURD, R.G. (1974). An analysis of the growth of young tomato plants in water culture at different light integrals and $CO_2$ concentrations. II. A mathematical model. *Annals of Botany* **38**, 389–400

WANN, M., RAPER, C.D. and LUCAS, H.L. (1978). A dynamic model for plant growth: a simulation of dry matter accumulation for tobacco. *Photosynthetica* **12**, 121–136

## Chapter 9
## Plant diseases and pests

### Plant diseases

BURLEIGH, J.R., ROELFS, A.P. and EVERSMEYER, M.G. (1972). Estimating damage to wheat caused by *Puccinia recondita tritiri. Phytopathology* **62**, 944–946

EVERSMEYER, M.G. and BURLEIGH, J.R. (1970). A method of predicting epidemic development of wheat leaf rust. *Phytopathology* **60**, 805–811

EVERSMEYER, M.G., BURLEIGH, J.R. and ROELFS, A.P. (1973). Equations for predicting wheat stem rust development. *Phytopathology* **63**, 348–351

HYRE, R.A. (1954). Progress in forecasting late blight of potato and tomato. *Plant Disease Reporter* **38**, 245–253

JAMES, W.C. (1971). An illustrated series of assessment keys for plant diseases, their preparation and usage. *Canadian Plant Disease Survey* **51**, 39–65

JAMES, W.C. (1974). Assessment of plant diseases and losses. *Annual Review of Phytopathology* **12**, 27–48

JAMES, W.C.. SHIH, C.S., HODGSON, W.A. and CALLBECK, L.C. (1972). The quantitative relationship between late blight of potato and loss of tuber yield. *Phytopathology* **62**, 92–96

KATSUBE, T. and KOSHIMIZU, Y. (1970). Influence of blast disease on harvests in rice plants. I: Effect of panicle infection on yield components and quality. *Bulletin Tohoku National Agricultural Experiment Station* **39**, 55–96
KRANZ, J. (ed.) (1974). *Epidemics of Plant Diseases: Mathematical Analysis and Modelling*. London: Chapman and Hall
KRAUSE, R.A. and MASSIE, L.B. (1975). Predictive systems: modern approaches to disease control. *Annual Review of Phytopathology* **13**, 31–47
KRAUSE, R.A., MASSIE, L.B. and HYRE, R.A. (1975). Blitecast: a computerized forecast of potato late blight. *Plant Disease Reporter* **59**, 95–98
LARGE, E.C. (1954). Growth stages in cereals: illustration of the Feekes scale. *Plant Pathology* **3**, 9–28
ROMIG, R.W. and CALPOUZOS, L. (1970). The relationship between stem rust and loss of yield of spring wheat. *Phytopathology* **60**, 1376–1380
VAN DER PLANK, J.E. (1963). *Plant Diseases: Epidemics and Control*. New York: Academic Press
WAGGONER, P.E. (1974). Simulation of epidemics. In *Epidemics of Plant Diseases: Mathematical Analysis and Modelling* (ed. J. Kranz), pp. 137–160. London: Chapman and Hall
WAGGONER, P.E., HORSFALL, J.G. and LUKENS, R.J. (1972). EPIMAY a simulator of southern corn blight. *Bulletin of the Connecticut Agricultural Experiment Station*, No. 729, 84 pp

### Pests, predators and parasites

CALTAGIRONE, L.E. (1981). Landmark examples in classical biological control. *Annual Review of Entomology* **26**, 213–232
CONWAY, G.R. (1977). Mathematical models in applied ecology. *Nature, London* **269**, 291–297
LESLIE, P.H. (1945). On the use of matrices in certain population mathematics. *Biometrika* **33**, 183–212
LEWIS, E.G. (1942). On the generation and growth of a population. *Sankhya* **2**, 93–96
LOTKA, A.J. (1925). *Elements of Physical Biology*. Baltimore: Williams and Wilkins
MACDONALD, N. (1978). *Time Lags in Biological Models*. Berlin: Springer
MAYNARD SMITH, J. (1974). *Models in Ecology*. Cambridge: Cambridge University Press
RABBINGE, R., ANKERSMIT, G.W. and PAK, G.A. (1979). Epidemiology and simulation of population development of *Sitobion avenae* in winter wheat. *Netherlands Journal of Plant Pathology* **85**, 197–220
STERN, V.M., SMITH, R.F., VAN DEN BOSCH, R. and HAGEN, K.S. (1959). The interaction of chemical and biological control of the spotted alfalfa aphid. The integrated control concept. *Hilgardia* **29**, 81–101
STREIFER, W. (1974). Realistic models in population biology. *Advances in Ecological Research* **8**, 199–266
VOLTERRA, V. (1926). Variazioni e fluttuazini del numero d'individui in specie animali conviventi. *Memorie dell'Academia Nazionale dei Lincei* **2**, 31–113
WILLIAMSON, M.H. (1972). *The Analysis of Biological Populations*. London: Arnold

## Chapter 10
## Animal processes

### Digestion

BALDWIN, R.L., KOONG, L.J. and ULYATT, M.J. (1977). A dynamic model of ruminant digestion for evaluation of factors affecting nutritive value. *Agricultural Systems* **2**, 255–288
BALDWIN, R.L., LUCAS, H.L. and CABRERA, R. (1970). Energetic relationships in the formation and utilization of fermentation end-products. In *Physiology of Digestion and Metabolism in the Ruminant* (ed. A.T. Phillipson), pp. 319–334. Newcastle upon Tyne: Oriel Press
BEEVER, D.E., BLACK, J.L. and FAICHNEY, G.J. (1981). Simulation of the effects of rumen function on the flow of nutrients from the stomach of sheep: part 2—assessment of computer predictions. *Agricultural Systems* **6**, 221–241
BLACK, J.L., BEEVER, D.E., FAICHNEY, G.J., HOWARTH, B.R. and GRAHAM, N.McC. (1981). Simulation of the effects of rumen function on the flow of nutrients from the stomach of sheep: part 1—description of a computer program. *Agricultural Systems* **6**, 195–219
FAICHNEY, G.J., BEEVER, D.E. and BLACK, J.L. (1981). Prediction of the fractional rate of outflow of water from the rumen of sheep. *Agricultural Systems* **6**, 261–268

FRANCE, J., THORNLEY, J.H.M. and BEEVER, D.E. (1982). A mathematical model of the rumen. *Journal of Agricultural Science, Cambridge* **99**, 343–353

HUNGATE, R.E. (1966). *The Rumen and its Microbes.* New York: Academic Press

McDONALD, I. (1981). A revised model for the estimation of protein degradability in the rumen. *Journal of Agricultural Science, Cambridge* **96**, 251–252

MERTENS, D.R. (1977). Dietary fiber components: relationship to the rate and extent of ruminal digestion. *Federation Proceedings of the Federation of American Societies for Experimental Biology* **36**, 187–192

ØRSKOV, E.R. and McDONALD, I. (1979). The estimation of protein degradability in the rumen from incubation measurements weighted according to rate of passage. *Journal of Agricultural Science, Cambridge* **92**, 499–503

REICHL, J.R. and BALDWIN, R.L. (1975). Rumen modeling: rumen input–output balance models. *Journal of Dairy Science* **58**, 879–890

ULLYAT, M.J., BALDWIN, R.L. and KOONG, L.J. (1976). The basis of nutritive value—a modelling approach. *Proceedings of the New Zealand Society of Animal Production* **36**, 140–149

WALDO, D.R., SMITH, L.W. and COX, E.L. (1972). Model of cellulose disappearance from the rumen. *Journal of Dairy Science* **55**, 125–129

## Metabolism

AGRICULTURAL RESEARCH COUNCIL (1980). *The Nutrient Requirements of Ruminant Livestock.* Slough: Commonwealth Agricultural Bureaux

BALDWIN, R.L. and BLACK, J.L. (1979). *Simulation of the Effects of Nutritional and Physiological Status on the Growth of Mammalian Tissue: Description and Evaluation of a Computer Program.* Animal Research Laboratories Technical Paper No. 6. Melbourne: CSIRO

BALDWIN, R.L., CRIST, K., WAGHORN, G. and SMITH, N.E. (1981). The synthesis of models to describe metabolism and its integration. *Proceedings of the Nutrition Society* **40**, 139–145

BLACK, J.L. and GRIFFITHS, D.A. (1975). Effects of liveweight and energy intake on nitrogen balance and total N requirements of lambs. *British Journal of Nutrition* **33**, 399–413

BROUWER, E. (1965). Report of the sub-committee on constants and factors. In *Energy Metabolism* (ed. K.L. Blaxter), pp. 441–443. London: Academic Press

DIXON, M. and WEBB, E.C. (1964). *Enzymes*, second edition. London: Longman

FORBES, J.M. (1977a). Interrelationships between physical and metabolic control of voluntary food intake in fattening, pregnant and lactating mature sheep: a model. *Animal Production* **24**, 91–101

FORBES, J.M. (1977b). Development of a model of voluntary food intake and energy balance in lactating cows. *Animal Production* **24**, 203–214

GEISLER, P.A. and NEAL, H. (1979). A model for the effects of energy nutrition on the pregnant ewe. *Animal Production* **29**, 357–369

GILL, E.M., THORNLEY, J.H.M., BLACK, J.L., OLDHAM, J.D. and BEEVER, D.E. (1984). Efficiency of utilization of absorbed energy in ruminants. To be submitted to the *British Journal of Nutrition*

GRAHAM, N.McC., BLACK, J.L., FAICHNEY, G.J. and ARNOLD, G.W. (1976). Simulation of growth and production in sheep—Model 1: a computer program to estimate energy and nitrogen utilization, body composition and empty liveweight change, day by day for sheep of any age. *Agricultural Systems* **1**, 113–138

KEENER, H.M. (1979). Simulation of energy utilisation of bovine animals. *Agricultural Systems* **4**, 79–100

KOONG, L.J., FALTER, K.H. and LUCAS, H.L. (1982). A mathematical model for the joint metabolism of nitrogen and energy in cattle. *Agricultural Systems* **9**, 301–324

LENG, R.A. and ANNISON, E.F. (1963). Metabolism of acetate, propionate and butyrate by sheep liver slices. *Biochemical Journal* **86**, 319–327

MAHLER, H.R. and CORDES, E.H. (1966). *Biological Chemistry.* New York: Harper and Row

McVEIGH, J.M. and TARRANT, P.V. (1982). Glycogen content and repletion rate in beef muscle, effect of feeding and fasting. *Journal of Nutrition* **112**, 1306–1314

MINISTRY OF AGRICULTURE, FISHERIES AND FOOD (1975). *Energy Allowances and Feeding Systems for Ruminants.* Technical Bulletin No. 33. London: HMSO

MURRAY, D.M. and SLEZACEK, O. (1976). Growth rate and its effect on empty body weight, carcass weight and dissected carcass composition of sheep. *Journal of Agricultural Science, Cambridge* **87**, 171–179

NATIONAL RESEARCH COUNCIL (1970). *Nutrient Requirements of Beef Cattle,* fifth edition. Washington DC: National Academy of Sciences
NATIONAL RESEARCH COUNCIL (1971). *Nutrient Requirements of Dairy Cattle,* fifth edition. Washington DC: National Academy of Sciences
NATIONAL RESEARCH COUNCIL (1975). *Nutrient Requirements of Sheep,* fifth edition. Washington DC: National Academy of Sciences
NEWTON, J.E. and EDELSTEN, P.R. (1976). A model on the effect of nutrition on litter size and weight in the pregnant ewe. *Agricultural Systems* **1**, 185–199
PETHICK, D.W., LINDSAY, D.B., BARKER, P.J. and NORTHROP, A.J. (1981). Acetate supply and utilization by the tissues of sheep in vivo. *British Journal of Nutrition* **46**, 97–110
SCHULZ, A.R. (1978). Simulation of energy metabolism in the simple-stomached animal. *British Journal of Nutrition* **39**, 235–254
SEARLE, T.W. and GRIFFITHS, D.A. (1976). The body composition of growing sheep during milk feeding, and the effect on composition of weaning at various body weights. *Journal of Agricultural Science, Cambridge* **86**, 483–493
SMITH, N.E., BALDWIN, R.L. and SHARP, W.M. (1980). Models of tissue and animal metabolism. In *Energy Metabolism* (ed. L.E. Mount), pp. 193–197. London: Butterworths
THORNLEY, J.H.M. (1976). *Mathematical Models in Plant Physiology*. London: Academic Press
WEBSTER, A.J.F., OSUJI, P.O., WHITE, F. and INGRAM, J.F. (1975). The influence of food intake on portal blood flow and heat production in the digestive tract of sheep. *British Journal of Nutrition* **34**, 125–139

# Chapter 11
# Animal products

## Milk production

ABRAMOWITZ, M. and STEGUN, I.A. (1965). *Handbook of Mathematical Functions*. New York: Dover
BHATTACHARJEE, G.P. (1970). The incomplete gamma integral. *Applied Statistics* **19**, 285–287
COBBY, J.M. and LE DU, Y.L.P. (1978). On fitting curves to lactation data. *Animal Production* **26**, 127–133
DHANOA, M.S. (1981). A note on an alternative form of the lactation model of Wood. *Animal Production* **32**, 349–351
DHANOA, M.S. and LE DU, Y.L.P. (1982). A partial adjustment model to describe the lactation curve of a dairy cow. *Animal Production* **34**, 243–247
GAINES, W.L. (1927). Persistency of lactation in dairy cows. *Bulletin of the Illinois Agricultural Experimental Station* No. 288, 355–424
GUEST, P.G. (1961). *Numerical Methods of Curve Fitting*. Cambridge: Cambridge University Press
KNIGHT, C.H. (1982). The mammary cell population in relation to milk yield. In *Hannah Research Institute Report 1981*, pp. 89–95. Ayr: Hannah Research Institute
MEPHAM, B. (1976). *The Secretion of Milk*. London: Arnold
MUMFORD, R.E. (1964). A review of anatomical and biochemical changes in the mammary gland with particular reference to quantitative methods of assessing mammary developments. *Dairy Science Abstracts* **26**, 293–304
NEAL, H.D.St.C. and THORNLEY, J.H.M. (1983). The lactation curve in cattle: a mathematical model of the mammary gland. *Journal of Agricultural Science, Cambridge* **101**, 389–400
NELDER, J.A. (1966). Inverse polynomials, a useful group of multi-factor response functions. *Biometrics* **22**, 128–141
PEARSON, K. (1922). *Tables of the Incomplete Γ-function*. London: HMSO
ROY, J.H.B. (1980) *The Calf*, fourth edition. London: Butterworths
TORRES-HERNANDEZ, G. and HOHENBOKEN, W.D. (1980). Biometric properties of lactation in ewes raising single or twin lambs. *Animal Production* **30**, 431–436
VUJIČIĆ, I. and BAČIĆ, B. (1961). New equation of the lactation curve [in Croatian]. *Letopis Naučnih Radova, Poljoprivredni Fakultet u Novum Sadu* No. 5. Novi Sad, Yugoslavia: Faculty of Agriculture, University of Novi Sad
WOOD, P.D.P. (1967). Algebraic model of the lactation curve in cattle. *Nature, London* **216**, 164–165

WOOD, P.D.P. (1969). Factors affecting the shape of the lactation curve in cattle. *Animal Production* **11**, 307–316

WOOD, P.D.P. (1979). A simple model of lactation curves for milk yield, food requirement and body weight. *Animal Production* **28**, 55–63

## Meat production

ADOLPH, E.F. (1949). Quantitative relations in the physiological constituents of mammals. *Science* **109**, 579–585

AGRICULTURAL RESEARCH COUNCIL (1980). *The Nutrient Requirements of Ruminant Livestock.* Slough: Commonwealth Agricultural Bureaux

BALDWIN, R.L. and BLACK, J.L. (1979). *Simulation of the Effects of Nutritional and Physiological Status on Growth of Mammal Tissues: Description and Evaluation of a Computer Program.* Animal Research Laboratories Technical Paper No. 6. Melbourne: CSIRO

BARTON, A.D. and LAIRD, A.K. (1969). Analysis of allometric and non-allometric differential growth. *Growth* **33**, 1–16

BRODY, S. (1945). *Bioenergetics and Growth.* New York: Reinhold

FOWLER, V.R. (1968). Body development and some problems in its evaluation. In *Growth and Development of Mammals* (ed. G.A. Lodge and G.E. Lamming), pp. 195–211. London: Butterworths

FOWLER, V.R. (1980). Growth in mammals for meat production. In *Growth in Animals* (ed. T.L.J. Lawrence), pp. 249–263. London: Butterworths

HARRIES, J.M., WILLIAMS, D.R. and POMEROY, R.W. (1975). Prediction of comparative retail values of beef carcasses. *Animal Production* **21**, 127–137

HARRIES, J.M., WILLIAMS, D.R. and POMEROY, R.W. (1976). Comparative retail value of beef carcasses from different groups of animals. *Animal Production* **23**, 349–356

LAIRD, A.K. (1965). Dynamics of relative growth. *Growth* **29**, 249–263

LAIRD, A.K., TYLER, S.A. and BARTON, A.D. (1965). Dynamics of normal growth. *Growth* **29**, 233–248

LAWRIE, R.A. (1975). Meat components and their variability. In *Meat* (ed. D.J.A. Cole and R.A. Lawrie), pp. 249–268. London: Butterworths

MUKHOTY, H. and BERG, R.T. (1971). Influence of breed and sex on the allometric growth patterns of major bovine tissues. *Animal Production* **13**, 219–227

MUNRO, H.N. (1969). An introduction to protein metabolism during the evolution and development of mammals. In *Mammalian Protein Metabolism* Volume III (ed. H.N. Munro), pp. 3–19. New York: Academic Press

ROUX, C.Z. (1976). A model for the description and regulation of growth and production. *Agroanimalia* **8**, 83–94

WHITTEMORE, C.T. and FAWCETT, R.H. (1974). Model responses of the growing pig to the dietary intake of energy and protein. *Animal Production* **19**, 221–231

WHITTEMORE, C.T. and FAWCETT, R.H. (1976). Theoretical aspects of a flexible model to simulate protein and lipid growth in pigs. *Animal Production* **22**, 87–96

WILLIAMS, D.R., POMEROY, R.W., HARRIES, J.M. and RYAN, P.O. (1974). Composition of beef carcasses. II. The use of regression equations to estimate total tissue components from observations on intact and quartered sides and partial dissection data. *Journal of Agricultural Science, Cambridge* **83**, 79–85

WILSON, B.J. (1980). Growth in birds for meat production. In *Growth in Animals* (ed. T.L.J. Lawrence), pp. 265–272. London: Butterworths

## Egg production

COMBS, G.F. (1960). Protein and energy requirements of laying hens. In *Proceedings of the Maryland Nutrition Conference for Feed Manufacturers 1960*, pp. 28–45

COMBS, G.F. (1968). Amino acid requirements of broilers and laying hens. In *Proceeding of the Maryland Nutrition Conference for Feed Manufacturers 1968*, pp. 86–96

CURNOW, R.N. (1973). A smooth population response curve based on an abrupt threshold and plateau model for individuals. *Biometrics* **29**, 1–10

FISHER, C. (1980). Protein deposition in poultry. In *Protein Deposition in Animals* (ed. P.J. Buttery and D.B. Lindsey), pp. 251–270. London: Butterworths

FISHER, C., MORRIS, T.R. and JENNINGS, R.C. (1973). A model for the description and prediction of the response of laying hens to amino acid intake. *British Poultry Science* **14**, 469–484
GILBERT, A.B. (1971). The female reproductive effort. In *Physiology and Biochemistry of the Domestic Fowl*, Volume 3 (ed. D.J. Bell and B.M. Freeman), pp. 1153–1162. London: Academic Press
HALLAM, D. (1981). *Econometric Forecasting in the UK Egg Market.* Tunbridge Wells: Eggs Authority
HURWITZ, S. and BORNSTEIN, S. (1973). The protein and amino acid requirements of laying hens: suggested models for calculation. *Poultry Science* **52**, 1124–1134
HURWITZ, S. and BORNSTEIN, S. (1977). The protein and amino acid requirements of laying hens: experimental evaluation of models of calculation. 1. Application of two models under various conditions. *Poultry Science* **56**, 969–978
KULSHRESHTA, S.N. (1971). A short-run model for forecasting monthly egg production in Canada. *Canadian Journal of Agricultural Economics* **19**, 36–46
ROY, S.K. and JOHNSON, P.N. (1973). Econometric models for quarterly shell egg prices. *American Journal of Agricultural Economics* **55**, 209–213
SMITH, W.K. (1978a). The amino acid requirements of laying hens: models for calculation. 1. Physiological background. *World's Poultry Science Journal* **34**, 81–96
SMITH, W.K. (1978b). The amino acid requirements of laying hens: models for calculation. 2. Practical application. *World's Poultry Science Journal* **34**, 129–136
SVENSSON, S.A. (1964). Composition and energy content of eggs, growing chicks and hens, with some notes on preparation and method of analysis. *Lantbrukshogskolaus Annalar* **30**, 405
THOMAS, O.P., TWINING, P.V. and BOSSARD, E.H. (1977). The available lysine requirement of 7–9 week old sexed broiler chicks. *Poultry Science* **56**, 57–60

# Chapter 12
# Farm planning and control: I

## Introduction

BARNARD, C.S. and NIX, J.S. (1973). *Farm Planning and Control.* Cambridge: Cambridge University Press
DALTON, G.E. (ed.) (1975). *Study of Agricultural Systems.* London: Applied Science Publishers
DENT, J.B. and ANDERSON, J.R. (ed.) (1971). *Systems Analysis in Agricultural Management.* Sydney: Wiley
DENT, J.B. and BLACKIE, M.J. (1979). *Systems Simulation in Agriculture.* London: Applied Science Publishers
SPEDDING, C.R.W. (1979). *An Introduction to Agricultural Systems.* London: Applied Science Publishers

## Resource allocation

AUDSLEY, E. (1979). Planning an arable farm's machinery needs—a linear programming application. In *Proceedings of an Operational Research Workshop 20–21 September 1978.* National Institute of Agricultural Engineering Report No. 32. Silsoe: NCAE/NIAE
AUDSLEY, E., DUMONT, S. and BOYCE, D.S. (1978). An economic comparison of methods of cultivating and planting cereals, sugar beet and potatoes and their interaction with harvesting, timeliness and available labour by linear programming. *Journal of Agricultural Engineering Research* **23**, 283–300
HARDAKER, J.B. (1979). A review of some farm management research methods for small-farm development in LDCs. *Journal of Agricultural Economics* **30**, 315–328
ROUMASSET, J.A., BOUSSARD, J.M. and SINGH, I. (1979). *Risk, Uncertainty and Agricultural Development.* New York: SEARCA and ADC
TAHA, H.A. (1971). *Operations Research—an Introduction.* New York: Macmillan

## Crop production

BURT, O.R. and ALLISON, J.R. (1963). Farm management decisions with dynamic programming. *Journal of Farm Economics* **45**, 121–136

## Grazing and conservation

AUDSLEY, E. (1974). A linear programming model of a high-temperature grass-drying enterprise. *Journal of the British Grassland Society* **29**, 291–298

BROUGHAM, R.W. (1956). The rate of growth of short-rotation ryegrass pastures in the late autumn, winter, and early spring. *New Zealand Journal of Science and Technology* **A38**, 78

CHRISTIAN, K.R., FREER, M., DONNELLY, J.R., DAVIDSON, J.L. and ARMSTRONG, J.S. (1978). *Simulation of Grazing Systems.* Wageningen: Pudoc

CORRALL, A.J., NEAL, H.D.St.C. and WILKINSON, J.M. (1982). *Silage in Milk Production.* Technical Report No. 29. Hurley, Maidenhead: Grassland Research Institute

DUMONT, A.G. and BOYCE, D.S. (1976). Leaf protein production and use on the farm: an economic study. *Journal of the British Grassland Society* **31**, 153–163

EDELSTEN, P.R. and CORRALL, A.J. (1979). Regression models to predict herbage production and digestibility in a non-regular sequence of cuts. *Journal of Agricultural Science, Cambridge* **92**, 575–585

FRANCE, J., BROCKINGTON, N.R. and NEWTON, J.E. (1981). Modelling grazed grassland systems: wether sheep grazing perennial ryegrass. *Applied Geography* **1**, 133–150

MORLEY, F.H.W. (1968). Pasture growth curves and grazing management. *Australian Journal of Experimental Agriculture and Animal Husbandry* **8**, 40–45

NOY-MEIR, I. (1976). Rotational grazing in a continuously growing pasture: a simple model. *Agricultural Systems* **1**, 87–112

NOY-MEIR, I. (1978). Grazing and production in seasonal pastures: analysis of a simple model. *Journal of Applied Ecology* **15**, 809–835

PARKE, D., DUMONT, A.G. and BOYCE, D.S. (1978). A mathematical model to study forage conservation methods. *Journal of the British Grassland Society* **33**, 261–273

RICHARDS, I.R. and HOBSON, R.D. (1979). Optimising land and fertiliser N usage in grassland systems where grazing and conservation areas are separate. *Agricultural Systems* **4**, 59–70

SIBBALD, A.R., MAXWELL, T.J. and EADIE, J. (1979). A conceptual approach to the modelling of herbage intake by hill sheep. *Agricultural Systems* **4**, 119–134

# Chapter 13
# Farm planning and control: II

## The dairy enterprise

BYWATER, A.C. (1976). Simulation of the intake and partition of nutrients by the dairy cow: part II—the yield and composition of milk. *Agricultural Systems* **1**, 261–279

BYWATER, A.C. and DENT, J.B. (1976). Simulation of the intake and partition of nutrients by the dairy cow: part I—management control in the dairy enterprise; philosophy and general model construction. *Agricultural Systems* **1**, 245–260

CHUDLEIGH, P. (1977). A model of the small scale dairying enterprise: an aid to resource allocation in agricultural development. *Agricultural Systems* **2**, 67–82

CRABTREE, J.R. (1970). Towards a dairy enterprise model. In *The Use of Models in Agricultural and Biological Research* (ed. J.G.W. Jones), pp. 35–42. Hurley, Maidenhead: Grassland Research Institute

FRANCE, J., HEAL, H.D.St.C., MARSDEN, S. and FROST, B. (1982). A dairy herd cash flow model. *Agricultural Systems* **8**, 129–142

KILLEN, L. and KEANE, M. (1978). A linear programming model of seasonality in milk production. *Journal of the Operational Research Society* **29**, 625–631

KIRK, J.H. (1981). Application of programmable calculators to mastitis control programs. *Journal of Dairy Science* **64**, 2048–2058

LINN, J.G. and SPIKE, P.L. (1980). Programmable calculators and their application to feeding and management of dairy cattle. *Journal of Dairy Science* **63**, 1390

REYES, A.A., BLAKE, R.W., SHUMWAY, C.R. and LONG, J.T. (1981). Multistage optimization model for dairy production. *Journal of Dairy Science* **64**, 2003–2016

TAHA, H.A. (1971). *Operations Research—an Introduction.* New York: Macmillan

WOOD, P.D.P. (1969). Factors affecting the shape of the lactation curve in cattle. *Animal Production* **11**, 307–316

WOOD, P.D.P. (1970). The relationship between the month of calving and milk production. *Animal Production* **12**, 253–259

WOOD, P.D.P. (1976). Algebraic models of the lactation curves for milk, fat and protein production, with estimates of seasonal variation. *Animal Production* **22**, 35–40

WOOD, P.D.P. (1979). A simple model of lactation curves for milk yield, food requirement and body weight. *Animal Production* **28**, 55–63

WOOD, P.D.P. (1980). Breed variations in the shape of the lactation curve of cattle and their implications for efficiency. *Animal Production* **31**, 133–141

WOOD, P.D.P. (1981). A note on regional variations in the seasonality of milk production in dairy cattle. *Animal Production* **32**, 105–108

## Beef systems

BECK, A.C. and DILLON, J.L. (1980). Brucellosis eradication planning: an application of simulation modelling. *Agricultural Systems* **5**, 165–180

BECK, A.C., HARRISON, I. and JOHNSTON, J.H. (1982). Using simulation to assess the risks and returns from pasture improvement for beef production in agriculturally underdeveloped regions. *Agricultural Systems* **8**, 55–71

CONGLETON Jr., W.R. and GOODWILL, R.E. (1980a). Simulated comparisons of breeding plans for beef production—part 1: a dynamic model to evaluate the effects of mating plan on herd age structure and productivity. *Agricultural Systems* **5**, 207–219

CONGLÈTON Jr., W.R. and GOODWILL, R.E. (1980b). Simulated comparisons of breeding plans for beef production—part 2: Hereford, Angus and Charolais sires bred to Hereford, Angus and Hereford–Angus dams to produce feeder calves. *Agricultural Systems* **5**, 221–232

CONGLETON Jr., W.R. and GOODWILL, R.E. (1980c). Simulated comparisons of breeding plans for beef production—part 3: systems for producing feeder-calves involving intensive culling and additional breeds of sire. *Agricultural Systems* **5**, 309–318

CONWAY, A.G. (1974a). A production function for grazing cattle. 1. Model of the relationship between liveweight gain and stocking rate. *Irish Journal of Agricultural Economics and Rural Sociology* **5**, 1–13

CONWAY, A.G. (1974b). A production function for grazing cattle. 4. An economic decision model for grazing steers. *Irish Journal of Agricultural Economics and Rural Sociology* **5**, 57–108

CURLL, M.L. (1978). Simulation: an aid to decisions on superphosphate use for beef production. *Agricultural Systems* **3**, 195–204

HALTER, A.N., MILLER, S.F., DE JONGH, R. and ESTRADA, H.R. (1976). Application of systems-simulation in the Venezuelan cattle industry. *Agricultural Systems* **1**, 139–162

JOANDET, G.E. and CARTWRIGHT, T.C. (1975). Modelling beef production systems. *Journal of Animal Science* **41**, 1238–1246

KELLEY WHITE, T., McCARL, B.A., MAY, R.D. and SPREEN, T.H. (1978). A systems analysis of the Guyanese livestock industry. *Agricultural Systems* **3**, 47–66

KENNEDY, J.O.S. (1972). A model for determining optimal marketing and feeding policies for beef cattle. *Journal of Agricultural Economics* **23**, 147–159

LOUW, A., GROSSKOPF, J.F.W. and GROENEWALD, J.A. (1979). Beef production systems and sales strategies in an extensive ranching region in South Africa. *Agricultural Systems* **4**, 101–114

MEYER, C.F. and NEWETT, R.J. (1970). Dynamic programming for feedlot optimization. *Management Science* **16**, 410–426

MILLER, W.C., BRINKS, J.S. and SUTHERLAND, T.M. (1978). Computer assisted management decisions for beef production systems. *Agricultural Systems* **3**, 147–158

MILLER, W.C., WARD, G.M., YORKS, T.P., ROSSITER, D.L. and COMBS, J.J. (1980). A mathematical model of the United States beef production systems. *Agricultural Systems* **5**, 295–307

MORLEY, F.H.W. (1977). In *Applications in Agricultural Modelling* (ed. A.J. de Boer and C.W. Rose), pp. 131–146. Brisbane: Australian Institute of Agricultural Science

MORRIS, C.A., PARKINS, J.J. and WILTON, J.W. (1976). Effects of creep feeding, mature cow weight and milk yield on farm gross margins in an integrated beef production model. *Canadian Journal of Animal Science* **56**, 87–97

RYAN, T.J. (1974). A beef feedlot simulation model. *Journal of Agricultural Economics* **25**, 265–276
SANDERS, J.O. and CARTWRIGHT, T.C. (1979a). A general cattle production systems model. I: Structure of the model. *Agricultural Systems* **4**, 217–227
SANDERS, J.O. and CARTWRIGHT, T.C. (1979b). A general cattle production systems model. Part 2—procedures used for simulating animal performance. *Agricultural Systems* **4**, 289–309
SULLIVAN, G.M., CARTWRIGHT, T.C. and FARRIS, D.E. (1981). Simulation of production systems in East Africa by use of interfaced forage and cattle models. *Agricultural Systems* **7**, 245–265
WILTON, J.W., MORRIS, C.A., JENSEN, E.A., LEIGH, A.O. and PFEIFFER, W.C. (1974). A linear programming model for beef cattle production. *Canadian Journal of Animal Science* **54**, 693–707
YORKS, T.P., MILLER, W.C., COMBS, J.J. and WARD, G.M. (1980). Energy minimised v. cost minimized alternatives for the US beef production system. *Agricultural Systems* **6**, 121–129

## Lamb and other livestock systems

BLACKIE, M.J. and DENT, J.B. (1974). The concept and application of skeleton models in farm business analysis and planning. *Journal of Agricultural Economics* **25**, 165–174
BLACKIE, M.J. and DENT, J.B. (1976). Analyzing hog production strategies with a simulation model. *American Journal of Agricultural Economics* **58**, 39–46
CURLL, M.L. and DAVIDSON, J.L. (1977). Lamb production: a case study of experimentation and simulation. *Agricultural Systems* **2**, 121–138
DEVINDAR, S., WILLIAMS, I.H. and TUNG, L. (1980). Simulation-aided development of the Hawaii swine research programme. *Agricultural Systems* **5**, 279–264
EDELSTEN, P.R. and NEWTON, J.E. (1977). A simulation model of a lowland sheep system. *Agricultural Systems* **2**, 17–32
GEISLER, P.A., NEWTON, J.E., SHELDRICK, R.D. and MOHAN, A. (1979). A model of lamb production from an autumn catch crop. *Agricultural Systems* **4**, 49–57
GEISLER, P.A., PAINE, A.C. and GEYTENBEEK, P.E. (1977). Simulation of an intensified lambing system incorporating two flocks and the rapid remating of ewes. *Agricultural Systems* **2**, 109–119
GREIG, I.D., HARDAKER, J.B., FARRELL, D.J. and CUMMING, R.B. (1977). Towards the determination of optimal systems of broiler production. *Agricultural Systems* **2**, 47–65
WALSINGHAM, J.M., EDELSTEN, P.R. and BROCKINGTON, N.R. (1977). Simulation of the management of a rabbit population for meat production. *Agricultural Systems* **2**, 85–98
WHITE, D.H. and MORLEY, F.H.W. (1977). Estimation of optimal stocking rate of Merino sheep. *Agricultural Systems* **2**, 289–304
WHITE, D.H., O'LEARY, G.J., BARTLETT, B.E. and ABU-SEREWA, S. (1978). Simulation of poultry egg production. *Agricultural Systems* **3**, 85–102

## Culling

GARTNER, J.A. and HERBERT, W.A. (1979). A preliminary model to investigate culling and replacement policy in dairy herds. *Agricultural Systems* **4**, 189–215
McARTHUR, A.T.G. (1973). Application of dynamic programming to the culling decision in dairy cattle. *Proceedings of the New Zealand Society of Animal Production* **33**, 141–147
SMITH, B.J. (1973). Dynamic programming of the dairy cow replacement problem. *American Journal of Agricultural Economics* **55**, 100–104
STEWART, H.M., BURNSIDE, E.B., WILTON, J.W. and PFEIFFER, W.C. (1977). A dynamic programming approach to culling decisions in commercial dairy herds. *Journal of Dairy Science* **60**, 602–617
THROSBY, C.D. (1964). Some dynamic programming models for farm management research. *Journal of Agricultural Economics* **16**, 98–110
WALSINGHAM, J.M., EDELSTEN, P.R. and BROCKINGTON, N.R. (1977). Simulation of the management of a rabbit population for meat production. *Agricultural Systems* **2**, 85–98
WHITE, W.C. (1959). The determination of an optimal replacement policy for a continually operating egg production enterprise. *Journal of Farm Economics* **41**, 1535–1542

## Rationing

AGRICULTURAL RESEARCH COUNCIL (1980). *The Nutrient Requirements of Ruminant Livestock.* Slough: Commonwealth Agricultural Bureaux
BATH, D.L. and BENETT, L.F. (1980). Development of a dairy feeding model for maximizing income above feed cost with access by remote computer terminal. *Journal of Dairy Science* **63**, 1379–1389
BENEKE, R.R. and WINTERBOER, R. (1973). *Linear Programming Applications to Agriculture.* Ames: Iowa State University Press
BLACK, J.R. and HLUBIK, J. (1980). Basics of computerized linear programs for ration formulation. *Journal of Dairy Science* **63**, 1366–1378
BROKKEN, R.F. (1971a). Programming models for use of the Lofgreen–Garrett net energy system in formulating rations for beef cattle. *Journal of Animal Science* **32**, 685–691
BROKKEN, R.F. (1971b). Formulating beef rations with varying levels of heat increment. *Journal of Animal Science* **32**, 692–703
CRABTREE, J.R. (1982). Interactive formulation system for cattle diets. *Agricultural Systems* **8**, 291–308
DENT, J.B. and CASEY, H. (1967). *Linear Programming and Animal Nutrition.* London: Crosby Lockwood
FRANCE, J. (1982). Using a programmable calculator for rationing beef cattle. *Journal of the Operational Research Society* **33**, 419–424
FRANCE, J., NEAL, H. and POLLOTT, G.E. (1982). Using a programmable calculator for rationing pregnant ewes. *Agricultural Systems* **9**, 267–279
GLEN, J.J. (1980). A parametric programming method for beef cattle ration formulation. *Journal of the Operational Research Society* **31**, 689–698
GUE, R.L. and LIGGETT, J.C. (1966). Mathematical programming models for hospital menu planning. *Journal of Industrial Engineering* **17**, 395–400
HARKINS, J., EDWARDS, R.A. and McDONALD, P. (1974). A new net energy system for ruminants. *Animal Production* **19**, 141–148
McDONOUGH, J.A. (1971). Feed formulation for least cost of gain. *American Journal of Agricultural Economics* **53**, 106–108
MINISTRY OF AGRICULTURE, FISHERIES AND FOOD (1975). *Energy Allowances and Feeding Systems for Ruminants.* Technical Bulletin No. 33. London: HMSO
SMITH, V.E. (1963). *Electronic Computation of Human Diets.* East Lansing: Michigan State University Business Studies

# Glossary

## Further reading

BOX, G.E.P. and JENKINS, G.M. (1976). *Time Series Analysis: Forecasting and Control.* San Francisco: Holden-Day
CARSLAW, H.S. and JAEGER, J.C. (1960). *Conduction of Heat in Solids*, second edition. Oxford: Oxford University Press
CHURCHILL, R.V. (1963). *Fourier Series and Boundary Value Problems.* New York: McGraw-Hill
CRANK, J. (1975). *Mathematics of Diffusion*, second edition. Oxford: Oxford University Press
DRAPER, N.R. and SMITH, H. (1981). *Applied Regression Analysis*, second edition. New York: Wiley
EDWARDS, A.W.F. (1972). *Likelihood.* Cambridge: Cambridge University Press
FELLER, W. (1968). *An Introduction to Probability Theory and Its Applications*, third edition. New York: Wiley
HAMMERSLEY, J.M. and HANDSCOMB, D.C. (1964). *Monte Carlo Methods.* London: Methuen
HILDEBRAND, F.B. (1965). *Methods of Applied Mathematics*, second edition. Englewood Cliffs, New Jersey: Prentice-Hall
HOEL, P.G. (1962). *Introduction to Mathematical Statistics*, third edition. New York: Wiley
KENDALL, M.G. (1973). *Time-series.* London: Griffin
NAYLOR, T.H., BALINTFY, J.L., BURDICK, D.S. and KONG CHU (1966). *Computer Simulation Techniques.* New York: Wiley
SIDDONS, A.W., SNELL, K.S. and MORGAN, J.B. (1952). *A New Calculus Part III.* Cambridge: Cambridge University Press

SNEDDON, I.A. (1957). *Elements of Partial Differential Equations*. New York: McGraw-Hill
SNEDECOR, G.W. and COCHRAN, W.G. (1980). *Statistical Methods*, seventh edition. Ames: Iowa State University Press
STEPHENSON, G. (1961). *Mathematical Methods for Science Students*. London: Longman
TAHA, H.A. (1971). *Operations Research—an Introduction*. New York: Macmillan
WAGNER, H.M. (1969). *Principles of Operation Research with Applications to Managerial Decisions*. Englewood Cliffs, New Jersey: Prentice-Hall
WEATHERBURN, C.E. (1968). *A First Course in Mathematical Statistics*. Cambridge: Cambridge University Press

# Index